REWARD AND DECISION MAKING IN CORTICOBASAL GANGLIA NETWORKS

ANNALS OF THE NEW YORK ACADEMY OF SCIENCES
Volume 1104

REWARD AND DECISION MAKING IN CORTICOBASAL GANGLIA NETWORKS

Edited by Bernard W. Balleine, Kenji Doya, John O'Doherty, and Masamichi Sakagami

Published by Blackwell Publishing on behalf of the New York Academy of Sciences
Boston, Massachusetts
2007

Library of Congress Cataloging-in-Publication Data

Reward and decision making in corticobasal ganglia networks /
edited by Bernard W. Balleine ... [et al.].
    p. ; cm. – (The annals of the New York Academy of Sciences,
ISSN 0077-8923 ; v. 1104)
    Includes bibliographical references.
    ISBN-13: 978-1-57331-674-3 (paper : alk. paper)
    ISBN-10: 1-57331-674-1 (paper : alk. paper)
    1. Reward (Psychology)–Physiological aspects–Congresses.
2. Decision-making–Physiological aspects–Congresses. 3. Basal
ganglia–Congresses. 4. Cerebral cortex–Congresses. 5. Neural
networks (Neurobiology)–Congresses. 6. Neural
circuitry–Congresses. I. Balleine, Bernard W. II. New York Academy
of Sciences. III. Series.
    [DNLM: 1. Basal Ganglia–Congresses. 2. Cerebral
Cortex–Congresses. 3. Decision Making–physiology–Congresses.
4. Reward–Congresses. W1 AN626YL v.1104 2007 / WL 307 R466
2007]

    QP395.R49 2007
    612.8–dc22

                                                    2007010302

The *Annals of the New York Academy of Sciences* (ISSN: 0077-8923 [print]; ISSN: 1749-6632 [online]) is published 28 times a year on behalf of the New York Academy of Sciences by Blackwell Publishing with offices at 350 Main St., Malden, MA 02148 USA; 9600 Garsington Road, Oxford, OX4 2ZG UK; and 600 North Bridge Rd, #05-01 Parkview Square, 18878 Singapore.

Information for subscribers: For new orders, renewals, sample copy requests, claims, changes of address and all other subscription correspondence please contact the Journals Department at your nearest Blackwell office (address details listed above). UK office phone: +44 (0)1865 778315, fax +44 (0)1865 471775; US office phone: 1-800-835-6770 (toll free US) or 1-781-388-8599; fax: 1-781-388-8232; Asia office phone: +65 6511 8000, fax; +44 (0)1865 471775, Email: customerservices@blackwellpublishing.com

Subscription rates:
Institutional Premium   The Americas: $4043   Rest of World: £2246
The Premium institutional price also includes online access to full-text articles from 1997 to present, where available. For other pricing options or more information about online access to Blackwell Publishing journals, including access information and terms and conditions, please visit www.blackwellpublishing. com/nyas
*Customers in Canada should add 6% GST or provide evidence of entitlement to exemption.
**Customer in the UK or EU: add the appropriate rate for VAT EC for non-registered customers in countries where this is applicable. If you are registered for VAT please supply your registration number.

Mailing: The *Annals of the New York Academy of Sciences* is mailed Standard Rate. Mailing to rest of world by DHL Smart & Global Mail. Canadian mail is sent by Canadian publications mail agreement number 40573520. **Postmaster:** Send all address changes to *Annals of the New York Academy of Sciences*, Blackwell Publishing Inc., Journals Subscription Department, 350 Main St., Malden, MA 02148-5020.

Membership information: Members may order copies of *Annals* volumes directly from the Academy by visiting www.nyas.org/annals, emailing membership@nyas.org, faxing 212-298-3650, or calling 800-843-6927 (US only), or 212-298-8640 (International). For more information on becoming a member of the New York Academy of Sciences, please visit www.nyas.org/membership. Claims and inquiries on member orders should be directed to the Academy at email: membership@nyas.org or Tel: 212-298-8640 (International) or 800-843-6927 (US only).

Printed in the USA. Printed on acid-free paper.

Annals are available to subscribers online at the New York Academy of Sciences and also at Blackwell Synergy. Visit www.blackwell-synergy.com or www.annalsnyas.org to search the articles and register for table of contents e-mail alerts. Access to full text and PDF downloads of *Annals* articles are available to nonmembers and subscribers on a pay-per-view basis at www.blackwell-synergy.com and www.annalsnyas.org.

The paper used in this publication meets the minimum requirements of the National Standard for Information Sciences Permanence of Paper for Printed Library Materials, ANSI Z39.48-1984.

ISSN: 0077-8923 (print); 1749-6632 (online)
ISBN-10: 1-57331-674-1 (paper); ISBN-13: 978-1-57331-674-3 (paper)

A catalogue record for this title is available from the British Library.

ANNALS OF THE NEW YORK ACADEMY OF SCIENCES

Volume 1104
May 2007

REWARD AND DECISION MAKING IN CORTICOBASAL GANGLIA NETWORKS

Editors
BERNARD W. BALLEINE, KENJI DOYA, JOHN O'DOHERTY, AND
MASAMICHI SAKAGAMI

This volume is the result of a conference entitled **Reward and Decision-Making in Cortico-basal Gangila Networks**, held at the UCLA Conference Center, Lake Arrowhead, California, on June 1–4, 2006.

CONTENTS

Financial assistance was received from:

- Brain Research Institute at the University of California, Los Angeles
- Okinawa Institute of Science and Technology
- California Institute of Technology
- Tamagawa University Research Institute

Introduction

Current Trends in Decision Making

Decision making—the field concerned with the processes by which animals choose between competing actions on the basis of the expected value, or utility, of their consequences—has long been a topic of interest in philosophy, psychology, economics, and ethology. Nevertheless, in recent years there has been considerable expansion in this field into new directions, particularly into neuroscience and machine learning, largely due to innovation, both in technology and in approach. Whereas classic theories tried to explain how a decision should be made under a stationary condition with sufficient knowledge of the problem, but without regard to the neural processes involved, recent decision-making research addresses how actual individuals learn to make decisions when faced with novel or uncertain situations, as well as which neural circuits and systems are involved during different types of decision and during different stages of learning. The maturation of neuroimaging techniques for studying brain functions in humans (e.g., functional magnetic resonance imaging) has played a key role in this development, but what characterizes the recent surge in decision-making research is convergence of different lines of research, including reinforcement learning theory, behavioral economics, focal lesions, and pharmacology.

The field as it has developed is very broad, crossing traditional boundaries between areas of science and neuroscience in its investigation of multiple species and brain areas, and through its use of multiple approaches to research, particularly in the degree to which previously distinct procedures, protocols, and paradigms are combined to solve specific problems. This has been an exciting time of rapid, sometimes frenetic, or even chaotic, advances. Nevertheless, some broad themes have begun to emerge in recent years, and it was recognition of this fact that first prompted the thought that a collection of papers surveying contemporary research in decision making might be possible. In fact, the articles in this volume grew out of a series of talks and presentations given at a meeting held at the Lake Arrowhead Conference Center at the University of California, Los Angeles in June 2006 (see FIG. 1). The decision to hold that meeting itself came from a series of discussions between the editors of this volume, initiated first at the Okinawa Computational Neuroscience Course in July 2005 (OCNC2005), in which each of us participated that year, and that were carried forward to the Society for Neurosciences meeting in Washington D.C. in November 2005. We felt that there had been sufficient advance in the

Ann. N.Y. Acad. Sci. 1104: xi–xv (2007). © 2007 New York Academy of Sciences.
doi: 10.1196/annals.1390.025

FIGURE 1. *Attendees at the meeting held at the UCLA Lake Arrowhead conference center in June 2006. Back row (L-R):* Robert Brown, Kazuyuki Samajima, Makoto Ito, Read Montague, Alan Hampton, David Reddish, Kersten Preuschoff, Betsy Murray, Mark Walton, Okihide Hikosaka, Andrew Delamater, Yasushi Kobayashi, Masamichi Sakagami, Bernard Balleine, Jeff Wickens, John O'Doherty, Geoffrey Schoenbaum. *Front row (L-R):* Yael Niv, Tony Brugier, Simon Killcross, Sean Ostlund, Peter Bossaerts, Kenji Doya, Greg Corrado, Brian Knutson, Daria Knoch, Mauricio Delgado, Rui Costa, Daeyeol Lee, Saori Tanaka, Masataka Watanabe. *Not pictured:* Greg Berns, Jon Horwitz, Nicolas Schweighofer.

understanding of decision processes in humans and in other animals, and that a sufficient number of active research labs were working on the problem to hold a meeting to explore common research themes and goals. Furthermore, we thought that we could detect beneath the diverse treatments, approaches, and methods in use in these labs, a common core of ideas and interests that we were all circling and that we felt could make communication across these disparate domains possible.

The meeting itself was structured to cover several broad themes, although we decided not to group these but to intermix them, in the hope of identifying common elements. For the same reason, the table of contents for this volume closely follows the original order of the talks given at the meeting. We have identified six themes around which substantial research efforts appear to be currently organized: (1) computational models of decision processes; (2) neuroeconomics and decision theory; (3) the executive functions of the prefrontal cortex; (4) the evaluative functions of the orbital cortex; (5) the integrative functions of corticostriatal networks; and (6) the role of neurotransmitter systems, particularly dopamine, in reward prediction. Of course, as is common in this research area, many of the articles in this volume cover more than one of these themes.

Computational Models

The emphasis on adaptive choice—the suggestion that animals select actions that maximize long-run future reward—has resulted in considerable interest in learning models that can capture these kinds of constraints, particularly reinforcement learning models. Several articles in the current volume describe the relevance of this approach to research in decision making, particularly with respect to imaging (O'Doherty), drug addiction (Redish and Johnson), dopamine function (Niv), and the function of the prefrontal cortex (Lee and Seo). In addition, Samejima and Doya extend this approach in contrasting model-based and model-free reinforcement learning in the context of the functions of discrete corticostriatal networks and their role in adaptive behavior. Similarly, on the basis of their discussion of temporal difference models, Knutson and Wimmer consider evidence for distinct error terms in fMRI data, involving the prefrontal cortex, the ventral striatum, and their potential functions.

Neuroeconomics and Decision Theory

Other computational approaches have been inspired by economics, particularly aspects of expected utility theory, and several articles here are concerned with these ideas as they relate to understanding the error signal associated with predictions of reward. In these cases, the computational benefit lies in establishing the means by which neural processes optimize reward predictions, whether applied to making decisions under risk (Preuschoff and Bossaerts), exerting control over impulsive choice (Knoch and Fehr), or to the binding dynamics of neurotransmitters, such as dopamine, that have been associated with reward prediction error (Berns, Capra, and Noussair). In addition, several articles in this volume are concerned with other influences on utility, specifically the discounting functions associated with delayed reward (Roesch, Calu, Burke, and Schoenbaum; Schweighofer, Tanaka, and Doya) and the influence of effort on action values (Walton, Rudebeck, Bannerman, and Rushworth; Niv).

Executive Functions

Other contributions focus on the role of specialized regions of the prefrontal cortex in response selection and other executive processes. For example, Knoch and Fehr describe the role of the human dorsolateral prefrontal cortex in regulating both impulsive choices and those choices based on the subjects' immediate self-interests. Transmagnetic stimulation interferes with activity in this region and increases selfish and risky choices, even when these reduce long-run future reward. Two articles, one by Lee and Seo and another by Sakagami and Watanabe, argue that much the same is true of primates: that the dorsolateral prefrontal cortex allows state-based predictions of reward to influence action selection, whereas connections of the lateral prefrontal cortex with motivational structures provide the basis for choice to be influenced by the subjects' basic desires. In addition, Haddon and Killcross review work on the rodent prefrontal cortex and the finding that the prelimbic and anterior cingulate areas—broadly speaking the rodent homologue of primate dorsolateral prefrontal cortex—resolve response conflict, particularly conflict based on contextual modulation of action selection.

Prediction and Value

The way that predictions of rewarding outcomes are translated from motivational imperatives into the more general predictions of value for association with potential actions is a topic of some concern in the decision-making literature, and two articles in this volume focus explicitly on this issue. Delamater and Oakeshott propose that multiple attributes of the reward, such as its identity, value, or anticipated time of occurrence, are encoded simultaneously but dissociably in the associative structure encoding predictive relations and that this information can be used flexibly to select actions that are dependent on current constraints. This implies that quite specific motivational factors could influence performance under some circumstances and that relative value might be more critical under others, whereas time to reward might be more important under still others. It is interesting to consider how these distinct predictions summate to determine net value. Roesch *et al.* propose that this is one of the specialized functions of the orbitofrontal cortex and describe distinct neural codes in this region in rats that combine factors that discount reward value with the influence that those factors have on action selection.

Corticostriatal Networks

However sophisticated the frontal lobe may be in parsing abstract relations and values, it does not act alone. Recent research suggests that the cortical influence on action selection and initiation is determined by its relationship with the basal ganglia, particularly the cortical inputs to dorsal striatum. Several articles in this volume argue this case based on evidence from rodents (Costa; Balleine and Ostlund), primates (Hikosaka; Samejima and Doya), and human subjects (Delgado). Although there is considerable evidence of corticostriatal involvement in motor learning, particularly in the acquisition of sensorimotor habits, recent evidence reviewed in each of these articles suggests that reward-related or goal-directed actions are mediated by discrete corticobasal ganglia circuits involving striatal processes distinct from those controlling habits. Hence, these studies also suggest that there is considerably more heterogeneity of function in the striatum than was previously anticipated.

Neurotransmitter Systems

Recent evidence from studies assessing the neurochemical inputs to the striatum suggests that there is heterogeneity in the function of specific neurotransmitter systems, most notably in that of dopamine, corresponding to the functional heterogeneity just described. Several articles in the current volume describe this developing and important area of research in decision making, focusing on neurophysiological studies in rodents (Costa; Wickens, Budd, Hyland, and Arbuthnott) and on behavioral studies manipulating dopaminergic signaling (Horvitz, Choi, Morvan, Eyny and Balsam; Haddon and Killcross). Other articles are concerned with the role of midbrain dopamine neurons as error signals from a computational (Berns *et al.*) and structural perspective (Kobayashi and Okada). Kobayashi and Okada, for example, consider the role of the pedunculopontine tegmentum, which is one of the main afferents on the dopamine-rich areas of the midbrain, in the calculation of the reward prediction error signal conveyed by the dopamine neurons. Other articles describe other neurotransmitter systems besides dopamine; for example, Schweighofer *et al.* consider the role of serotonin in the evaluation of delayed rewards, whereas Redish and Johnson consider the claim that opioid signaling provides hedonic evaluation of rewarding events and evaluate whether the opioid system is involved in computing the value of expected rewards.

It is notable that each of the articles in this volume is as much concerned with future directions as with current research. There are, of course, many important problems to be solved. Some, like the future of the marriage between neuroscience and economics, may only be clearly resolved with considerable passage of time. Answers to others, for example, whether the error signal generated by midbrain dopamine neurons is actually critical for predictive learning, might optimistically come sooner. There are, however, many questions that researchers are only just starting to ask and that will form the basis for future research, such as the possibility, given the multiple neurotransmitter systems involved in learning, that there are several error signals that contribute to predictive learning—for example, signals predicting reward versus punishment and with different discount factors. Decision-making research is currently rich with possibilities and, although, naturally, we believe that this volume will make a substantive contribution to this subject, we hope that it will also serve to encourage future development in this field.

—BERNARD W. BALLEINE
University of California Los Angeles
Los Angeles, California, USA

—KENJI DOYA
Okinawa Institute of Science and Technology
Okinawa, Japan

—JOHN O'DOHERTY
California Institute of Technology
Pasadena, California, USA

—MASAMICHI SAKAGAMI
Tamagawa University Research Institute
Machida, Tokyo 194-8610, Japan

Learning about Multiple Attributes of Reward in Pavlovian Conditioning

ANDREW R. DELAMATER AND STEPHEN OAKESHOTT

Psychology Department, Brooklyn College, City University of New York, Brooklyn, New York, USA

ABSTRACT: The nature of the reward representation in Pavlovian conditioning has been of perennial interest to students of associative learning theory. We consider the view that it consists of a range of different attributes, each of which may be governed by different learning rules. We investigated this issue through a series of experiments using a time-sensitive Pavlovian-to-instrumental transfer procedure, aiming to dissociate learning about temporal and specific sensory features of a reward. Our results successfully demonstrated that learning about these different features appears to be dissociable, with learning about the specific sensory features of a Pavlovian unconditioned stimulus (US) occurring very rapidly across a wide range of experimental procedures, while learning about the temporal features of the US occurred slightly less quickly and was more sensitive to parametric disruption. These results are discussed with regard to the potential independence or interdependence of the relevant learning processes, and to some recent neurophysiological recording and brain lesion work, which provide additional means to investigate these dissociations.

KEYWORDS: CS–US interval; temporal learning; Pavlovian–instrumental transfer; trace conditioning; sensory-specific learning

INTRODUCTION

Questions related to the structure of learning have long occupied the attention of learning theorists. Recent advances in the application of behavioral and neuroscience techniques have permitted a more complete analysis to this problem than has been historically possible. The purpose of this article is to provide a framework for asking new questions about the structure of learning in Pavlovian conditioning, and to also present some recent findings from our lab, which illustrate a potential difference in how organisms learn about qualitative and temporal aspects of reinforcers.

Address for correspondence: Andrew R. Delamater, Psychology Department, Brooklyn College, CUNY, 2900 Bedford Avenue, Brooklyn, NY 11210. Voice: +718-951-5000, ext: 6026; fax: +718-951-4814.
andrewd@brooklyn.cuny.edu

Ann. N.Y. Acad. Sci. 1104: 1–20 (2007). © 2007 New York Academy of Sciences.
doi: 10.1196/annals.1390.008

THE PROBLEM OF STRUCTURE IN THE STANDARD
ASSOCIATIVE MODEL

When an organism is exposed to a positive contingency between conditioned and unconditioned stimuli (CS and US, respectively) in a Pavlovian learning task, the organism indicates through its behavior that it has learned something about this contingent relation. The classic example comes from Pavlov's own observations that repeated pairings between an initially neutral auditory stimulus and food result in the animal developing new salivary-conditioned responses (or CRs) to the auditory stimulus.[1] However, the nature of that learning is not directly obvious from the behavior itself. For instance, these changes could have occurred because new connections developed between the auditory CS and the salivary response to the food US. More properly speaking, new connections could have developed between the neural representations of the CS and UR (the unconditioned response), connections that have historically been referred to as stimulus–response (S–R) associations, such that presentation of the CS leads directly to production of the CR. Alternatively, the emergence of CRs could be explained by the development of new connections between the neural representations of the CS and the US (stimulus–stimulus, or S–S, associations). In this case, the CR is generated indirectly, as CS presentation evokes a representation of the US, which in turn leads to responding. Since both of these explanations can account for the same change in behavior, special methods are required to probe the structure of the underlying learning.

Whereas the issue of whether learning was best construed in terms of S–R or S–S associations dominated early discussions of learning theory,[2] today the question of the structure of learning is considerably more complex. One level of analysis concerns questions about the nature of the link between the elements of the association, or, put differently, about the computational rules whereby new connections are established.[3–5] A second, more recently addressed, question at this level concerns determining whether new associative links are binary or hierarchical in character, that is, whether they "connect" two or more than two elements.[6] Another level of analysis of the issue of structure of learning addresses questions concerned with the representational content of learning. In other words, exactly what do the elements that become connected as a result of learning represent?

Displayed in FIGURE 1 is a simple schematic depiction of some of the possible answers to this question. It is intended to portray an elaboration of what may be referred to as the standard associative model, which simply involves a connection between neural representations of the CS and US. The elaborated model acknowledges that there may exist multiple features of the US, any set of which can enter into associations with the CS. It assumes that a US is "processed" in multiple ways, that is, in terms of its specific sensory qualities, its hedonic tone, its more general activating or "motivational" properties, its temporal occurrence, as well as in terms of its specific response-eliciting

Associations with Different Reinforcer Attributes

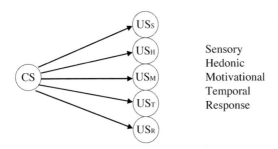

FIGURE 1. Possible components of Pavlovian US representations, and possible associative links between the CS and those US components.

characteristics, and that each of these can participate in the association process. Perhaps the most central question arising from this framework concerns the nature of any interactions between learning about these multiple reinforcer attributes. It seems possible that different neural systems or structures underlie each one of these different reinforcer "representations," and that learning involving these different neural systems or structures is independent of one another. Alternatively, there may exist important interdependencies in learning about different reinforcer attributes such that learning of one attribute in some way either facilitates or competes with learning about other attributes.

Important empirical and theoretical challenges naturally arise from this perspective. In particular, with separate measures of learning involving any one of the attributes shown in FIGURE 1, one can, in theory, determine to what extent learning involving different pairs of attributes is independent or interdependent of one another. Here we will begin this undertaking by reviewing some recent data from our laboratory concerning the possible independence of learning about sensory and temporal attributes.

AN ILLUSTRATIVE EXAMPLE

Some of the issues raised above can better be appreciated with reference to the results from a rather basic conditioning experiment examining appetitive conditioning in laboratory rats where subjects were given presentations of a 20-sec CS with a US occurring at the termination of the stimulus. For reasons that will become clear shortly, subjects in this experiment were trained with two different CS–US pairs (e.g., flashing light–pellet and noise–sucrose, counterbalanced) over 24 sessions in each of which there occurred two trials of each type. Our basic measure of conditioning in this setting is the rate of magazine approach responding elicited by presentation of the CSs. Both pellet (2, 45-mg pellets) and liquid sucrose (0.1 mL of a 20% solution) USs were delivered to the same food magazine, and magazine entry responses were recorded with an infrared photodetection circuit. We also included a random control group that received equivalent numbers of CS and US presentations in each session, but with the USs presented randomly within the intertrial intervals, rather than at CS offset.

FIGURE 2 shows a common data pattern found in our lab with this preparation. The acquisition of conditioned magazine responding is shown for both groups collapsed over both types of trials and presented in successive 4-sec intervals within the CSs in separate four-session blocks of training. There are a few noteworthy features of these data. First, whereas initial responding in the two groups was similar, a clear difference emerged over training, with the rats who had received CS–US pairings (Gp Paired) displaying increased levels of responding relative to those rats who had received random CS, US presentations (Gp Random). Second, by the end of training Gp Paired subjects displayed a clearly timed CR profile, as their level of magazine responding steadily increased across the CS interval (as the time of US delivery approached). Third, this timed CR profile emerged gradually over training, appearing to require a greater number of CS–US pairings than did simple magazine approach. Specifically, although paired subjects displayed higher levels of responding than the random controls as early as block 2 of training, the distribution of responding across the CS at this point in training was flat. Temporal control of responding started to emerge only in block 3 and became progressively sharper with further training.

What does this acquisition data tell us about the underlying structure of learning? It seems clear from these data that subjects come to represent, in some sense, the temporal features of the US. In addition, it appears as though some nontemporal form of learning develops prior to learning about these temporal features. However, the nature of this nontemporal learning process cannot be determined from these data alone. It is clear that associations between the CSs and *some* feature of the USs occur earlier in training than associations between the CSs and the temporal components of the USs, but it is impossible to tell from these data whether these associations involve the sensory, hedonic, motivational, or response properties of the USs. The next challenge, then, is determining which of these nontemporal components of learning best describes the results. To answer this question we need to move onto other, more selective assays of learning.

THE PAVLOVIAN-TO-INSTRUMENTAL TRANSFER TEST (PIT)

The PIT has been used extensively and successfully as a measure of learning involving both motivational and sensory features of different USs.[7–12] In this test procedure, CSs are tested for their effects on instrumental responses with which they have never been trained. The CS has been shown in this procedure

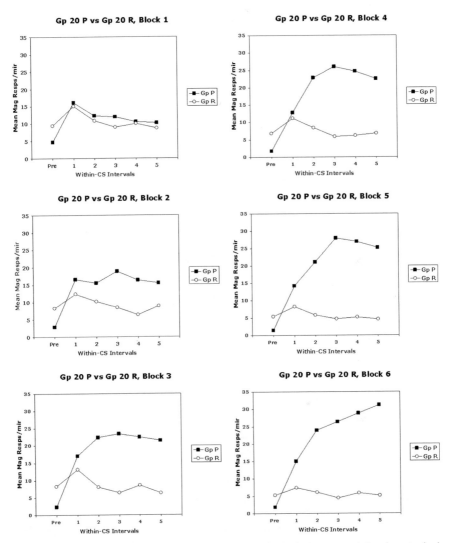

FIGURE 2. Conditioned magazine response topography across training in a typical experiment. One group received CS–US pairings over six blocks of training, while a second group received random presentations of CS and US. Responding is shown during the pre-CS period as well as in five successive 4-sec bins during the CS.

to selectively elevate the rate of instrumental responding when the CS and instrumental response were previously reinforced with the same US. Although it is not well understood at present precisely why this selective effect occurs, it is commonly assumed that this effect could only occur if both the CS and the instrumental response were to associatively activate a specific sensory representation of the US. In addition, under some circumstances the CS has also been shown in this procedure to nonselectively elevate instrumental responses when the CS and instrumental response were previously reinforced with different USs. This nonselective effect is commonly taken as evidence of learning about some general activating or "motivational" attribute of the US, which might be shared with other, similar, USs.

Thus, the PIT test presents us with a potentially powerful tool to dissect the nature of learning. The studies reported below used this test procedure to investigate the effects of different behavioral treatments on learning about different reinforcer attributes. The general strategy is to assume that, if different behavioral treatments affect different components of learning in different ways, then this would begin to suggest that dissociations may exist at an underlying process level.

ACQUISITION OF SENSORY AND TEMPORAL LEARNING

Returning to the data reported above, it appears that some nontemporal form of learning occurs prior to the development of temporally specific learning. This study investigated the nature of this nontemporal learning by using the PIT test procedure after varying amounts of Pavlovian training. In addition to assessing the overall presence or absence of sensory- and motivationally based CS–US associations in the PIT tests, we also assessed the timing of these effects across the CS interval.

Rodent subjects initially were given instrumental training with two different response-US pairs (e.g., lever press-pellet and chain pull-sucrose, counterbalanced). At first, the animals were trained on a continuous reinforcement schedule and were then switched to progressively leaner variable interval schedules (10 sec, 30 sec, and 60 sec) in subsequent sessions, until stable responding was recorded. Following this training, subjects were split into six groups, each of which received different amounts of Pavlovian training, either 4, 8, 16, 24, 56, or 112 presentations of each stimulus, with two distinct CS–US pairs (e.g., noise-pellet, flashing light-sucrose). Training was arranged such that each group completed their Pavlovian training phase on the same day. The CSs used in this study were 60 sec in duration with the appropriate US occurring at the offset of the CS, and the average intertrial interval was 5 min. On the day following Pavlovian training all subjects were given one retraining session with each instrumental response. The PIT tests occurred over the next 2 days. In each session, subjects were given a choice between the lever and chain, in the absence of any reinforcement, and each CS was presented a total of eight times in a pseudorandom order. Stimulus periods alternated with 60-sec prestimulus periods. We monitored levels of magazine directed responding as well as instrumental responding throughout these test sessions, such that the no-stimulus periods provided a measure of baseline responding.

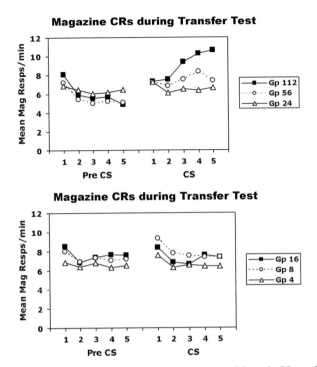

FIGURE 3. Conditioned magazine approach in pre-CS and CS periods during Pavlovian-to-instrumental transfer tests (PIT) conducted after varying amounts of Pavlovian training were given to separate groups of subjects.

FIGURE 3 displays the magazine approach CR data collected during these PIT sessions. It shows, not surprisingly, that groups given more training performed more magazine approach CRs, and that timed magazine responding was most clearly seen in the group given the most Pavlovian training.

The data from the PIT tests are shown in FIGURE 4. These data show that, in spite of the large differences in magazine responding and timing across the CS interval, all the groups display selective PIT, preferring the instrumental response that was previously reinforced with the same US as that signaled by the CS, to the response paired with the different US. In the group given only four trials with each CS–US pair, this effect did not interact with time, with the preference appearing relatively stable across the CS. All of the other groups not only displayed an overall bias toward the "same" response over the "different" response, but the effect was more pronounced toward the end of the CS, the period where the US would normally have occurred. Thus, it appears from these data that these rats learned to associate the CS with sensory specific attributes of the US early in training, and that they learned to time the specific US's arrival with relatively little additional training. It is of some interest that both

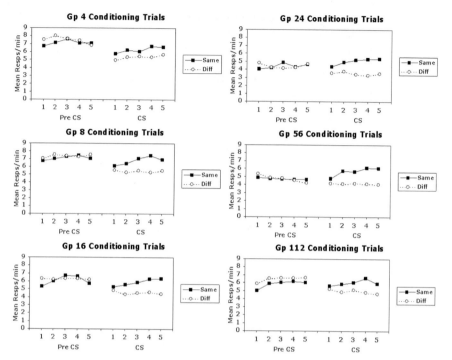

FIGURE 4. Instrumental responding during Pavlovian-to-instrumental test sessions is displayed in pre-CS and CS periods for separate groups given different amounts of Pavlovian training prior to the test. The data are segregated according to whether the CS and response were reinforced previously with the same or different US.

of these effects developed before the onset of measurable magazine approach behavior, although the present study also provided no evidence for acquisition of general "motivational" effects of the CS upon performance of "different" outcome responses.

Thus, it appears plausible that learning about sensory and temporal qualities of the US proceed at different rates. This begins to suggest a separation in the processes underlying these two components of learning. The next experiment examined this idea further by investigating another variable shown to importantly influence CR development, CS–US interval.

EFFECTS OF CS–US INTERVAL ON SENSORY AND TEMPORAL LEARNING

A well-established fact about conditioning is that CS–US interval has a strong influence on the development of CRs as well as their topography.[13,14] However, the question of how this variable influences the development of sensory-specific associations and US timing is less clear. Some authors have suggested that sensory-specific learning ought to be optimal with short CS–US intervals because similarity between short duration CSs and the relatively short duration sensory components of the US might favor association formation.[15] In addition, other authors have suggested that similar interval timing processes should be engaged regardless of the CS–US interval.[16] Recently, we have examined CS–US interval effects using PIT test procedures in an effort to more directly investigate potential effects on sensory-specific learning as well as the timing of sensory-specific control.[17]

In one experiment we gave subjects separate instrumental and Pavlovian training like that described above, with the exception that different groups of subjects were trained with different CS–US intervals during Pavlovian training sessions. Animals were trained with either 20-sec, 60-sec, or 180-sec CS–US intervals, and they were all given enough training sessions to ensure asymptotic magazine CR responding prior to the PIT tests. The total number of trials during training (48) and the total session time (50 min) were held constant in each group. To avoid stimulus generalization problems, the PIT test sessions were conducted with each group tested at their training stimulus durations. Each CS was presented enough times in the PIT test to total 9 min of sampling time.

The magazine CR data over successive blocks of training are shown for each group in FIGURE 5. There were large differences in magazine approach responding in the three groups, with short CS–US intervals supporting more CRs than longer CS–US intervals. In spite of these large differences in magazine responding, the stimuli exerted comparable reinforcer-specific control over instrumental responding in the PIT tests in the three groups (FIG. 6). In other words, equivalent sensory-specific learning occurred in these three groups despite rather large differences in CS–US interval and magazine approach conditioning.

Acquisition of Magazine Approach

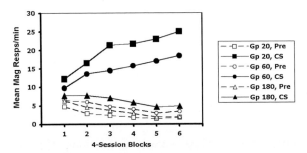

FIGURE 5. Acquisition of magazine approach responding in separate groups trained with different CS–US intervals. Responding is shown in six 4-session blocks in pre-CS and CS periods.

Choice Test: Paired Groups

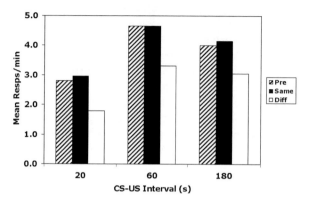

FIGURE 6. Results from PIT tests conducted after separate groups were given equivalent amounts of training with different CS–US intervals. Instrumental responding is shown in the pre-CS period as well as during the CS when the response and CS were reinforced with the same or different US.

The data presented in FIGURES 5 AND 6 display the results collapsed across the entire CS, which differed in duration across the three groups. To assess timing processes, the data were also examined during the PIT tests broken down in terms of successive one-fifths of the interval, as in FIGURE 7. Timing research has shown that conditioned responding across different intervals often superimposes when plotted in this way.[18] It is clear that groups trained with 20-sec and 60-sec CS–US intervals display greater sensory-specific Pavlovian control over instrumental response choice toward the end of the CS–US interval. However, the group trained with 180-sec CS–US intervals did not display any temporal control over the specific PIT effect. In other words, these data suggest that good sensory-specific associations were learned across all CS–US intervals, but that good timing occurred only with the relatively shorter CS–US intervals. Once again, it appears as though learning about sensory-specific features of the US can occur in the absence of good interval timing.

SENSORY AND TEMPORAL LEARNING
IN TRACE CONDITIONING

Another manipulation frequently found to alter the course of Pavlovian learning is the introduction of a temporal gap between CS offset and delivery of the US, in a procedure described as trace conditioning. The presence of this gap, or trace interval, typically results in a reduction of learned behavior across training, with the degree of reduction depending on the particular procedure employed. In the context of the present discussion, it is of interest whether

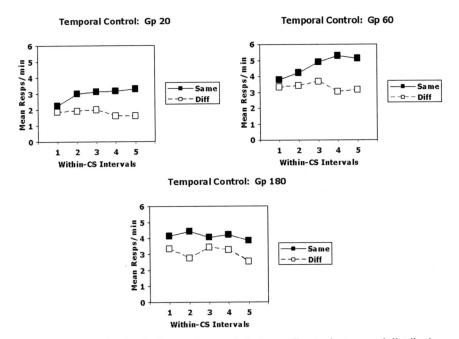

FIGURE 7. The data in FIGURE 6 are replotted according to the temporal distribution of same and different responding across the CSs in groups trained with different CS–US intervals.

the disruptive impact of the trace interval might differentially affect learning about stimulus timing and about sensory-specific features. A recent study in our laboratory investigated this issue with a PIT procedure similar to that outlined above, where hungry rats were initially given instrumental training with two response–outcome pairs, and then received Pavlovian training with the same two outcomes. Thirty-second, CSs were employed, and the novel feature of this study was the presence of a 30-sec trace interval, in which no stimuli were present, between the offset of each CS and delivery of the relevant US. Following sufficient training to allow asymptotic magazine approach, the animals received PIT testing, and their behavior was monitored across both CS presentation and the subsequent trace interval, broken down into 10-sec bins. As can be seen from FIGURE 8A, these rats showed more interest in the food magazine during the trace interval than during the CS itself, which might indicate some degree of response timing, but showed no tendency to time their responding appropriately within either interval, responding more slowly at the end than at the beginning. The most straightforward way to interpret this pattern of results is probably that the animals come to associate the offset of the CS with reinforcement, but that they are unable to learn about the temporal features of the trace interval, possibly due to generalization from the intertrial interval (ITI).[19]

(A)   Trace Conditioning PIT test: Mag CRs

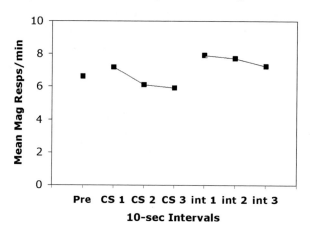

(B)   Trace Conditioning PIT test

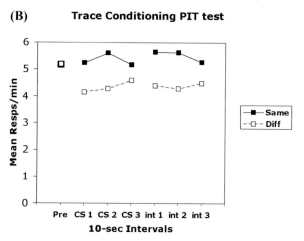

FIGURE 8. (A) Magazine approach responding during PIT tests conducted following Pavlovian trace conditioning. Responding is shown in the pre-CS period as well as in three, 10-sec bins during the CS and the trace interval. **(B)** Instrumental responding during the PIT conducted following Pavlovian trace conditioning. Responding is shown during the pre-CS period as well as in three, 10-sec bins during the CS and trace interval. Responses are segregated according to whether the CS and response were reinforced previously with the same or different US.

However, these rats clearly were able to learn about the specific sensory aspects of the US across the trace interval, as is illustrated by FIGURE 8B, showing a significant preference for the "same" response during both the CS and the trace interval. Indeed, the observed same versus different effect during CS presentation is of particular interest given the lack of magazine approach responding, suggesting a dissociation between the processes underlying these behaviors. It is also worth noting that there is no sign that the animals are timing their sensory-specific behavior any more than their magazine approach behavior, with the same/different effect appearing relatively consistent across both the CS and trace intervals.

Overall, the results of this study supply further evidence to dissociate learning about stimulus timing from that about sensory-specific stimulus features, since the introduction of the trace interval seems to prevent the former, at least with the amount of training we employed here, but not the latter. Accordingly, the present data support those described earlier, indicating that learning about sensory-specific qualities of the US precedes learning to time the US, implying a dissociation between the two.

EXTINCTION AND SENSORY VERSUS TEMPORAL LEARNING

One additional manipulation we have explored concerns the possible effects of Pavlovian extinction training upon different components of learning. Earlier research indicates that extinction has no impact on the status of sensory-specific CS–US associations,[10,20–22] and that while extinction reduces conditioned responding overall it has little impact on CR timing.[23] However, no study has examined the impact of extinction on the timing of sensory-specific learning. It seems possible that extinction might have dissociable effects on sensory-specific PIT and on the timing of this effect. We explored this possibility next.

Four groups of subjects were given instrumental training with two response–outcome pairs, prior to being trained with two separate stimulus–outcome pairs, as described above. One pair of groups (groups Late-Ext and Late-No Ext) were trained with 60-sec CSs in which the USs occurred in the final 10 sec of the stimuli while, for a second pair of groups (groups Early–Ext and Early–No Ext), US delivery instead occurred 10 sec after the onset of the 60-sec CSs. Following training, groups Late-Ext and Early-Ext were given 10 days of extinction training during which each CS was presented without the US. The other two groups (groups Late-No Ext and Early-No Ext) were placed in the conditioning chambers for an equivalent time, but received no CS or US presentations. The effects of extinction on learning about sensory and temporal features of the US were assessed in PIT tests. The data from these tests are presented in FIGURE 9.

The data in the top panel indicate that presenting the USs early versus late in the CSs resulted in appropriately timed magazine CRs in the groups, as indicated by those groups not given extinction, but that extinction reduced the level of magazine CRs in each of these different training conditions. The instrumental response data from the PIT test are of more interest. These data (in the middle and bottom rows) suggest that instead of extinction hindering

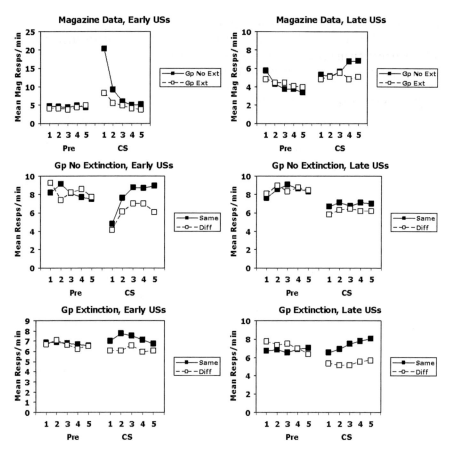

FIGURE 9. Magazine (*upper row*) and instrumental (*middle* and *lower rows*) responding is shown during the PIT tests conducted after separate groups were given extinction or
not following Pavlovian training in which the USs occurred in separate groups either early
(*left column*) or late (*right column*) in the 60-sec CSs.

one type of learning or the other, it actually *enhanced* sensory-specific temporal control. The No Extinction control groups displayed sensory-specific
PIT that was not temporally specific (see middle row). However, the Extinction groups trained with the USs presented early in the CS or late in the CS
displayed appropriately timed sensory-specific PIT (see bottom row). Thus,
it appears that far from undermining sensory or temporal learning, extinction
had a beneficial effect on the retrievability of these attributes, at least relative to context-exposed controls. In the absence of an extinction treatment, the
CSs failed to exert temporal control over the sensory-specific PIT effect. It
is of interest, however, that Group Early/No Extinction and Group Late/No
Extinction both displayed sensory-specific PIT, even though this effect was

not timed appropriately. These data possibly suggest that temporal information of the US is more subject to forgetting than sensory information over a 10-day retention interval. If this is so, then why would subjects given extinction show improved temporal control over sensory-specific PIT? Extinction may alleviate this forgetting through a process akin to "reconsolidation."[24]

Though these comments are highly speculative, it seems clear that training with USs at different times within the CS does produce sensory-specific learning that also comes under temporal control. Extinction does little to undermine both of these forms of learning, although they may be differentially susceptible to forgetting. If so, this would provide us with additional evidence to support the more general claim that learning about sensory and temporal features of the US are dissociable processes.

SUMMARY AND CONCLUSIONS

We began by noting that a key issue for understanding the structure of Pavlovian learning is to explore the extent to which learning about different components of the reinforcer involves interdependent or independent processes. While this issue can be addressed, in theory, in a number of different ways, we have explored it by focusing on learning about one pair of reinforcer attributes portrayed in FIGURE 1—reinforcer quality and reinforcer time. Our studies suggest that learning involving these reinforcer attributes involves different underlying systems. This conclusion follows from our findings that (1) although sensory and temporal learning both develop rapidly, sensory learning seems to occur earlier, (2) training with a long CS–US interval, or with a trace interval between CS and US, results in sensory learning in the apparent absence of US timing, and (3) learning about sensory and temporal features of the US is differentially affected by forgetting over a delay interval.

looseness-1Contrary to our results, some investigators working at the behavioral level have suggested that learning to time the US may be primary, and that learning additional information about the US is secondary to this temporal learning.[25] This claim is based on findings that learning to time when the US occurs and learning to respond to a CS occur together,[26] and that when conditioned responses first start to appear their latency is close to the time at which the US occurs.[25,27] The data we present here are not entirely consistent with these results. We believe that, by using the selective PIT test together with magazine approach CRs, our results can perhaps more easily separate between learning about reinforcer quality and time than is possible by looking at different measures of conditioned magazine approach responding by itself. In other words, having multiple measures that are specific in which aspects of learning they measure seem absolutely critical to the sort of undertaking suggested here.

Collectively our data support the view that learning about sensory and temporal features of the US involves different underlying systems, such that it should be possible to find evidence of this distinction at the neural systems level. Perhaps the clearest way to do so would be to find the neural systems underlying each form of learning and demonstrate a double dissociation in the effects of different lesions.

Currently, we do not understand how the brain codes temporal aspects of the US well enough to ask whether a double dissociation between the effects of lesions could be demonstrated for sensory and temporal learning. Recent neurophysiological recording work, however, suggests that learning about reinforcer timing and magnitude may differ. Roesch and colleagues have identified cells in the orbitofrontal cortex (OFC), which are sensitive to differences in reinforcer delay but not magnitude, and others that are sensitive to reinforcer magnitude but not delay.[28] The finding that different populations of cells are involved in coding different aspects of the reinforcer seems minimally required in order to claim that learning involving those different aspects involves different systems, and are independent of one another. However, to more fully address these issues, one needs to ask not only if learning about one attribute of the US can occur when another attribute has been functionally removed, a task made difficult by the fact that cells coding for each type of attribute may coexist within the same neural structure (e.g., OFC), but also whether learning about one attribute may facilitate or hinder learning about another attribute when both are available. In the case of reinforcer timing and quality, for instance, will learning about the sensory attributes of the US facilitate or impair learning about the temporal features, or will they proceed independently? As yet, we have no direct evidence either way, but lesion experiments involving structures known to code sensory attributes of the US could begin to provide an answer to this question.

Double dissociations have been made in studies examining learning about sensory and motivational attributes of the US. Corbit and Balleine have demonstrated that the central nucleus (CN) and basolateral amygdala (BLA) code different aspects of the US.[8] They observed that lesions of the CN eliminated general, but not sensory-specific, PIT effects, while lesions of the BLA eliminated sensory-specific, but not general, PIT effects. These results are consistent with the view that motivational encoding of the US depends upon the CN, while sensory-specific encoding depends upon the BLA. Further, these results begin to suggest that learning involving these two components of the US can proceed independently. However, two qualifications should be noted. First, while this sort of double dissociation suggests a distinction in the processes giving rise to motivational and sensory learning, it remains to be determined to what extent learning about these different reinforcer attributes might interact with one another in an intact animal. For instance, there is some data to suggest that prior learning of the motivational aspects of the US can under some circumstances facilitate and under other circumstances block learning about sensory-specific aspects of the US.[29] If these two forms of learning were truly independent, then such interactions would not occur.

Second, additional brain structures have been identified as playing a role in these effects, rendering the overall story more complex. For instance, Ostund and Balleine[30] have shown that BLA and OFC seem to interact in mediating selective PIT effects (see also Refs. 31,32). Whereas lesions of the BLA conducted before or after Pavlovian training eliminate selective PIT, lesions of the OFC conducted after training, but not before, eliminate selective PIT. Thus, sensory learning likely involves multiple structures at different points in training. Accordingly, the process of eliminating one form of learning to determine if this impairs another form of learning will necessarily be complex, requiring a comprehensive understanding of which brain structures participate in what aspect of conditioning.

By adopting the perspective noted above, a variety of additional experiments suggest themselves concerning the possible interactions or independence among different forms of learning. We have focused on sensory and temporal learning, and, as just noted, other work has begun to explore possible differences between sensory and motivational learning. However, if we can also devise good measures to assess associations with hedonic and response components of the US, then it will be possible to explore a variety of additional questions. Research directed toward an analysis of learning about the response properties of the US is, surprisingly, only in its formative stage. In this regard, studies of the acquisition of stimulus–response habits[33,34] provide an important step toward such investigations. Indeed, the distinction between dopamine-dependent and dopamine-independent conditioned responding,[35] described as reflecting a transition from goal-directed to stimulus–response habit-based responding, has a ready interpretation in terms of a transition from sensory- to response-based associations.

Other research has begun to identify differences in the neural basis of what may be described as learning about the motivational and hedonic aspects of the US. The distinction between "wanting" and "liking," for example, makes sense from this perspective.[36] While research in this area points to a distinction in the neural basis of these two forms of conditioning, it is less clear that the behavioral measures distinguishing between these effects uniquely assess learning about general motivational and hedonic components of learning, respectively. For instance, the typical measure of "wanting" involves assessing the effects of a CS upon instrumental responding,[37] while the most commonly used measure of "liking" is the taste reactivity test.[38] For the present purposes, it will be important to determine if the CS elevates instrumental responding through its association with the sensory attribute of the US or the more general motivational attribute. As noted above, this distinction can readily be made with the PIT test by looking for selective or nonselective effects. Furthermore, earlier research has demonstrated that an auditory CS for sucrose (or quinine) can increase (or decrease) the number of ingestive taste reactivity responses displayed to plain water when it is presented coincident with the CS.[39] It is tempting to conclude from this observation that the CS is increasing "liking" for water through its association with the hedonic aspects of the US. However, it could also be exerting its effects through its association with the sensory aspects of the US.[40] In short, before it can be determined whether learning about motivational and hedonic aspects of the US is independent or interdependent, highly specific response measures and procedures will need to be employed.

In summary, we have made explicit an assumption that we believe has been implicitly guiding research in several areas. This assumption is that when a US is presented to an organism, it evokes a host of potentially dissociable processes that can be described as sensory, hedonic, motivational, temporal, or response based. If we acknowledge that simple associative learning may actually entail differential learning about each or all of these different processes, then a variety of questions present themselves. The most basic questions, it seems to us, involve determining if learning about different components of the US is dissociable, and if so whether learning of these different components is independent or interdependent of one another. We have tried to illustrate one approach to addressing these issues by focusing on learning about sensory and temporal attributes of the US. Our data suggest that sensory and temporal learning can be behaviorally distinguished, but the degree to which these forms of learning are independent or interdependent of one another must await future research.

REFERENCES

1. PAVLOV, I.P. 1927. Conditioned Reflexes. Oxford University Press. Oxford.
2. ROZEBOOM, W.W. 1958. "What is learned?"—An empirical enigma. Psychol. Rev. **65:** 22–33.
3. MACKINTOSH, N.J. 1975. A theory of attention: variations in the associability of stimuli with reinforcement. Psychol. Rev. **82:** 276–298.
4. PEARCE, J.M. & G. HALL. 1980. A model for Pavlovian learning: variations in the effectiveness of conditioned but not of unconditioned stimuli. Psychol. Rev. **87:** 532–552.
5. RESCORLA, R.A. & A.R. WAGNER. 1972. A theory of Pavlovian conditioning: variations in the effectiveness of reinforcement and nonreinforcement. In Classical Conditioning II: Current Research and Theory. A.H. Black & W.F. Prokasy, Eds.: 64–99. Appleton-Century-Crofts. New York.
6. HALL, G. 2002. Associative structures in Pavlovian and instrumental conditioning. In Steven's Handbook of Experimental Psychology. Third edition. Vol 3: Learning, Motivation, and Emotion. H. Pashler & R. Gallistel, Eds.: 1–45. John Wiley & Sons. New York.
7. BLUNDELL, P., G. HALL & S. KILLCROSS. 2001. Lesions of the basolateral amygdala disrupt selective aspects of reinforcer representation in rats. J. Neurosci. **21:** 9018–9026.
8. CORBIT, L.H. & B.W. BALLEINE. 2005. Double dissociation of basolateral and central amygdala lesions on the general and outcome-specific forms of Pavlovian-instrumental transfer. J. Neurosci. **25:** 962–970.

9. DELAMATER, A.R. 1995. Outcome-selective effects of intertrial reinforcement in a Pavlovian appetitive conditioning paradigm with rats. Anim. Learn. Behav. **23:** 31–39.
10. DELAMATER, A.R. 1996. Effects of several extinction treatments upon the integrity of Pavlovian stimulus-outcome associations. Anim. Learn. Behav. **24:** 437–449.
11. HOLLAND, P.C. 2004. Relations between Pavlovian-instrumental transfer and reinforcer devaluation. J. Exp. Psychol. Anim. Behav. Process **30:** 104–117.
12. KRUSE, J.M. *et al.* 1983. Pavlovian conditioned stimulus effects upon instrumental choice behavior are reinforcer specific. Learn. Motiv. **14:** 165–181.
13. SCHNEIDERMAN, N., Ed. 1972. Response System Divergences in Aversive Classical Conditioning. Appleton-Century-Crofts. New York.
14. TIMBERLAKE, W., G. WAHL & D. KING. 1982. Stimulus and response contingencies in the misbehavior of rats. J. Exp. Psychol. Anim. Behav. Process **8:** 62–85.
15. KONORSKI, J. 1967. Integrative Activity of the Brain. University of Chicago Press. Chicago.
16. GIBBON, J. & P.D. BALSAM. 1981. The spread of association in time. *In* Autoshaping and Conditioning Theory. C.M. Locurto, H.S. Terrace & J. Gibbon, Eds.: 219–254. Academic Press. New York.
17. DELAMATER, A.R. & P.C. HOLLAND. The influence of CS-US interval on several different indices of learning in appetitive conditioning. J. Exp. Psychol. Anim. Behav. Processes. In press.
18. CHURCH, R.M. 2002. Temporal learning. *In* Steven's Handbook of Experimental Psychology Third edition. Vol 3: Learning, Motivation, and Emotion. H. Pashler & R. Gallistel, Eds.: 365–393. John Wiley & Sons. New York.
19. ZENTALL, T.R., E.D. KLEIN & R.A. SINGER. 2004. Evidence for detection of one duration sample and default responding to others by pigeons may result from an artifact of retention-test ambiguity. J. Exp. Psychol. Anim. Behav. Processes **30:** 129–134.
20. DELAMATER, A.R. 2004. Experimental extinction in Pavlovian conditioning: behavioural and neuroscience perspectives. Q. J. Exp. Psychol. Sect. B Comp. Physiol. Psychol. **57:** 97–132.
21. RESCORLA, R.A. 1996. Preservation of Pavlovian associations through extinction. Q. J. Exp. Psychol. Sect. B Comp. Physiol. Psychol. **49:** 245–258.
22. RESCORLA, R.A. 2001. Experimental extinction. *In* Handbook of Contemporary Learning Theories. R.R. Mowrer & S.B. Klein, Eds.: 119–154. Lawrence Erlbaum Associates. Mahwah, NJ.
23. OHYAMA, T.J. *et al.* 1999. Temporal control during maintenance and extinction of conditioned keypecking in ring doves. Anim. Learn. Behav. **27:** 89–98.
24. NADER, K. 2003. Memory traces unbound. Trends. Neurosci. **26:** 65–72.
25. BALSAM, P.D., M.R. DREW & C. YANG. 2002. Timing at the start of associative learning. Learn. Motiv. **33:** 141–155.
26. KIRKPATRICK, K. & R.M. CHURCH. 2000. Independent effects of stimulus and cycle duration in conditioning: the role of timing processes. Anim. Learn. Behav. **28:** 373–388.
27. DREW, M.R. *et al.* 2005. Temporal control of conditioned responding in goldfish. J. Exp. Psychol. Anim. Behav. Process **31:** 31–39.
28. ROESCH, M.R., A.R. TAYLOR & G. SCHOENBAUM. 2006. Encoding of time-discounted rewards in orbitofrontal cortex is independent of value representation. Neuron **51:** 509–520.

29. GEWIRTZ, J.C., S.E. BRANDON & A.R. WAGNER. 1998. Modulation of the acquisition of the rabbit eyeblink conditioned response by conditioned contextual stimuli. J. Exp. Psychol. Anim. Behav. Process. **24(1)**: 106–117.
30. OSTLUND, S.B. & B.W. BALLEINE. 2005. Lesions of the orbitofrontal cortex disrupt Pavlovian, but not instrumental, outcome-encoding. Program No. 71.20. Abstract Viewer/Itinerary Planner. Washington, DC: Society for Neuroscience.
31. SADDORIS, M.P., M. GALLAGHER & G. SCHOENBAUM. 2005. Rapid associative encoding in basolateral amygdala depends on connections with orbitofrontal cortex. Neuron **46**: 321–331.
32. SCHOENBAUM, G. & M. ROESCH. 2005. Orbitofrontal cortex, associative learning, and expectancies. Neuron **47**: 633–636.
33. HADDON, J.E. & S. KILLCROSS. Contextual control of choice performance: behavioral, neurobiological and neuro-chemical influences. This volume.
34. HORVITZ, J.C. et al. A "Good Parent" Function for Dopamine in Response Acquisition and Expression. This volume.
35. CHOI, W.Y., P.D. BALSAM & J.C. HORVITZ. 2005. Extended habit training reduces dopamine mediation of appetitive response expression. J. Neurosci. **25**: 6729–6733.
36. BERRIDGE, K.C. 2007. The debate over dopamine's role in reward: the case for incentive salience. Psychopharmacology **191**: 391–431.
37. WYVELL, C.L. & K.C. BERRIDGE. 2000. Intra-accumbens amphetamine increases the conditioned incentive salience of sucrose reward: enhancement of reward "wanting" without enhanced "liking" or response reinforcement. J. Neurosci. **20**: 8122–8130.
38. GRILL, H.J. & R. NORGREN. 1978. The taste reactivity test. I. Mimetic responses to gustatory stimuli in neurologically normal rats. Brain. Res. **143**: 263–279.
39. DELAMATER, A.R., V.M. LoLORDO & K.C. BERRIDGE. 1986. Control of fluid palatability by exteroceptive Pavlovian signals. J. Exp. Psychol. Anim. Behav. Process **12**: 143–152.
40. HOLLAND, P.C. 1990. Event representations in Pavlovian conditioning: image and action. Cognition. **37**: 105–131.

Should I Stay or Should I Go?

Transformation of Time-Discounted Rewards in Orbitofrontal Cortex and Associated Brain Circuits

MATTHEW R. ROESCH,[a] DONNA J. CALU,[b] KATHRYN A. BURKE,[b] AND GEOFFREY SCHOENBAUM[a,b,c]

[a]Department of Anatomy and Neurobiology, University of Maryland School of Medicine, Baltimore, Maryland, USA

[b]Program in Neuroscience, University of Maryland School of Medicine, Baltimore, Maryland, USA

[c]Department of Psychology, University of Maryland Baltimore County, Baltimore, Maryland, USA

ABSTRACT: Animals prefer a small, immediate reward over a larger delayed reward (time discounting). Lesions of the orbitofrontal cortex (OFC) can either increase or decrease the breakpoint at which animals abandon the large delayed reward for the more immediate reward as the delay becomes longer. Here we argue that the varied effects of OFC lesions on delayed discounting reflect two different patterns of activity in OFC; one that bridges the gap between a response and an outcome and another that discounts delayed reward. These signals appear to reflect the spatial location of the reward and/or the action taken to obtain it, and are encoded independently from representations of absolute value. We suggest a dual role for output from OFC in both discounting delayed reward, while at the same time supporting new learning for them.

KEYWORDS: reward; orbitofrontal cortex; delay; time discounting; value

INTRODUCTION

Should I stay or should I go? Every year, at the annual Society For Neuroscience (SfN) meeting, we ponder this question while waiting in an endless line

Address for correspondence: Matthew Roesch, Department of Anatomy and Neurobiology, University of Maryland School of Medicine, 20 Penn Street HSF-2, Rm S251 Baltimore, MD 21201. Voice: 410-706-8910; fax: 410-706-2512.
mroes001@umaryland.edu

Ann. N.Y. Acad. Sci. 1104: 21–34 (2007). © 2007 New York Academy of Sciences.
doi: 10.1196/annals.1390.001

for a cup of coffee. Late to the conference and in desperate need of caffeine, we wonder: is it really worth the wait (FIG. 1A)?

Starbucks has a lot riding on the answer to this question; but beyond that, it also addresses a fundamental issue underlying how neural systems value different rewards that differ in how quickly they can be obtained. In the lab, the neural mechanisms underlying this aspect of decision making are studied in tasks that ask animals to choose between a small reward delivered immediately and large reward delivered after some delay. As the delay to the large reward becomes longer subjects usually discount the value of the large reward, biasing their choices toward the small, immediate reward. Interestingly, lesions of the orbitofrontal cortex (OFC) alter the breakpoint at which animals abandon the large delayed reward for the more immediate reward, effectively influencing whether they "stay" or "go."

In this article, we review these studies in the light of recent data collected from our lab examining neural correlates of time discounting in OFC. We will argue that the varied effects of OFC lesions, which sometimes increase and other times decrease this breakpoint, reflect two patterns of neural activity in OFC, one that maintains representations of reward across a delay and the other that discounts delayed rewards. We will show that these representations of the discounted reward are maintained independently of representations of absolute reward value. This is consistent with the finding that lesions to OFC disrupt delay discounting but often do not affect preference between differently sized rewards. The independent encoding of different aspects of reward value contradicts recent hypotheses that OFC neurons signal the value of outcomes in a kind of "common currency." Finally, we will discuss findings that suggest that OFC might represent the spatial location and/or the action associated with delayed reward.

What Does Neural Activity Tell Us about the Role of OFC in Time Discounting?

Several studies have reported abnormal behavior in OFC-lesioned rats forced to choose between small, immediate and larger, delayed rewards. Some studies report that OFC lesions make animals more impulsive, that is, less likely to wait for a delayed reward.[1–3] These results suggest that OFC is critical for responding to rewards when they are delayed. Other studies report that OFC lesions make animals less impulsive; that is, more likely to wait for the delayed reward,[3,4] suggesting that OFC is critical for discounting or devaluing the delayed reward. We have recently found two different patterns of neural activity in OFC, which appear to map on to these roles.

One pattern is evident in neurons that fire in anticipation of the delayed reward; activity in these neurons is similar to outcome-expectant activity seen in other settings.[5–13] In our study, we trained animals to respond to one of two wells located under a central odor port (FIG. 1B–D). We then manipulated how long the animal had to wait to receive reward after responding. FIGURE 2 shows neuronal activity when the reward was delayed from 1–4 sec (gray). FIGURE 2A shows an OFC neuron that fires after the response into the fluid well and during the anticipation of the delayed reward. FIGURE 2B illustrates population activity of 27 OFC neurons sharing the same characteristic. When rewards were delivered after a short delay, activity rose after the response but quickly declined after reward delivery (black). However, when reward was delayed by 3 sec (gray) activity continued to rise until the delayed reward was delivered, resulting in higher levels of activity for rewards that were delayed compared to those that were delivered immediately.

These signals appear to maintain a representation of the imminent delivery of reward during the delay. Such a representation would facilitate the formation of associative representations in other brain regions. For example, we have recently shown that input from OFC is important for rapid changes in cue-outcome encoding in basolateral amygdala.[14] This deficit may be due to the loss of these expectancy signals, normally generated in OFC.[7] Similarly, the effects of pretraining OFC lesions on delayed discounting may reflect the absence of these expectancies when associations with immediate or delayed rewards must be learned. Rats lacking this signal during training would encode associations with the large, delayed reward more weakly than associations with the small, immediate reward. As a result, rats with pretraining lesions would exhibit apparently "impulsive" responding for the more strongly encoded small, immediate reward. This interpretation is consistent with reports that lesions of OFC before training cause impulsive responding.

Interestingly, experience with the large, delayed reward can reduce impulsive choice in OFC-lesioned rats.[2] In this experiment, a large, delayed reward was initially pitted against a small, immediate reward. OFC-lesioned rats were more impulsive than controls. However, after a period of training in which both rewards were experienced at equal delays, the OFC-lesioned rats were able to perform normally when they were returned to a setting in which the small reward was no longer delayed. These results are consistent with the idea that OFC facilitates initial learning and that lesions can make rats more impulsive due to the loss of expectancy signals. Recovery of function under these circumstances may reflect the formation of more normal strength associations with the large reward, due to the additional experience in the symmetrically delayed variant of the task. Alternatively, associations with the small reward may be weakened in OFC-lesioned animals when the delay is imposed, for the same reason that associations with the large reward are weakened during initial training. These hypotheses might be distinguished by examining the strength of downstream encoding in areas such as basolateral amygdala and nucleus accumbens in lesioned rats.

However, OFC lesions can also result in lower levels of impulsivity, as if signals involved in discounting the delayed reward are lost.[4] Interestingly,

(A)

(B)

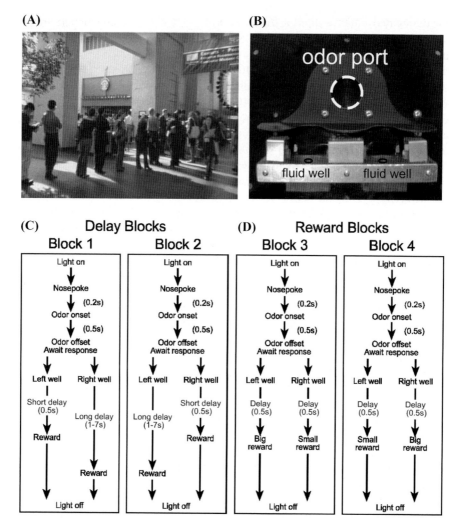

(C) Delay Blocks   **(D)** Reward Blocks

FIGURE 1. (**A**) Photograph taken at the annual neuroscience meeting in 2005 illustrates the all too familiar situation of waiting in an endless line for coffee. Late to the conference and in desperate need of caffeine you have to decide, do I go left and wait for coffee, or go right and see that last poster on my itinerary. (**B**) Apparatus used in our lab to study this type of decision making. (**C–D**). Choice task during which we varied (**C**) the delay preceding reward delivery and (**D**) the size of reward. Figure shows sequence of events in each trial in four blocks in which we manipulated the time to reward or the size of reward. Trials were signaled by illumination of the panel lights inside the box. When these lights were on, nosepoke into the odor port resulted in delivery of the odor cue to a small hemicylinder located behind this opening. One of three different odors was delivered to the port on each trial, in a pseudorandom order. At odor offset, the rat had 3 sec to make a response at one of the two fluid wells located below the port. One odor instructed the rat to go to the left to get reward, a second odor instructed the rat to go to the right to get

the majority of OFC neurons recorded in well-trained rats in our study seemed to perform this function. Unlike the example in FIGURE 2A, activity in these neurons did not bridge the gap between the response and reward delivery, but instead declined as the delay to reward increased (FIG.2C). FIGURE 2D represents the average firing rate of 65 neurons that showed this characteristic. Under short delay conditions, these neurons fired in anticipation of and during delivery of reward (black), however, when the reward was delayed (gray), activity declined until reward was delivered. Remarkably, this activity was correlated with a decreased tendency of the rat to choose the long delay on future free choice trials (FIG. 3; chi-square, $P < 0.05$). Thus, activity in this population biased decision making toward immediate gratification.

The effects of posttraining OFC lesions on delayed discounting may reflect the absence of these discounting signals, since posttraining lesions would not affect formation of the associative representations, but would cause the rats to be unable to discount the value of the large reward during actual task performance. As a result, rats with posttraining lesions would be less impulsive.[4] This interpretation is consistent with reports that lesions of OFC after training cause lower levels of impulsive response. Of course, it should be noted that one study has reported that posttraining lesions can induce elevated levels of impulsivity.[2] The amount of training in this study, though substantial, was still far less than that reported in Winstanley *et al.*[4] suggesting the rats may have not formed strong associations prior to surgery.

In summary, the effects of OFC lesions on delayed discounting may reflect the nature of these two signals and their role at different stages during learning. Rats lesioned after being fully trained have normal associative representations but are simply unable to discount the value of the large reward due to the loss of OFC. This reflects the loss of the discounting signal illustrated in FIGURE 2 C–D, which was the predominant signal in our well-trained rats. In contrast, rats lesioned before any training may have an associative learning deficit that renders meaningless the loss of any delayed discounting function in OFC; these rats fail to normally encode the reward associations due to the loss of expectancy signals during learning. As a result, they exhibit apparently "impulsive" responding, selecting the small reward lever even at very short delays for the simple reason that the associative representations of value for this response are better encoded.

---

reward, and a third odor indicated that the rat could obtain reward at either well. (**C**) One well was randomly designated as short and the other long at the start of the session (block 1). In the second block of trials these contingencies were switched (block 2). (**D**) In later blocks we held the delay preceding reward delivery constant while manipulating the size of the expected reward (adapted from Roesch *et al.*[33]).

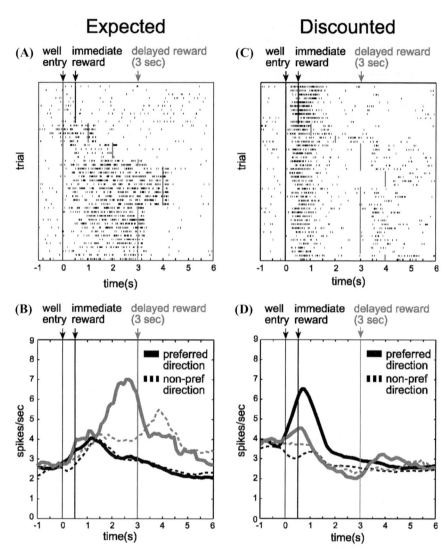

FIGURE 2. (**A**) Single cell example of an expectancy neuron. Activity is plotted for the last 10 trials in a block in which reward was delivered in the cell's preferred direction after 500 msec (black) followed by trials in which the reward was delayed by 1 to 4 sec (gray). Each row represents a single trial, each tick mark represents a single action potential and the black/gray lines indicate when the reward was delivered. (**B**) Population histogram ($n = 27$) representing firing rate as a function of time during the trial for neurons that fired in anticipation of a reward delayed by 3 sec. Activity is aligned on well entry. Preferred direction refers to the spatial location that elicited the stronger response for each neuron. Black: short. Gray: long. Solid: preferred direction. Dashed: nonpreferred direction. (**C**) Single cell example of a neuron that discounts delayed rewards (gray), but fires strongly for rewards delivered immediately (black). (**D**) Population histogram representing firing rate as a function of time

Do Neurons in OFC Signal Time-Discounted Rewards in a Common "Value" Currency?

The results described above indicate that OFC is critical for discounting the value of the delayed reward after learning.[4] This is conceptually similar to the role OFC plays for devaluation in other settings.[15–18] The proposal that OFC is performing the same function in discounting and devaluation is consistent with the idea that output from OFC provides a context-free representation of value. This hypothesis is supported by single-unit recording work[6,8,10,11,13,19–23] and functional imaging studies,[24–29] which show that activity in OFC seems to encode the value of different goals or outcomes in a common currency. For example, Tremblay *et al.*[8] showed that OFC neurons fire selectively after responding in anticipation of the different rewards and that this selective activity is influenced by the monkey's reward preference. Activity in anticipation of a particular reward differed according to whether the monkey valued it more or less than the other reward available within the current block of trials. It has been proposed that this signal integrates available information that impacts this judgment, providing a context-free representation of a thing's value.[25,26]

If this hypothesis is correct, then neural activity that encodes the delay to reward should also be influenced by changes in reward magnitude, either at a single-unit or population level. Such covariance has been reported when delay and reward size are manipulated at the same time.[23,30] However, in the study described above, we found that when delay and reward size were manipulated across different blocks of trials, analogous to the manipulations of reward preference made by Schultz and colleagues,[8] OFC neurons maintained dissociable representations of the value of delayed and differently sized rewards. Thus neurons that fired more (or less) for immediate reward did not fire more (or less) for a larger reward when the delay was held constant. This is illustrated for the 65 neurons that fired more for rewards delivered immediately (FIG. 4C), as well as for those ($n = 27$) that fired more strongly in anticipation of the delayed reward (FIG. 4A). Neither population showed any relation to reward size (FIG. 4B, D).

As a result, expectancy signals in rat OFC in our study did not track relative reward preference. This is evident in the activity elicited by the small immediate reward in the trial blocks that differed by delay versus magnitude of reward. Neurons encoding relative value in a common currency should have responded less for the small immediate reward when it was pitted against the

---

during the trial for neurons that fired more strongly after short delays during the reward epoch ($n = 65$). Conventions are the same as in A and B. "Preferred direction" refers to the spatial location of the well for which higher firing rate was elicited when other variables were held constant (i.e., reward size and delay length). Population activity is indexed to each neuron's preferred direction to average across cells (adapted from Roesch *et al.*[33]).

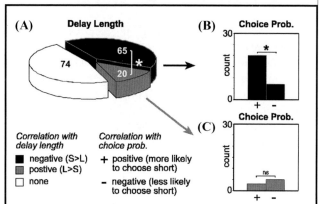

FIGURE 3. Dependency of firing rate during the reward epoch on delay length and future choice probability as revealed by multiple regression analysis. (**A**) Black and gray represent cases in which the correlation between firing rate and delay length was negative (stronger firing for short) or positive (stronger firing for long), respectively. (**B, C**). Each bar represents the number of neurons in which the correlation between firing rate and future choice probability was positive (more likely to choose direction associated with short) or negative (less likely to choose direction associated with short) for those cells that also showed (**B**) a negative or (**C**) a positive correlation with delay length. $*P < 0.05$; chi-square. (adapted from Roesch et al.[33]).

large reward at the same delay (i.e., when it was nonpreferred), but more for this same small reward when it was pitted against a delayed reward of equal size (i.e., when it was preferred). Yet the relative value of the small reward was not reflected in the activity of either population (FIG. 2, black vs. FIG. 4, gray; t-test, $P > 0.7380$) or in the counts of single neurons (chi-square; $P > 0.31$).

　　The fact that we were able to dissociate the effects of reward and delay on single-unit activity in OFC indicates that encoding of these different types of value information may involve different neural processes. This dissociation is perhaps not surprising considering recent behavioral data that support the view that learning about sensory and temporal features of stimuli involves different underlying systems (see Delamater's article, this volume). In addition, many mathematical models of value typically treat size and delay as separate variables in their equations.[3] However, our data indicate that, despite several reports to the contrary,[8,22,23] OFC neurons do not always provide a generic value signal. Our ability to detect this difference may reflect a species difference; however, a more interesting explanation is that the difference may reflect the amount of training typically required in primate studies, which is usually much greater than the training given to a rat. It is possible that with

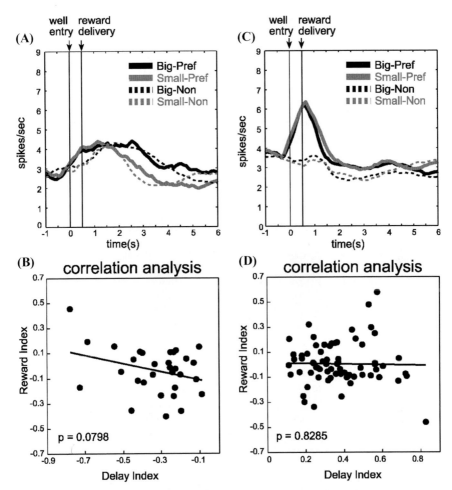

FIGURE 4. (**A**) Population histogram of same 27 neurons (shown in FIG. 2B) during trials when delay was held constant but reward size varied. Black: big. Gray: small. Solid: preferred direction. Dashed: nonpreferred direction. (**B**) Relation of firing dependent on delay length to firing dependent on reward size for those neurons that fired more strongly after long delays (shown in FIG. 2B). The delay index and reward index are computed on the basis of firing during the reward epoch. Delay index $= (S - L) / (S + L)$ where S and L represent firing rates on short- and long-delay trials, respectively. Reward index $= (B - S) / (B + S)$ where B and S represent firing rates on big- and small-reward trials, respectively. (**C**) Population histogram of same 65 neurons (shown in FIG. 2D) during trials when delay was held constant but reward size varied. Black: big. Gray: small. Solid: preferred direction. Dashed: nonpreferred direction. (**D**) Relation of firing dependent on delay length to firing dependent on reward size for those neurons that fired more strongly after short delays (shown in FIG. 2D). Conventions are the same as in A and B. (adapted from Roesch et al.[33]).

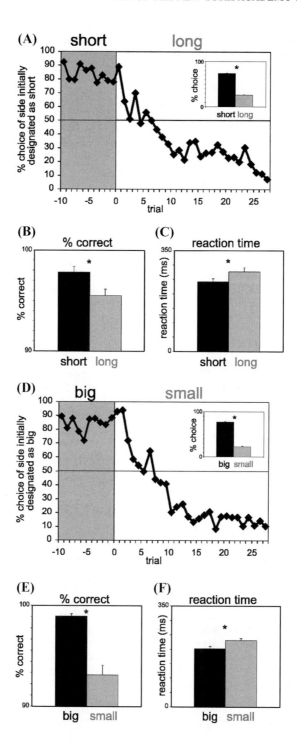

extended training, OFC neurons become optimized to provide these generic value representations. This would have interesting implications as it would suggest that OFC might do a good job integrating commonly encountered variables into a common currency and would do less well integrating variables that are unique or rarely encountered into these calculations.

Interestingly, on a population level, there was relatively little impact of reward size on activity in OFC, even though the rats responded similarly to size and delay manipulations (FIG. 5). Thus, OFC may have a particularly fundamental role in discounting delayed rewards that is not necessary for encoding the absolute value of a reward. This is supported by the finding that lesions to corticolimbic structures disrupt delay discounting but typically do not alter size preference.[1,4,31]

Do Neurons in OFC Signal Time-Discounted Rewards Dependent on the Response Required to Obtain Them?

So how is OFC representing responses that lead to immediate versus delayed rewards? Until recently, the involvement of OFC in encoding the action taken to receive reward has been largely neglected. This is in large part due to the finding that task-related activity in primate OFC is generally not dependent on the direction of the motor response, suggesting that OFC encodes the value of rewards independent of the actions required to obtain them. However, in these studies, response direction is typically not a predictor of reward; instead monkeys are generally highly trained to associated reward qualities with the visual properties of conditioned stimuli.[5,8–11,13,22,23]

In contrast, recent rodent work has explicitly paired reward with direction. These studies have found OFC neurons to be directionally selective, suggesting that OFC is involved in monitoring either the spatial goal or the action taken to achieve that goal.[32–34] For example, in a study by Feierstein *et al.*,[32] rats made responses to either a left or right well based on odor-direction contingencies. Remarkably, nearly half of the neurons recorded in OFC were directionally selective, firing more for one direction but not the other across multiple task epochs. They also found that neurons sensitive to the outcome (rewarded or not rewarded) were modulated by the side the animal had gone to, integrating

---

FIGURE 5. Impact of delay length (**A–C**) and reward size (**D, E**) on behavior. (**A**) Average choice rate, collapsed across direction, for all sessions for trials before and after the switch from short to long. Inset: The height of each bar indicates the percentage choice of short delay and long delay taken over all choice trials. (**B, C**) The height of each bar indicates the percentage correct (**B**) and reaction time (**C**) across all recording sessions in all rats on short-delay (black) and long-delay (gray) forced-choice trials. (**D, E**) Impact of reward size on the same behavior measures described in (**A–C**). Asterisks: *t*-test, $P <$ 0.05. Error bars: standard errors. (adapted from Roesch *et al.*[33]).

directional and outcome information that may be used to guide future decisions. Similarly, in our discounting study, we found that discounting signals were directionally specific (FIG. 2; solid vs. dashed). Moreover, this directional signal did indeed impact choice behavior (FIG. 3).

The fact that signals in OFC are directional is important because most discounting tasks typically increase delays for only one response. Indeed, OFC does not appear to be necessary when delays are increased for both responses at the same time.[2] The presence of correlates reflecting the value of different directional responses in OFC is interesting in light of recent data from Balleine and colleagues showing that OFC lesions do not affect changes in instrumental responding after devaluation.[35] Such behavioral changes are presumed to require action–outcome associations. A lack of effect of OFC lesions on instrumental devaluation suggests that directional correlates, such as those demonstrated by us or the Mainen group,[32] are either not critical to behavior or are not encoding action–outcome contingencies. One alternative is that they may encode associations between sensory features of the action and the outcome. In the case of our report, such features might include the spatial position of the well, for example. This is supported by connections with areas that carry spatial information[36] and by a number of lesion studies that implicate OFC in spatial tasks.[37,38] Future studies that minimize these unique sensory aspects of the response would help elucidate the nature of this representation and its role in impulsive behavior.

CONCLUSION

We conclude that the varied effects of OFC lesions on delayed discounting reflect two different patterns of activity in OFC; one that bridges the gap between a response and an outcome and another that discounts delayed reward. These signals appear to reflect the spatial location of the reward and/or the action taken to obtain it, and are encoded independently from representations of absolute value. This suggests a dual role for output from OFC in both discounting delayed reward, while at the same time supporting new learning for them. Output from OFC may impact reward representations in downstream areas shown to be involved in time discounting, such as basolateral amygdala and nucleus accumbens. Of course these predictions remain to be validated; the impact of delay on neuronal activity in these areas is unknown and remains critical to our understanding of how we decide to "stay" or "go" when rewards are delayed.

REFERENCES

1. MOBINI, S. *et al.* 2002. Effects of lesions of the orbitofrontal cortex on sensitivity to delayed and probabilistic reinforcement. Psychopharmacology (Berl.) **160:** 290–298.

2. RUDEBECK, P.H., M.E. WALTON, A.N. SMYTH, *et al.* 2006. Separate neural pathways process different decision costs. Nat. Neurosci. **9:** 1161–1168.

3. KHERAMIN, S. *et al.* 2002. Effects of quinolinic acid-induced lesions of the orbital prefrontal cortex on inter-temporal choice: a quantitative analysis. Psychopharmacology (Berl.) **165:** 9–17.

4. WINSTANLEY, C.A., D.E. THEOBALD, R.N. CARDINAL & T.W. ROBBINS. 2004. Contrasting roles of basolateral amygdala and orbitofrontal cortex in impulsive choice. J. Neurosci. **24:** 4718–4722.

5. ROLLS, E.T. 2000. The orbitofrontal cortex and reward. Cereb. Cortex **10:** 284–294.

6. SCHOENBAUM, G., A.A. CHIBA & M. GALLAGHER. 1998. Orbitofrontal cortex and basolateral amygdala encode expected outcomes during learning. Nat. Neurosci. **1:** 155–159.

7. SCHOENBAUM, G. & M. ROESCH. 2005. Orbitofrontal cortex, associative learning, and expectancies. Neuron **47:** 633–666.

8. TREMBLAY, L. & W. SCHULTZ. 1999. Relative reward preference in primate orbitofrontal cortex. Nature **398:** 704–708.

9. TREMBLAY, L. & W. SCHULTZ. 2000. Reward-related neuronal activity during go-nogo task performance in primate orbitofrontal cortex. J. Neurophysiol. **83:** 1864–1876.

10. WALLIS, J.D. & E.K. MILLER. 2003. Neuronal activity in primate dorsolateral and orbital prefrontal cortex during performance of a reward preference task. Eur. J. Neurosci. **18:** 2069–2081.

11. ROESCH, M.R. & C.R. OLSON. 2004. Neuronal activity related to reward value and motivation in primate frontal cortex. Science **304:** 307–310.

12. HIKOSAKA, K. & M. WATANABE. 2004. Long- and short-range reward expectancy in the primate orbitofrontal cortex. Eur. J. Neurosci. **19:** 1046–1054.

13. HIKOSAKA, K. & M. WATANABE. 2000. Delay activity of orbital and lateral prefrontal neurons of the monkey varying with different rewards. Cereb. Cortex **10:** 263–271.

14. SADDORIS, M.P., M. GALLAGHER & G. SCHOENBAUM. 2005. Rapid associative encoding in basolateral amygdala depends on connections with orbitofrontal cortex. Neuron **46:** 321–331.

15. SCHOENBAUM, G. & B. SETLOW. 2005. Cocaine makes actions insensitive to outcomes but not extinction: implications for altered orbitofrontal-amygdalar function. Cereb. Cortex **15:** 1162–1169.

16. PICKENS, C.L. *et al.* 2003. Different roles for orbitofrontal cortex and basolateral amygdala in a reinforcer devaluation task. J. Neurosci. **23:** 11078–11084.

17. IZQUIERDO, A.D., R.K. SUDA & E.A. MURRAY. 2004. Bilateral orbital prefrontal cortex lesions in rhesus monkeys disrupt choices guided by both reward value and reward contingency. J. Neurosci. **24:** 7540–7548.

18. BAXTER, M.G., A. PARKER, C.C. LINDNER, *et al.* 2000. Control of response selection by reinforcer value requires interaction of amygdala and orbital prefrontal cortex. J. Neurosci. **20:** 4311–4319.

19. ROLLS, E.T. 1996. The orbitofrontal cortex. Philos. Trans. R. Soc. Lond. B. Biol. Sci. **351:** 1433–1443; discussion 1443–4.

20. CRITCHLEY, H.D. & E.T. ROLLS. 1996. Olfactory neuronal responses in the primate orbitofrontal cortex: analysis in an olfactory discrimination task. J. Neurophysiol. **75:** 1659–1672.

21. SCHOENBAUM, G., A.A. CHIBA & M. GALLAGHER. 1999. Neural encoding in orbitofrontal cortex and basolateral amygdala during olfactory discrimination learning. J. Neurosci. **19:** 1876–1884.

22. PADOA-SCHIOPPA, C. & J.A. ASSAD. 2006. Neurons in the orbitofrontal cortex encode economic value. Nature **441:** 223–226.
23. ROESCH, M.R. & C.R. OLSON. 2005. Neuronal activity in primate orbitofrontal cortex reflects the value of time. J. Neurophysiol. **94:** 2457–2471.
24. GOTTFRIED, J.A., O'J. DOHERTY & R.J. DOLAN. 2003. Encoding predictive reward value in human amygdala and orbitofrontal cortex. Science **301:** 1104–1107.
25. KRINGELBACH, M.L. 2005. The human orbitofrontal cortex: linking reward to hedonic experience. Nat. Rev. Neurosci. **6:** 691–702.
26. MONTAGUE, P.R. & G.S. BERNS. 2002. Neural economics and the biological substrates of valuation. Neuron **36:** 265–284.
27. O'Doherty J., H. CRITCHLEY, R. DEICHMANN & R.J. DOLAN. 2003. Dissociating valence of outcome from behavioral control in human orbital and ventral prefrontal cortices. J. Neurosci. **23:** 7931–7939.
28. O'Doherty J., M.L. KRINGELBACH, E.T. ROLLS, et al. 2001. Abstract reward and punishment representations in the human orbitofrontal cortex. Nat. Neurosci. **4:** 95–102.
29. ARANA, F.S. et al. 2003. Dissociable contributions of the human amygdala and orbitofrontal cortex to incentive motivation and goal selection. J. Neurosci. **23:** 9632–9638.
30. KALENSCHER, T. et al. 2005. Single units in the pigeon brain integrate reward amount and time-to-reward in an impulsive choice task. Curr. Biol. **15:** 594–602.
31. CARDINAL, R.N., C.A. WINSTANLEY, T.W. ROBBINS & B.J. EVERITT. 2004. Limbic corticostriatal systems and delayed reinforcement. Ann. N. Y. Acad. Sci. **1021:** 33–50.
32. FEIERSTEIN, C.E., M.C. QUIRK, N. UCHIDA, et al. 2006. Representation of spatial goals in rat orbitofrontal cortex. Neuron **51:** 495–507.
33. ROESCH, M.R., A.R. TAYLOR & G. SCHOENBAUM. 2006. Encoding of time-discounted rewards in orbitofrontal cortex is independent of value representation. Neuron **51:** 509–520.
34. LIPTON, P.A., P. ALVAREZ & H. EICHENBAUM. 1999. Crossmodal associative memory representations in rodent orbitofrontal cortex. Neuron **22:** 349–359.
35. OSTLUND, S.B. & B.W. BALLEINE. 2005. Lesions of the orbitofrontal cortex disrupt pavlovian, but not instrumental, outcome-encoding. Soc. Neurosci. Abstracts **71.2**
36. REEP, R.L., J.V. CORWIN & V. KING. 1996. Neuronal connections of orbital cortex in rats: topography of cortical and thalamic afferents. Exp. Brain Res. **111:** 215–232.
37. CORWIN, J.V., M. FUSSINGER, R.C. MEYER, et al. 1994. Bilateral destruction of the ventrolateral orbital cortex produces allocentric but not egocentric spatial deficits in rats. Behav. Brain Res. **61:** 79–86.
38. VAFAEI, A.A. & A. RASHIDY-POUR. 2004. Reversible lesion of the rat's orbitofrontal cortex interferes with hippocampus-dependent spatial memory. Behav. Brain Res. **149:** 61–68.

Model-Based fMRI and Its Application to Reward Learning and Decision Making

JOHN P. O'DOHERTY,[a,b] ALAN HAMPTON,[a] AND HACKJIN KIM[b]

[a]Computation and Neural Systems Program, California Institute of Technology, Pasadena, California, USA

[b]Division of Humanities and Social Sciences, California Institute of Technology, Pasadena, California, USA

ABSTRACT: In model-based functional magnetic resonance imaging (fMRI), signals derived from a computational model for a specific cognitive process are correlated against fMRI data from subjects performing a relevant task to determine brain regions showing a response profile consistent with that model. A key advantage of this technique over more conventional neuroimaging approaches is that model-based fMRI can provide insights into how a particular cognitive process is implemented in a specific brain area as opposed to merely identifying where a particular process is located. This review will briefly summarize the approach of model-based fMRI, with reference to the field of reward learning and decision making, where computational models have been used to probe the neural mechanisms underlying learning of reward associations, modifying action choice to obtain reward, as well as in encoding expected value signals that reflect the abstract structure of a decision problem. Finally, some of the limitations of this approach will be discussed.

KEYWORDS: computational models; neuroimaging; prediction error; expected value; conditioning; striatum; ventromedial prefrontal cortex

INTRODUCTION

In this review, we will highlight a recent development in neuroimaging research, the "model-based" approach, which involves the application of computational models in the design and analysis of neuroimaging experiments, particularly those involving functional magnetic resonance imaging (fMRI). We will outline the advantages of this method over more traditional neuroimaging approaches as a means of advancing psychological or computational theories

Address for correspondence: John P. O'Doherty, California Institute of Technology, M/C 228-77, 1200 E. California Boulevard, Pasadena, CA 91125, USA. Voice: +1-626-395-5981; fax: +1-626-793-8580.

jdoherty@hss.caltech.edu

Ann. N.Y. Acad. Sci. 1104: 35–53 (2007). © 2007 New York Academy of Sciences.
doi: 10.1196/annals.1390.022

of brain function. It will be argued that the model-based approach provides a means of using neuroimaging data to discriminate between existing theories of brain function as well as to advance new novel theories, in a way that could not be achieved with more conventional neuroimaging techniques or in traditional behavioral studies. This argument will be supported with reference to the field of reward learning and decision making, where the model-based approach has up to now been most profitably employed.

MODEL-BASED fMRI

The central approach behind model-based imaging is that one needs to start out with a quantitative computational model that describes a mapping or transformation between a set of stimulus inputs, and a set of behavioral responses. The specific "internal" operations required to effect such a transformation are the variables of interest in the neuroimaging study, as it is these variables that will ultimately be correlated with the neuroimaging data. Perhaps one of the simplest examples of quantitative models in psychology, are those developed to account for the acquisition of conditioned responses during Pavlovian conditioning, the most widely cited example being the Rescorla–Wagner (RW) model.[1] In the RW model, the process by which a conditioned stimulus (CS) comes to produce a conditioned response, is represented by two scalar variables: v, which is the strength of the conditioned response elicited, or more abstractly the value of the CS, and u, which is the value of the unconditioned stimulus (UCS). For conditioning to occur, through repeated contingent presentations of the CS and US, the variable v (which may be initially zero) should converge toward the value of u. At the core of the RW model is that this convergence is accomplished by means of a prediction error δ, which is the difference between the current value of u and v, on each conditioning trial. The value of v is then updated in proportion to δ, so that over the course of learning, v eventually converges to u, δ tends to zero, and learning is complete. Here, in this simple model there are three variables of interest, v, δ, and u. In traditional behavioral psychology, the only directly observable variable here is v, because it can be measured by assessing the strength of the conditioned response in a human or an animal undergoing conditioning. The other variables, such as δ, while not directly observable can be inferred indirectly from the behavioral data, by for example, analyzing the behavioral learning curves.[2] However, the ability to directly measure such variables as they are implemented in the brain, is arguably a much more compelling approach in establishing the validity of such a model. In other words, the neural data become a rich source of evidence, which in addition to the behavior, can be used to constrain the computational models.[3]

Once one moves beyond simple models, such as the RW model, to models accounting for more complex cognitive or behavioral phenomena with many internal variables, then the ability to validate such models on the basis of be-havioral observation alone becomes an ever more difficult proposition as the more complex the model (and hence the more associated free parameters), the more unconstrained the behavioral fitting becomes. In this event, the use of neurophysiology or neuroimaging techniques to measure such internal vari-ables and thus impose additional constraints, becomes even more critical. In the latter part of this review, we will give a specific example of how neuroimag-ing data can be used to impose just such constraints and permit discrimination between competing models.

A RECIPE FOR MODEL-BASED fMRI

The standard approach in model-based MRI is to first fit the computational model to subjects' actual behavior to find specific values for the free parameters in the model, which minimize the difference between the model predictions and the behavioral data. In the case of the RW model example outlined previously, this would involve finding a value for the one free parameter in the simplest variant of this model, the learning rate α, which is used to scale the updates to v in proportion to the prediction error δ on each trial t: ($v_t = v_{t-1} + \alpha\delta$), such that the trial by trial values for v are as close as possible to the actual behavioral conditioned responses elicited in a particular subject or group of subjects. Once the best-fitting model parameters have been found, then the different model components can be regressed against the fMRI data. For the RW model, there are two variables that change on a trial by trial basis as a function of learning: v and δ. Both of these variables can be entered into the fMRI analysis as a time series and convolved with a canonical hemodynamic response function or other basis function(s), to capture the effects of hemodynamic lag, as is standard in many fMRI analysis approaches. These signals are then regressed against the fMRI data using a general linear model,[4] to identify areas where the model-predicted time series show significant correlations with the actual changes in blood oxygenation level-dependent (BOLD) signal over time. While the RW model only accounts for signal changes on a trial by trial basis, model time series predictions can be much more fine-grained, capturing dynamic changes in activity within as well as between trials, on a second by second basis. A schematic of the model-based fMRI procedure is provided in FIGURE 1.

MODEL VALIDATION AND COMPARISON

Establishing correlations between a given model and a time series of fMRI activity in a given region does not of course demonstrate conclusively that this region is implementing that specific model. It is possible that some aspects

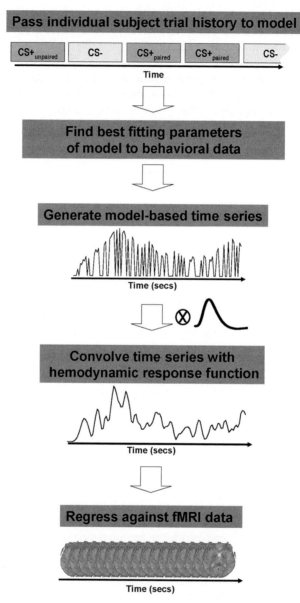

FIGURE 1. Illustration of model-based fMRI approach. Each individual subject's trial history is passed to the model, and the parameters of the model are fit so as to minimize the difference between the model predictions and an external behavioral measure, which in the conditioning example could be an external measure of conditioning, such as galvanic skin conductance responses or pupil dilation. Next, the best model-fitting parameters are used to generate a time series for each trial in the fMRI, which are then convolved with basis function(s) to account for the effects of hemodynamic lag, such as the canonical hemodynamic response function, and then regressed against the fMRI data.

of the time series data correlate well with the model while other parts of the data do not, yet a significant correlation still prevails when the model is fitted to the fMRI data overall. To address this possibility it is useful to plot the model predictions against the observed data to evaluate whether there are any systematic deviations of the model against the data. If evidence is found for systematic deviations of the data from the model predictions (i.e., structure in the residuals), this could indicate that the computational model needs to be modified to account better for the neural activity in a specific area.

A more principled approach is to compare the explanatory power of different competing models against both the behavioral and fMRI data. Usually this is done first for the behavioral data, such that the fit of a variety of models may be compared against the behavioral data to select the subset of models that provide the best account of behavior. An important consideration when performing model comparison is that in the trivial case, models with more free parameters will tend to provide a better fit than models with fewer parameters, thus the addition of extra free parameters needs to be penalized appropriately during the fitting procedure to take this into account.[5] However, in the event that two or more competing models might provide qualitatively similar predictions about behavior, such that there is no way to use the behavioral data alone to discriminate them, it is necessary to turn to the fMRI data to impose additional constraints. In this event such models can be entered into a regression analysis against the fMRI data and allowed to compete against each to determine which model provides a better fit to the fMRI data, for a given brain region.

INFERENCE IN MODEL-BASED fMRI

Model-based fMRI exploits the spatial and temporal resolution of event-related fMRI to characterize dynamic changes in neural activity over time in terms of a particular computational process. Moreover, this technique can be used as a means of discriminating between competing computational models of cognitive and neural function. Thus, model-based fMRI provides insight into "how" a particular cognitive function might be implemented in the brain, not only "where" it is implemented. The model-based approach described here can be contrasted with the approach taken in many conventional neuroimaging studies in which a set of brain areas are merely reported to be "activated" in a particular task or condition, which contribute knowledge about spatial localization of specific cognitive processes, but arguably provide little or no insight into how these processes are being implemented in the activated areas.

MODEL-BASED fMRI OF REWARD LEARNING AND DECISION MAKING

Model-based fMRI will now be illustrated with reference to the problem of how humans can learn to make adaptive decisions under uncertainty to maximize rewards. Decision making in this context can usefully be fractionated into two distinct components, a reward prediction component in which the expected future reward associated with a particular set of stimuli or actions is learned through experience, and an action selection component in which over the course of learning an agent learns to bias its action choice to favor those action(s) that yield the greatest reward.

LEARNING REWARD PREDICTIONS

Reward prediction in its simplest form can be studied through the phenomenon of classical or Pavlovian conditioning, whereby an arbitrary initially affectively neutral or CS takes on reward value through repeated contingent presentation with an appetitive UCS, such as a food reward. Error correcting learning rules, such as the RW model described earlier, have proved to be successful in accounting for much though by no means all behavioral phenomena in classical conditioning. The first evidence for reward prediction error signals in the brain emerged from the work of Wolfram Schultz and colleagues who observed such signals by recording from the phasic activity of dopamine neurons in awake, behaving, nonhuman primates undergoing simple instrumental or classical conditioning tasks.[6-8] The response profile of these neurons does not correspond to a simple RW rule but rather a real-time extension of this rule called temporal difference learning in which predictions of future reward are computed at each discrete time interval t within a trial, such that the error signal is generated by computing the difference in successive predictions.[9,10] This specific model provides a good approximation to the temporal profile of activity of these neurons during classical conditioning in which the dopamine neurons first respond at the time of the UCS before learning is established but shift back in time within a trial to respond instead at the time of presentation of the cue once learning is established. To test for evidence of a temporal difference prediction error signal in the human brain, O'Doherty and colleagues[11] scanned human subjects while they underwent a classical conditioning paradigm in which associations were learned between arbitrary visual fractal stimuli and a pleasant sweet taste reward (glucose). One cue was followed most of the time by the taste reward, whereas another cue was followed most of the time by no reward. However, in addition, subjects were exposed to low frequency "error" trials in which the cue associated with reward was presented but the reward was omitted, and the cue associated with no reward was presented but a reward was unexpectedly delivered. The specific trial history that each subject experienced was next fed into a temporal difference model to generate a time series that specified the model-predicted prediction error signal at three different time points in a trial from the time at which the CS is presented until the time at which the reward is delivered (3 sec later) (FIG. 2A, B).

This time series was then convolved with a canonical hemodynamic response function and regressed against the fMRI data for each individual subject, to identify brain regions correlating with the model-predicted time series. This analysis revealed significant correlations with the model-based predictions in a number of brain regions, most notably the ventral striatum (ventral putamen bilaterally) (FIG. 2C) and orbitofrontal cortex, both prominent target regions of dopamine neurons.[12] These results suggest that prediction error signals are present in the human brain during reward learning, and that these signals conform to a response profile consistent with a specific computational model: temporal difference learning. Another study by McClure and colleagues also revealed activity in ventral striatum consistent with a reward prediction error signal using an event-related trial-based analysis.[13]

ACTION SELECTION FOR REWARD: THE ACTOR/CRITIC IN THE HUMAN BRAIN

While classical conditioning is a useful paradigm for studying the passive learning of reward predictions to gain insight into the process by which humans can learn to perform actions to obtain reward, it is necessary to turn to instrumental conditioning, which involves learning of stimulus-response or stimulus-response-outcome associations. Insight into how humans or other animals might implement instrumental conditioning has come from the application of a family of models collectively known as reinforcement learning, originally developed in computer science.[14] In one such model, the actor/critic,[15,16] action selection is conceived as involving two distinct components: a critic, which learns to predict future reward associated with particular states in the environment, and an actor, which chooses specific actions to move the agent from state to state according to a learned policy. The critic encodes the value of particular states in the world and as such has the characteristics of a Pavlovian reward prediction signal described above. The actor stores a set of probabilities for each action in each state of the world, and chooses actions according to those probabilities. The goal of the model is to modify the policy stored in the actor such that over time, those actions associated with the highest predicted reward are selected more often. This is accomplished by means of a prediction error signal, which computes the difference in predicted reward as the agent moves from state to state. This signal is then used to update value predictions stored in the critic for each state, but also to update action probabilities stored in the actor such that if the agent moves to a state associated with greater reward (and thus generates a positive prediction error), then the probability of choosing that action in future is increased. Conversely, if the agent moves to a state associated with less reward, this generates a negative prediction error and the probability of choosing that action again is decreased.

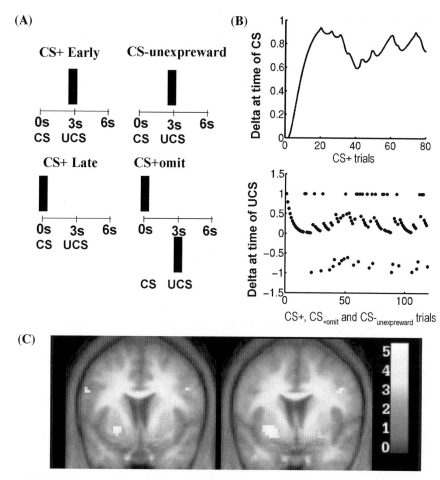

FIGURE 2. Model-based fMRI of stimulus-reward learning. (**A**) Properties of the temporal difference prediction error signal during reward learning in which a cue (CS+) is paired repeatedly with a reward (UCS) presented 3 sec later. During the initial stages of learning (CS + early trials), the error signal responds at the time of presentation of the UCS, but over the course of learning transfers back to the time of presentation of the CS (CS + late trials). On trials in which the CS+ is not presented but the reward is delivered anyway (CS–unexp. reward), the signal shows a positive response at the time the reward is delivered, whereas on trials in which the CS is presented but the reward is unexpectedly omitted the signals show a negative response at the time of outcome. (**B**) Plot of model-generated prediction error signals at the time of presentation of the CS, and the time of presentation of the UCS, over the course of the experiment for a typical subject. (**C**) Area of bilateral ventral striatum (ventral putamen bilaterally) showing significant correlations with the temporal difference prediction error signal while subjects underwent classical conditioning with sweet taste reward (1M glucose). Data from O'Doherty *et al.*[11]

Some computational neuroscientists have drawn analogies between the anatomy and connections of the basal ganglia, and possible neural architectures for implementing reinforcement learning models including the actor/critic. Houk and colleagues[17] have proposed that the actor and critic could be implemented within patch/striosome and matrix compartments distributed throughout the striatum. Montague and colleagues[10] proposed that the ventral and dorsal striatum implemented the critic and actor respectively, on the grounds of extant knowledge of the putative functions of these structures at the time, derived primarily from animal lesion studies. To test these hypotheses, O'Doherty and colleagues[18] scanned hungry human subjects with fMRI while they performed a simple instrumental conditioning task in which they were required to choose one of two actions leading to juice reward with either a high or low probability (FIG. 3A). Neural responses corresponding to the generation of prediction error signals during performance of the instrumental task were compared to that elicited during a control Pavlovian task in which subjects experienced the same stimulus-reward contingencies but did not actively choose which action to select. This comparison was designed to isolate the actor, which was hypothesized to be engaged only in the instrumental task, from the critic, which was hypothesized to be engaged in both the instrumental and Pavlovian control tasks. Consistent with the proposal of a dorsal versus ventral actor/critic architecture, activity in dorsal striatum was found to be specifically correlated with prediction error signals when subjects were actively performing instrumental responses to obtain reward. In contrast, ventral striatum was found to be active in both the instrumental and Pavlovian tasks (FIG. 3B).

These results suggest a dorsal–ventral distinction within the striatum whereby ventral striatum is more concerned with Pavlovian or stimulus-outcome learning, while the dorsal striatum is more engaged during learning of stimulus-response or stimulus-response-outcome associations. The suggestion that human dorsal striatum is specifically involved under situations when subjects need to select actions to obtain reward has received support from a number of other fMRI studies, both model based[19] and trial based.[20]

EXPECTED VALUE

This raises the question as to where in the brain expected values are represented. A number of model-based fMRI studies have consistently implicated ventromedial prefrontal cortex in encoding the value of chosen actions. Kim and colleagues,[21] used a variant of the actor/critic algorithm to generate expected value signals as subjects made decisions between which of two possible actions to chose to obtain monetary reward, as well as in a different condition, to avoid losing money. In this study, different available actions were associated

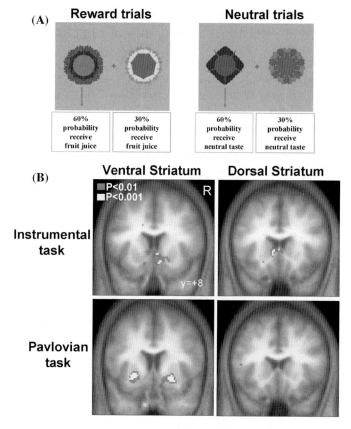

FIGURE 3. Model-based fMRI of action selection for reward. (**A**) Schematic of instrumental choice task used by O'Doherty *et al.*[18] On each trial of the reward condition subject chooses between two possible actions, one associated with a high probability of obtaining juice reward (60%), the other a low probability (30%). In a neutral condition subjects also choose between actions with similar probabilities but in this case they receive an affectively neutral outcome (tasteless solution). Prediction error responses during the reward condition of the instrumental choice task were compared to prediction error signals during a yoked Pavlovian control task. (**B**) Significant correlations with the reward prediction error signal generated by an actor/critic model were found in ventral striatum (ventral putamen extending into nucleus accumbens proper) in both the Pavlovian and instrumental tasks, suggesting that this region is involved in stimulus-outcome learning. In contrast, a region of dorsal striatum (anteromedial caudate nucleus) was found to be correlated with prediction error signals only during the instrumental task, suggesting that this area is involved in stimulus-response or stimulus-response-outcome learning. Data from O'Doherty *et al.*[18]

with distinct probabilities of either winning or losing money, such that in the reward condition one action was associated with a 60% probability of winning money, and the other action with only a 30% probability of winning. To maximize their cumulative reward, subjects should learn to choose the 60%

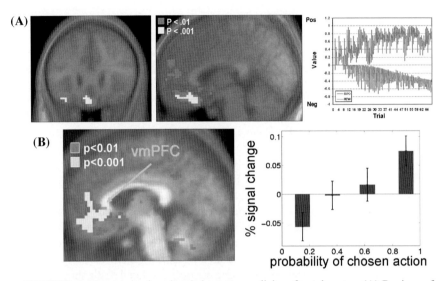

FIGURE 4. Expected value signals in ventromedial prefrontal cortex. (**A**) Regions of ventromedial prefrontal cortex (medial and central orbitofrontal cortex extending into medial prefrontal cortex) correlating with expected value signals generated by a variant of the actor/critic model during an fMRI study of instrumental choice of reward and avoidance (left and middle panels). The model-predicted expected value signals are shown for one subject in the right panel for both the reward (top line) and avoidance (bottom line) conditions. Data from Kim *et al.*[21] (**B**) Similar regions of ventromedial prefrontal cortex correlating with model-predicted expected value signals during performance of a four-armed bandit task with nonstationary reward distributions (left panel). BOLD signal changes in this region are shown plotted against model predictions (right panel), revealing an approximately linear relationship between expected value and BOLD signal changes in this region. Data from Daw *et al.*[22] The figure appears in color online.

probability action. In the avoidance condition, subjects were presented with a choice between the same probabilities, except in this context, 60% of the time after choosing one action they avoided losing money, whereas this only occurred 30% of the time after choosing the alternate action. To minimize their losses, subjects should learn to choose the action associated with the 60% probability of loss avoidance. Model-generated expected value signals for the action chosen were found to be correlated on a trial-by-trial basis with BOLD responses in bilateral orbitofrontal cortex and adjacent medial prefrontal cortex in both the reward and avoidance conditions, such that activity in these areas increased in proportion to the expected future reward associated with the specific action chosen (FIG. 4A).

Similar results were obtained by Daw and colleagues,[22] who used a four-armed bandit task in which "points" (that would later be converted into money) were paid out on each bandit. However, unlike the studies described previously, in this case, the mean payoff available on each bandit drifted over time, such that no one bandit was consistently paying out more than the others, but at any one time some bandits paid out more than the others. As a consequence, subjects had to keep track of the mean payoffs available on each bandit over the course of the experiment, to work out which one paid out the most at any one moment. Model-predicted expected value signals specific to the action chosen were found to be correlated on a trial by trial basis with fMRI activity in ventromedial prefrontal cortex (FIG. 4B). Tanaka and colleagues also showed significant correlations between value signals derived from reinforcement learning models and activity in medial prefrontal cortex during an fMRI study where subjects performed a Markov decision task involving choices to obtain immediate or delayed rewards.[23]

Taken together, these findings suggest that orbital and medial prefrontal cortex are involved in keeping track of the expected future reward associated with chosen actions, and that these areas show a response profile consistent with an expected value signal generated by reinforcement learning models. The fact that these regions were found to be correlated with the "chosen" action, raises the question as to how the chosen action comes to be chosen in the first place. Presumably somewhere in the brain it is necessary to encode the value of each available action before a specific action is chosen to compare between them and ultimately select a particular action. In future studies it will be important to establish the presence of prechoice action values to distinguish between regions actively computing the decision itself from those merely reporting its consequences. Another open question is whether state values, corresponding to the average reward available in a particular state (usually signaled by a specific combination of stimuli) irrespective of the actions subsequently chosen, are represented separately from action values, which correspond to the future reward associated with selection of a specific action. So far, neuroimaging studies have failed to distinguish between state and action values, most probably because such signals are usually highly correlated with each other in the experimental designs that have been used to date.

ENCODING ABSTRACT RULES IN A DECISION PROBLEM

Although standard RL models can account for a wide range of human and animal choice behavior, these models do have important limitations. One such limitation is a failure to account for higher order structure in a decision problem, such as rules describing interdependencies between actions, rewards, and other exogenous variables, such as time. Yet, many real-life decision problems do incorporate such structure.[24–26] Simple RL models assume the actions available in the world and the rewards that can be obtained from choice of those actions are independent from each other, such that in a given state, information gained about the rewards available from choice of one action provides no information about the rewards available from choice of another action. However, in many situations rules describing interdependencies between different actions do exist, and if subjects can exploit these rules, this will lead to greater reward than otherwise. One of the simplest examples of a decision task with such an abstract rule is probabilistic reversal learning.[25,26] In this task, subjects can choose between one of two actions, which give out monetary rewards or losses on a probabilistic basis, similar to the instrumental choice tasks outlined previously. However, in this case, at any one time, one of the actions pays out more reward than the other, such that if the subject continues to choose the high paying action they will obtain the greatest reward. After a time the contingencies reverse, and the subject has to switch their choice of action to continue to maximize their reward. The structure in this task is the anticorrelation between the distribution of rewards available on the two actions: when one action is "good" the other is "bad" and *vice versa*, as well as the rule that after a time the contingencies will reverse. A standard RL model could be used to learn such a task, but in this case the model would not incorporate the abstract rules in the reversal task but would instead simply learn about the values of the two actions independently.

However, what if a model were to incorporate such rules? To establish whether human subjects do use an abstract representation of the task structure to guide their choices, Hampton and colleagues[27] scanned subjects with fMRI while they underwent probabilistic reversal learning. A computational model that incorporated the structure of the task was then constructed and fitted to both the behavioral and fMRI data. This structure-based model was implemented as a hidden Markov model (HMM), where the hidden state to be estimated was the probability that on a given trial the "correct" (or currently high reward value action) is being chosen. This model also incorporated the fact that the identity of the correct action would reverse from time to time. The structure-based model was found to provide a good fit to subjects' behavior, and the signal derived from the model representing the probability that subjects were choosing the correct action (prior correct) was found to be significantly correlated at the time of action choice with fMRI activity in ventromedial prefrontal cortex (FIG. 5A). The areas thus identified overlap markedly with those regions found to correlate with expected value derived from standard RL models outlined previously. This is perhaps not surprising as the prior correct signal from the HMM model is strongly colinear with expected value signals derived from RL.

MODEL COMPARISON: STRUCTURE-BASED INFERENCE VERSUS STANDARD RL

However, the key question is whether the model that incorporates the rules of the decision task can account *better* for subjects' behavioral and fMRI data

(A)

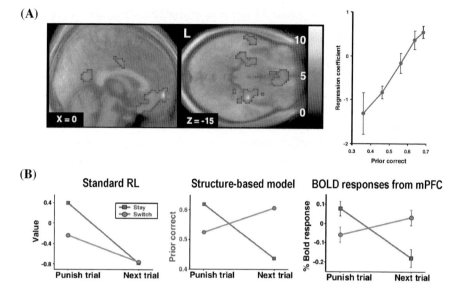

(B)

FIGURE 5. Expected value signals in ventromedial prefrontal cortex reflect abstract task structure. **(A)** Regions of ventromedial prefrontal cortex correlating with expected value signals while subjects performed a decision task with abstract structure: probabilistic reversal learning (left panel). Expected signals were generated by a model that incorporates the structure of the task. Plot of responses in this region against expected value signals reveal a strongly linear relationship (similar to that shown in FIG. 4B). **(B)** Dissociating a model that incorporates the structure of the task (structure-based model) from standard RL models that do not incorporate the rules of the task. The predictions from standard RL and the structure-based model are plotted for expected value signals before and after switching action choice following a reversal, where the subject switches back to an action that was chosen previously (shown in left and middle panels as the circled points). The structure-based model predicts that expected value signals jump up following a reversal, whereas the RL models predict no such increase in reward expectation. The actual BOLD signal in medial PFC shows a response profile consistent with the structure-based model and not simple RL (right panel). Data from Hampton et al.[27] Copyright 2006 Society for Neuroscience. The figure appears in color online.

than does standard RL, as this would provide evidence that human subjects *do* use knowledge of task structure to guide their choices as opposed to merely learning action values independently. In this study we therefore compared the goodness-of-fit of the HMM model to subjects' behavioral data to the fit achieved by a family of different RL models. The structure-based model was found to provide a significantly better fit to the behavioral data than did the best-fitting RL model, even after adjusting for the number of free parameters in the model. Thus, even at the behavioral level, there is evidence to suggest that subjects are using knowledge of the structure of the reversal task to guide their choices.

To determine whether neural coding of expected values also reflected knowledge of this structure, we looked specifically at those times in the experiment when the predictions of the structure-based model and of the standard RL model would be maximally divergent. This happens to be immediately after subjects switch their choice of action. According to both RL and the structure-based model, subjects should switch their choice of action once they have a low expected value for that action, presumably after having received a string of nonrewarding outcomes after selecting that action on previous trials. Where the models diverge, is when subjects switch back to an action that they previously switched away from after a prior contingency reversal. According to standard RL, once subjects switch back to an action they should still have a low expected reward for that action because the last time they chose that action it had a low expected value (hence they switched away from it). However, according to the structure-based model, once subjects switch back to an action they previously switched away from, this time they should have a high expected reward value, because they understand that the contingencies have reversed, and that therefore this action must now have a high reward value. FIGURE 5B shows the predictions of both the structure-based model and standard RL models alongside the actual signal at the time of choice extracted from the medial prefrontal cortex. As can be seen from this figure the actual fMRI data mirror the predictions of the structure-based model, by showing that subjects' expectations of reward jumps up once they switch back to a previously chosen action following a reversal. In a further analysis, model predictions from the structure-based model and standard RL model were both entered into a regression analysis against the fMRI data and the fit of both models to the fMRI data was directly compared. Once again, the structure-based model was found to provide a better fit to the fMRI data than the best-fitting RL model.

These results suggest that subjects can take into account the abstract rules of a decision problem and that knowledge of such rules can modulate expected value signals in prefrontal cortex that in turn may be used to guide behavioral choice. These results do not invalidate RL models *per se*, but rather highlight the need to incorporate additional inference mechanisms that may exist alongside standard RL either cooperatively or competitively.[28] Future studies will be needed to establish whether both structure-based and standard RL signals are present in the brain at the same time, and if so, whether different regions of the brain are involved in implementing these different learning processes. Alternatively, known rules of the task could be incorporated into the structure of the state space in a given task, and then standard RL mechanisms could be used to learn the values in this state space from then on. More generally, this particular study provides an example of how fMRI data can be used to inform about the specific computational functions being implemented in a particular brain region, in this case ventromedial prefrontal cortex. By comparing the degree to which activity in this region can be accounted for by different competing computational models, it was possible to show that signals in this region are accounted better by one model than another. Thus, fMRI has provided a means to discriminate between different computational models of brain function.

LIMITATIONS OF MODEL-BASED IMAGING

In this review we have argued that model-based fMRI provides a powerful means with which to test computational models of brain function, and argue that this approach overcomes many of the limitations of more conventional approaches to neuroimaging in which a set of regions are merely identified as being "active" in a given task or condition. Unlike these previous approaches that permit inferences about where a given cognitive function is implemented, model-based fMRI provides insights not only into where but also as to how a specific function might be carried out.

However, it should be noted that this technique does have many limitations. Use of the model-based approach does entail testing for highly specific signals in the brain, which permits testing of specific computational hypotheses, but at the same time this approach may limit the possibility of detecting unexpected findings that would not conform to a specific *a priori* hypothesis. Consequently, it is probably a reasonable policy to continue to employ more conventional trial-based analyses alongside the model-based approach. A complementary model-free approach that has been used to explore relationships between objective stimulus events and behavioral and neural data is to examine correlations between current behavior or neural activity on a given trial against previous experience, such as, for example, regressing current choice against the history of rewarding outcomes received. By analyzing how correlations between these variables change over time, it is in principle possible to extract information about the type of computational process being used by the subject to drive behavior as a function of experience, without imposing a specific computational model at the outset. For example, by using this approach Sugrue *et al.*, were able to gain insight into the manner by which reward information was integrated over time to drive performance on a matching task.[29] A similar approach could in principle be adopted in human imaging or behavioral studies, with the caveat that while this approach has proven useful in monkey neurophysiology studies where thousands of trials are typically available for analysis, it is less clear whether this technique will have sufficient statistical power in human studies that usually involve far fewer trials.

Furthermore, model-based fMRI just as with any other direct or indirect technique for recording neural activity, can only provide insight into correlations between neural activity and behavior but cannot establish causal links between activity in a particular brain area and subsequent behavior. For this, in humans at least, it is necessary to study the effects of disruption to a region either by means of assessing the effects of a permanent lesion to this area, or by inducing temporary inactivation in an area through transcranial magnetic stimulation (TMS). The model-based approach can of course be applied to lesion and TMS data just as it can be applied to fMRI data. Combining the results of model-based analyses of fMRI data with data derived from these other approaches will likely be an important direction for future research.

Another limitation of model-based fMRI as with any other use of fMRI data, is the poor spatiotemporal resolution of this technique compared to single or multiunit neurophysiology recording techniques available in rats and nonhuman primates. Computational signals observed with fMRI are likely to reflect cases where a large population of neurons essentially conveys the same signal, as is thought to be the case, for example, in the phasic activity of dopamine neurons. Naturally, model-based fMRI cannot provide insight into more fine-grained computational signals conveyed at the single neuron level, especially where different interleaved populations of neurons within a region may be conveying distinct computational signals. Model-based fMRI is only useful under situations where computational signals are conveyed by means of a change in mean firing rate in a population of neurons such that variation in the computational signal is reflected in variation in metabolic demand and hence blood oxygenation within a particular region. For some types of computational processes, changes in computations within a region may not result in a change in overall activity level, but rather may reflect a change in a distributed pattern of neural activity without any overall change in mean firing rates. Such signals would not be detectable with the model-based fMRI approach.

On account of these limitations, more so than ever perhaps, it will be necessary to combine the results from model-based fMRI studies in humans with other techniques, such as single- or multiunit neurophysiology in other animals, as well as with imaging methods in humans that afford greater temporal resolution at the expense of spatial precision, such as MEG or EEG. Just as the case has been made here for model-based approaches to fMRI, the case can be made for model-based approaches in these other methodologies. Indeed, it is important to note that the model-based approach to fMRI data parallels a similar approach that is increasingly being adopted in the animal neurophysiology literature, whereby the activity of single neurons are correlated against specific computational models.[24,30–33] As a consequence, the tendency toward incorporating computational models into fMRI studies of neural activity may be seen as part of a wider trend in the field toward a more quantitative and theory-driven approach to experimental neuroscience.

ACKNOWLEDGMENTS

This work was funded by grants from the Gimbel Discovery Fund for Neuroscience, the Gordon and Betty Moore Foundation, and a Searle Scholarship to JOD. We would like to thank Nathaniel Daw, Peter Dayan, Ray Dolan, Karl Friston, and Ben Seymour at UCL, and Peter Bossaerts and Shin Shimojo at Caltech, who were major collaborators on some of the research studies described here.

REFERENCES

1. RESCORLA, R.A. & A.R. WAGNER. 1972. A theory of Pavlovian conditioning: variations in the effectiveness of reinforcement and nonreinforcement. *In* Classical Conditioning II: Current Research and Theory. A.H. Black & W.F. Prokasy, Eds.: 64–99. Appleton Crofts. New York.
2. RESCORLA, R.A. 2002. Comparison of the rates of associative change during acquisition and extinction. J. Exp. Psychol. Anim. Behav. Process. **28:** 406–415.
3. HENSON, R. 2005. What can functional neuroimaging tell the experimental psychologist? Q. J. Exp. Psychol. A **58:** 193–233.
4. FRISTON, K.J. *et al.* 1995. Statistical parametric maps in functional imaging: a general linear approach. Hum. Brain Map. **2:** 189–210.
5. SCHWARZ, G. 1978. Estimating the dimension of a model. Ann. Stat. **6:** 461–464.
6. SCHULTZ, W. 1998. Predictive reward signal of dopamine neurons. J. Neurophysiol. **80:** 1–27.
7. MIRENOWICZ, J. & W. SCHULTZ. 1994. Importance of unpredictability for reward responses in primate dopamine neurons. J. Neurophysiol. **72:** 1024–1027.
8. HOLLERMAN, J.R. & W. SCHULTZ. 1998. Dopamine neurons report an error in the temporal prediction of reward during learning. Nat. Neurosci. **1:** 304–309.
9. SCHULTZ, W., P. DAYAN & P.R. MONTAGUE. 1997. A neural substrate of prediction and reward. Science **275:** 1593–1599.
10. MONTAGUE, P.R., P. DAYAN & T.J. SEJNOWSKI. 1996. A framework for mesencephalic dopamine systems based on predictive Hebbian learning. J. Neurosci. **16:** 1936–1947.
11. O'DOHERTY, J.P. *et al.* 2003. Temporal difference models and reward-related learning in the human brain. Neuron **38:** 329–337.
12. JOEL, D. & I. WEINER. 2000. The connections of the dopaminergic system with the striatum in rats and primates: an analysis with respect to the functional and compartmental organization of the striatum. Neuroscience **96:** 451–474.
13. MCCLURE, S.M., G.S. BERNS & P.R. MONTAGUE. 2003. Temporal prediction errors in a passive learning task activate human striatum. Neuron **38:** 339–346.
14. SUTTON, R.S. & A.G. BARTO. 1998. Reinforcement Learning. MIT Press. Cambridge, MA.
15. BARTO, A.G. 1992. Reinforcement learning and adaptive critic methods. *In* Handbook of Intelligent Control: Neural, Fuzzy, and Adaptive Approaches. D.A. White & D.A. Sofge, Eds.: 469–491. Van Norstrand Reinhold. New York.
16. BARTO, A.G. 1995. Adaptive critics and the basal ganglia. *In* Models of Information Processing in the Basal Ganglia. J.C. Houk, J.L. Davis & B.G. Beiser, Eds.: 215–232. MIT Press. Cambridge, MA.
17. HOUK, J.C., J.L. ADAMS & A.G. BARTO. 1995. A model of how the basal ganglia generate and use neural signals that predict reinforcement. *In* Models of Information Processing in the Basal Ganglia. J.C. Houk, J.L. Davis & B.G. Beiser, Eds.: 249–270. MIT Press. Cambridge.
18. O'DOHERTY, J. *et al.* 2004. Dissociable roles of ventral and dorsal striatum in instrumental conditioning. Science **304:** 452–454.
19. HARUNO, M. *et al.* 2004. A neural correlate of reward-based behavioral learning in caudate nucleus: a functional magnetic resonance imaging study of a stochastic decision task. J. Neurosci. **24:** 1660–1665.
20. TRICOMI, E.M., M.R. DELGADO & J.A. FIEZ. 2004. Modulation of caudate activity by action contingency. Neuron **41:** 281–292.

21. KIM, H., S. SHIMOJO & J.P. O'DOHERTY. 2006. Is avoiding an aversive outcome rewarding? Neural substrates of avoidance learning in the human brain. PLoS Biol. **4:** e233.
22. DAW, N.D. *et al.* 2006. Cortical substrates for exploratory decisions in humans. Nature **441:** 876–879.
23. TANAKA, S.C. *et al.* 2004. Prediction of immediate and future rewards differentially recruits cortico-basal ganglia loops. Nat. Neurosci. **7:** 887–893.
24. SUGRUE, L.P., G.S. CORRADO & W.T. NEWSOME. 2004. Matching behavior and the representation of value in the parietal cortex. Science **304:** 1782–1787.
25. O'DOHERTY, J. *et al.* 2003. Dissociating valence of outcome from behavioral control in human orbital and ventral prefrontal cortices. J. Neurosci. **23:** 7931–7939.
26. COOLS, R. *et al.* 2002. Defining the neural mechanisms of probabilistic reversal learning using event-related functional magnetic resonance imaging. J. Neurosci. **22:** 4563–4567.
27. HAMPTON, A.N., P. BOSSAERTS & J.P. O'DOHERTY. 2006. The role of the ventromedial prefrontal cortex in abstract state-based inference during decision making in humans. J. Neurosci. **26:** 8360–8367.
28. DAW, N.D., Y. NIV & P. DAYAN. 2005. Uncertainty-based competition between prefrontal and dorsolateral striatal systems for behavioral control. Nat. Neurosci. **8:** 1704–1711.
29. SUGRUE, L.P., G.S. CORRADO & W.T. NEWSOME. 2005. Choosing the greater of two goods: neural currencies for valuation and decision making. Nat. Rev. Neurosci. **6:** 363–375.
30. PLATT, M.L. & P.W. GLIMCHER. 1999. Neural correlates of decision variables in parietal cortex. Nature **400:** 233–238.
31. BARRACLOUGH, D.J., M.L. CONROY & D. LEE. 2004. Prefrontal cortex and decision making in a mixed-strategy game. Nat. Neurosci. **7:** 404–410.
32. SAMEJIMA, K. *et al.* 2005. Representation of action-specific reward values in the striatum. Science **310:** 1337–1340.
33. DAW, N.D. & K. DOYA. 2006. The computational neurobiology of learning and reward. Curr. Opin. Neurobiol. **16:** 199–204.

Splitting the Difference

How Does the Brain Code Reward Episodes?

BRIAN KNUTSON AND G. ELLIOTT WIMMER

Department of Psychology, Stanford University, Stanford, California, USA

> So, nat'ralists observe, a flea
> Hath smaller fleas that on him prey;
> And these have smaller still to bite 'em;
> And so proceed *ad infinitum.*
>
> —Jonathan Swift, *On Poetry: A Rhapsody*, 1733

ABSTRACT: Animal research and human brain imaging findings suggest that reward processing involves distinct anticipation and outcome phases. Error terms in popular models of reward learning (such as the temporal difference [TD] model) do not distinguish between the updating of expectations in response to reward cues and outcomes. Thus, correlating a single error term with neural activation assumes recruitment of similar neural substrates at each update. Here, we split the error term to separately model reward prediction and prediction errors, and compare the fit of single versus split error terms to functional magnetic resonance imaging (FMRI) data acquired during a monetary incentive delay task. We speculate and find that while the nucleus accumbens computes gain prediction in response to cues, the mesial prefrontal cortex (MPFC) computes gain prediction errors in response to outcomes. In addition to offering a more comprehensive and anatomically situated view of reward processing, split error terms generate novel predictions about psychiatric symptoms and lesion-induced deficits.

KEYWORDS: reward; anticipation; FMRI; human; computation; accumbens; prefrontal

MOTIVATION

Reminiscent of the infinite tower of fleas in Jonathan Swift's sardonic ode to poetry, the image of brain as an ascending hierarchy stands as a lasting

Address for correspondence: Brian Knutson, Department of Psychology, Bldg. 420, Jordan Hall, Stanford, CA 94305. Voice: +650-724-2965; fax: +650-725-5699.
knutson@psych.stanford.edu

Ann. N.Y. Acad. Sci. 1104: 54–69 (2007). © 2007 New York Academy of Sciences.
doi: 10.1196/annals.1390.020

contribution of neurologist John Hughlings Jackson's prolific but scattered writings.[1] Based on this hierarchical organization, Jackson predicted that lesions of outermost brain regions should produce not only "negative symptoms" or attenuation of critical faculties, but also "positive symptoms" or accentuation of previously inhibited faculties localized further down. Neurophysiologist Paul MacLean extended the notion of neural hierarchy in his sketch of the "triune brain," by stacking socioemotional concerns atop survival programs, which were in turn crowned by a higher level of symbolic representation.[2] While the triune brain concept has fallen out of favor due to ambiguous specification of subcortical circuitry and the challenge of distinguishing emotional from cognitive function,[3] recent innovations in brain imaging offer new hope for testing hierarchical localization schemes.

Functional magnetic resonance imaging (FMRI) methods have advanced rapidly since the technique's inception in the early 1990s. Conceptual advances in design and analysis soon followed after physical improvements in image acquisition. While initial FMRI experiments were modeled after positron emission tomography (PET) studies, with relatively reduced spatiotemporal resolution (i.e., ~ 8 mm^3/120 sec), event-related designs have dramatically enhanced FMRI's spatiotemporal resolution (i.e., ~ 4 mm^3/2 sec), potentially allowing investigators to acquire images at the speed of phenomenology. These new methods call for new models and modes of analysis.

Reward processing represents an evolutionarily conserved, yet environmentally flexible phenomenon that could benefit from temporally precise analysis. To promote survival and reproduction, subjective evaluation often must supersede and direct the processing of other types of information.[4] Thus, flexible evaluators must predict as well as respond to incentive outcomes. If rewarding stimuli are defined as those that an organism will work to obtain, reward processing minimally refers to the unfolding of reward anticipation and outcome phases over time,[5] consistent with a historic ethological distinction between appetitive and consummatory motivation.[6] In prior FMRI research, analyses have suggested that while reward anticipation primarily activates the subcortical nucleus accumbens (NAcc), reward outcomes primarily activate a region of the mesial prefrontal cortex (MPFC).[7] These analyses relied upon simple statistical contrasts that highlighted regions in which one experimental condition elicited greater local oxygen utilization (or "activation") than another. But beyond the critical first step of localization, how can investigators best model the dynamic flow of activity coursing through different brain regions? For instance, when does a signal deviate from baseline, in which direction (up or down), and to what degree? Models that can address these specific questions promise not only the practical benefit of increasing sensitivity to detect activation, but also the theoretical benefit of improving functional understanding of what a given region computes.

Computational models that generate temporally specific predictions have already yielded elegant and profound insights about brain regions implicated in reward processing. One of the most popular of these models is the temporal difference (TD) model.[8] A key term in TD models is the reward prediction error (i.e., "delta"), or the difference between expected and actual reward. The reward prediction error term can be used to update expectations. For instance, reward outcomes that are initially unexpected create a positive reward prediction error, since such an event exceeds expectations. However, once a cue begins to predict an uncertain reward, the reward prediction error shifts to the unexpected cue onset rather than reward delivery. Thus, like a dog salivating at the sound of a dinner bell, the model learns that specific cues predict eventual rewards.

TD models have recently been used not only to describe behavior, but also the firing rates of midbrain dopamine neurons.[9,10] Specifically, while midbrain dopamine neurons initially increase firing in response to an unexpected juice squirt, after monkeys learn that a cue predicts juice delivery, dopamine firing shifts to cue onset, implying a positive reward prediction error. Importantly, when juice does not follow the cue, dopamine neurons briefly stop firing at the time of expected juice delivery, implying a negative reward prediction error, which rules out surprise or novelty as alternative explanations for changes in firing rate.[10,11] Although the TD error term mimics key features of dopamine neuron firing, it does not indicate whether other brain regions modulate the activity of dopamine neurons, or how they do so.

Based on the temporal distinction between reward anticipation and outcome, as well as a spatial distinction between subcortical and cortical brain regions, we propose an "ascending differences" (AD) split of the TD error term (see TABLE 1). This modification assumes that the brain distinguishes between uncertain and certain events.[12] While uncertain events imply that something may occur in the future (i.e., probability falls between 1 or 0), certain events imply that something has occurred (i.e., probability collapses to either 1 or 0). Thus, uncertainty implies anticipation of an outcome, while certainty implies the outcome itself (which might include either the occurrence or nonoccurrence of an

TABLE 1. AD and TD regressor computation

	Anticipation	Outcome
Temporal difference (TD)		
Gain prediction error (GPE)	CV–EV*	OV–CV
Ascending differences (AD)		
Gain prediction	CV–EV	0
Gain prediction error	0	OV–CV

EV = average(CV) [or average V(t)].
CV = cue magnitude X cue probability [delta(t) at cue or V(t)].
OV = outcome magnitude X outcome probability (i.e., either 0 or 1) [delta(t) at outcome or r(t)].
*Omitting EV from this term does not improve the fit of TD GPE. A future reward prediction term (as found in typical TD error terms) is not included because monetary incentive delay (MID) task trial outcomes are independent (i.e., the current outcome carries no information about future outcomes).

event). Once learning (and thus anticipation) has stabilized, cue presentation elicits a prediction, which can be computed as the cue value minus the baseline expected value (or averaged outcome value up to that point). On the other hand, outcome presentation elicits a prediction error, which can be computed as the outcome value minus the predicted cue value. Thus, the AD modification uses the same error term as TD models, but "splits" reward prediction ("how good might it be?") and reward prediction error ("how good is it?") into two separate terms (see TABLE 1). This split allows for the possibility that different brain regions compute reward prediction and reward prediction error. Here, we name these split terms gain prediction (GP) and gain prediction error (GPE), since the data to be modeled were collected in humans undergoing FMRI as they anticipated and received monetary gains in a monetary incentive delay task.[a] Substitution of the word "gain" for "reward" retains a positive connotation, but allows for gain to be coded relative to some neutral reference point, rather than absolute zero.[13]

Research that initially applied TD models to FMRI data in dynamic tasks involving reward learning revealed that striatal activation correlated with TD reward prediction error,[14,15] a finding borne out by later research.[16,17] After stabilization of learning, however, spatially distinct neural substrates may separately represent GP versus GPEs.[18,19] It is important to extend computational models from dynamic to stable incentive processing tasks, since some psychiatric disorders may involve stable deficits in incentive processing, as suggested by emerging FMRI findings.[20,21] Thus, our goals in this article were to generalize learning models to FMRI data acquired during a stable incentive processing task, and to directly compare the fit of TD versus AD error terms to these data.

MODEL COMPARISON

Based on more than a century of behavioral research,[22] our laboratory has devised a monetary incentive delay (MID) task in which people anticipate and respond to monetary incentives while undergoing FMRI. Monetary incentives provide experimental flexibility because they are nearly universally valued, can be either positive or negative, and can be scaled (features that also facilitate computational modeling). Prior to scanning, subjects are trained on the MID task without pay, and then are shown the cash that they can make while playing in the scanner. In a typical MID task trial, subjects see a cue indicating potential gain or loss of varying magnitudes ($0.00, $0.20, $1.00, $5.00), wait for a brief period (anticipation: 2–3 sec), respond to a rapidly presented target with a

---

[a]TD models also include a "reward prediction" term (v). However, AD GP as defined here corresponds both with TD (delta(t)) at cue presentation, as well as TD reward prediction (v(t)). These TD terms are collinear in the MID task due to temporal overlap of cues and the short anticipation period that follows. AD GPE corresponds to TD delta(t) at outcome. (see TABLE 1 legend for mappings of AD to TD terms).

button press ($\sim$160–260 msec), and receive feedback indicating whether they have either gained or avoided losing money in addition to their cumulative total (outcome: 2 sec), based on the previous cue and whether they pushed the button before the target disappeared.[23] Based on the subject's reaction time, target speed can be adjusted to elicit a desired range of performance (here, an average hit rate of 66%). However, the outcome of each trial is determined independent of the outcomes of previous or subsequent trials.

The analyses presented below focus on changes in brain activation of 26 subjects during gain anticipation (2 sec following cue presentation) and in response to gain outcomes (2 sec during feedback presentation). These data have been previously presented in two reports, but were analyzed with simple unit-weighted contrasts rather than computationally derived regressors.[7,24] To compare TD and AD difference terms, continuous regressors were derived using the formulas described in TABLE 1, with the restriction that each regressor model equals numbers of deviations from baseline, thereby equating the power of each to detect correlated activation. Regressors were convolved with a gamma function to model the lag in hemodynamic response,[25] and entered into otherwise identical multiple regression models (i.e., models also included nuisance regressors that covaried out baseline, linear, and second-order trends for each session, as well as six motion parameter estimates) using Analysis of Functional Neural Images software.[26] Thus, in terms of regressors of interest, the TD model contained gain and loss prediction error regressors, while the AD model contained gain and loss prediction regressors as well as gain and loss prediction error regressors.

Analyses included three stages. First, in localization analyses, group maps were constructed in which model coefficients were tested against the null hypothesis of no activation using t-tests ($P < 0.0001$ uncorrected; corrected for the approximate total volume of striatal and mesial frontal gray matter volumes of interest at $P < 0.05$). For the AD model, AD GP and GPE coefficient maps were conjoined to confirm the predicted regional disjunction of activation ($P < 0.0001$ uncorrected). Second, in comparison analyses, coefficients from different models were directly compared using within subjects paired t-tests. The key comparisons specifically contrasted AD GP versus TD GPE and AD GPE versus TD GPE, with a focus on NAcc and MPFC volumes of interest ($P < 0.01$ uncorrected). An additional comparison contrasted AD GPE with simple outcome value (i.e., r(t); included in an otherwise identical AD model instead of AD GPE) in the MPFC volume of interest ($P < 0.01$, uncorrected). Third, in verification analyses, activation was averaged and extracted from NAcc and MPFC volumes of interest (based on Knutson et al.[7]) and plotted against model predictions for high incentive gain trials (+$5.00 hit and miss, which produce the strongest signal changes).

In localization analyses, statistical maps indicated that TD GPE significantly correlated with NAcc activation and, less robustly, MPFC activation. However, AD regressor maps indicated that while GP maximally correlated with

NAcc activation, GPE instead maximally correlated with MPFC activation (see FIG. 1). For both TD and AD models, loss-related regressors did not positively correlate with activation in these regions (see also TABLE 2). Conjunction of AD GP and AD GPE revealed no conjoint activation

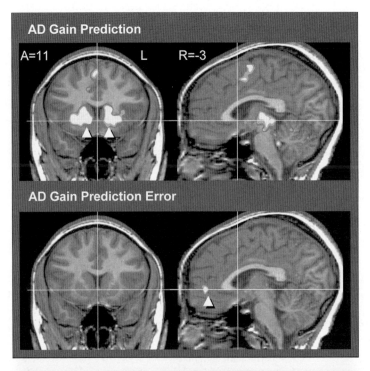

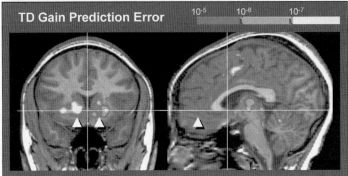

FIGURE 1. Ascending difference (AD) and temporal difference (TD) regressor maps ($n = 26$; $P < 0.0001$, uncorrected).

TABLE 2. AD and TD regressor foci

	Region	Z-score	R	A	S
AD GP (Gain)					
	R anterior cingulate	4.43	9	33	22
		4.83	7	16	36
		5.47	3	7	51
		4.59	4	−13	48
	L middle frontal gyrus	4.27	−31	28	32
	R anterior insula	5.60	27	15	0
	L anterior insula	5.93	−30	19	0
	R nucleus accumbens	6.65	10	7	−1
	L nucleus accumbens	5.85	−11	11	−1
	R caudate	6.08	13	9	10
		5.08	11	1	12
	L caudate	6.50	−12	11	9
		4.61	−7	2	16
	R putamen	6.55	20	7	0
		5.53	27	−5	8
	L putamen	6.23	−18	8	0
		5.98	−25	−2	5
	R thalamus	6.10	6	−20	4
	L thalamus	6.09	−7	−18	4
	R SNc/midbrain	5.36	9	−18	−6
	L SNc/midbrain	4.81	−5	−17	−9
	PAG/midbrain	6.08	0	−24	−7
	R BA 6	4.56	18	2	54
	R SMA/BA 31	4.65	6	−26	49
	R sup. frontal gyrus/BA 6	5.02	25	−10	48
	L sup. frontal gyrus/BA 6	4.63	−17	−7	60
	R precentral gyrus/BA6	4.42	30	−18	48
	L precentral gyrus/BA6	5.18	−33	−15	55
		4.83	−25	−11	63
	R paracentral lobule/BA 6	4.65	6	−25	49
	L postcentral gyrus/BA 3	5.19	−26	−31	58
	R superior parietal lobule	4.93	23	−60	44
	L superior parietal lobule	5.61	−22	−59	56
	R inferior parietal lobule	4.61	27	−48	39
	L inferior parietal lobule	4.66	−29	−48	37
	L precuneus	5.00	−18	−71	19
	R cuneus	4.39	7	−75	7
AD LP (Loss)					
	R superior frontal gyrus	−4.70	2	34	42
		−4.43	4	4	56
	R anterior cingulate	−4.73	6	17	26
	R caudate	−5.34	11	13	9
	L caudate	−4.28	−11	9	11
	R putamen	−5.16	19	10	0
		−5.36	24	1	−2
	L putamen	−4.59	−18	10	0
		−4.50	−23	−4	5

Continued.

TABLE 2. (Continued)

	Region	Z-score	R	A	S
	R thalamus	−5.29	8	−19	12
	L thalamus	−5.28	−5	−17	12
	L precentral gyrus/BA 6	−5.15	−27	−27	50
	R SNc/midbrain	−4.59	6	−18	−5
	PAG/midbrain	−4.89	2	−26	−16
AD GPE (Gain)					
	R MPFC	4.49	4	50	0
	L MPFC	5.50	−4	50	0
	L middle frontal gyrus	4.34	−27	29	37
	R NAcc	4.34	12	12	−7
	Posterior cingulate	4.83	0	−37	30
	R posterior cingulate	4.78	7	−52	15
	L posterior cingulate	4.74	−3	−57	15
	R paracentral lobule	4.31	2	−38	54
	R cuneus	5.10	25	−87	9
AD LPE (Loss)	(No regions survive threshold)				
	and cluster criteria*				
TD GPE (Gain)					
	R anterior cingulate	4.42	4	25	30
		5.64	7	3	46
		4.89	2	−3	31
	L anterior cingulate	5.56	−4	1	44
		4.44	−4	−15	45
	R cingulate	4.59	6	−23	49
	L cingulate	4.30	−8	−13	38
	R superior frontal gyrus	4.12	7	44	21
	L superior frontal gyrus	4.26	−10	34	20
	L genual cingulate	5.29	−7	41	−3
	R middle frontal gyrus	4.56	30	−3	56
		4.63	22	0	47
	R anterior insula	5.52	31	16	1
	L anterior insula	4.34	−33	18	2
	R nucleus accumbens	5.17	11	10	−4
	L nucleus accumbens	4.68	−10	12	−2
	R caudate	5.51	13	8	10
		5.03	11	−1	−15
	L caudate	4.50	−14	10	10
		4.52	−15	1	19
	R putamen	5.50	17	10	0
		5.87	23	−1	4
	L putamen	5.04	−18	7	0
	R thalamus	5.88	8	−18	4
	L thalamus	5.01	−7	−12	6
	R SNc/midbrain	5.15	3	−15	−7
	PAG/midbrain	4.97	−3	−30	−13
	R precentral gyrus/BA6	4.10	30	−22	56

Continued

TABLE 2. (Continued)

	Region	Z-score	R	A	S
	L precentral gyrus/BA6	4.26	−30	−9	57
	R precuneus	4.27	12	−42	50
		4.55	8	−68	23
	R inferior parietal lobule	4.60	22	−59	41
	L inferior parietal lobule	4.74	−30	−54	42
		5.34	−25	−62	37
		4.74	−12	−68	42
	R cuneus	4.88	11	−60	9
TD LPE (loss)	No regions survive threshold and cluster criteria*				

$P < 0.0001$, uncorrected; cluster = 3 voxels.

in the NAcc or MPFC (or any other region). Relaxing the cluster criterion revealed a single conjointly activated voxel in the NAcc, but this region was more than two orders of magnitude smaller (50 mm^3) than the region activated by AD GP alone (>15,000 mm^3). Comparison of AD GPE with simple outcome value revealed that AD GPE more robustly correlated with MPFC activation than outcome value (left TC: −4,50,−3; Z = 3.02; right TC: 4,50,3; Z = 3.51; ps < 0.01). This result is consistent with MPFC activation time course plots (see FIG. 2), which indicate not only that gain outcomes increase MPFC activation (predicted by both AD GPE and outcome), but also that nongain outcomes decrease MPFC activation (predicted only by AD GPE).

In comparison analyses, AD GP correlated with NAcc activation more robustly than TD GPE, while AD GPE correlated with MPFC activation marginally more than TD GPE, as predicted ($P < 0.01$, warm colors, see FIG. 3). Conversely, and consistent with less spatial specificity for the TD error term, TD GPE correlated with MPFC activation more robustly than AD GP, while TD GPE correlated with striatal activation more robustly than AD GPE (particularly in the lateral putamen and dorsal caudate; cool colors; see FIG. 3 and TABLE 3). Intriguingly, TD GPE correlated more robustly with anterior cingulate activation than both AD GP and AD GPE, implying that the TD GPE term may more closely model activation in regions other than the NAcc or MPFC.

In verification analyses, visual inspection of activation time course plots for high gain (+$5.00) conditions (hits and misses) confirmed findings from direct comparisons. Specifically, AD GP fit NAcc time course data more closely than did TD GPE, while AD GPE fit MPFC time course data more closely than did TD GPE (see FIG. 2). The AD model fits appeared closest for hits, with miss data falling somewhere between the predictions of AD and TD error terms in the NAcc (see TABLE 3 for direct statistical contrasts of models—plots provide visual comparison only).

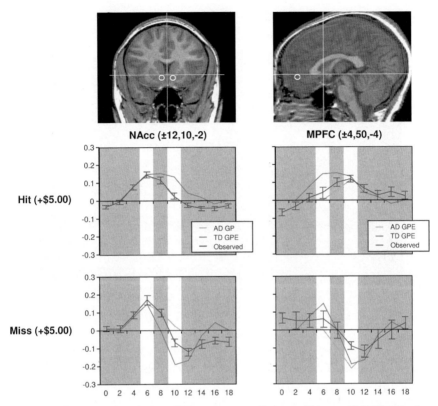

FIGURE 2. Relation of AD and TD regressor predictions to NAcc and MPFC activation time course data for +$5.00 hit and miss trials (mean ± SEM; $n = 26$; white bars indicate predicted peak activation for anticipation and outcome phases of trials lagged by 6 sec).

As suspected, AD error terms highlighted a dissociation undetected by the TD error term: while NAcc activation correlated with AD GP after cue presentation, MPFC activation correlated instead with AD GPE in response to outcomes. Both direct within-subject comparisons and visualization of activation time courses confirmed this dissociation. While the present AD split of the TD error term is conceptually simple and easy to implement, it represents more of a beginning than an end. Since the AD error terms are optimized for the MID task, which is stable and involves minimal learning, the present AD modification might require further elaboration to generalize to more dynamic learning scenarios. For instance, a parameter representing memory decay over time could be added to the GPE term. Also, rather than computing baseline expected value as an average, baseline expected value could be biased toward recent experiences. Finally, outcomes might also elicit uncertainty (i.e., which might serve as both outcomes and cues for future rewards) when they

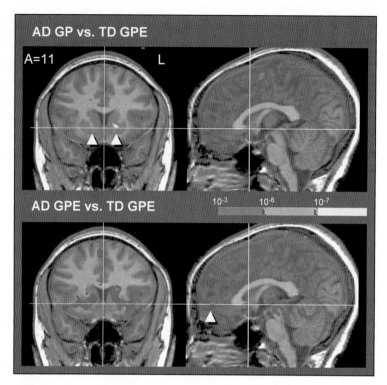

FIGURE 3. AD versus TD regressor comparison ($n = 26$, within-subjects; warm colors indicate better fit for AD terms while cool colors indicate better fit for TD terms; $P < 0.01$, uncorrected).

substantially deviate from the range of predictions suggested by a prior cue, or when they are correlated with outcomes of neighboring trials. Nonetheless, even the present simple AD split yields novel and testable predictions that extend beyond those generated by a single TD error term.

IMPLICATIONS

In the context of a stable incentive processing task, a temporal differences reward prediction error term correlated with activation in both NAcc and, to a lesser extent, MPFC regions, replicating a pattern of findings reported in a growing number of reports.[14,19,27,28] However, the currently proposed split of the TD error term revealed a spatiotemporal dissociation in which GP correlated most closely with NAcc activation, while GPE correlated more closely with MPFC activation.

In each region, AD GP and GPE terms predicted both increases and decreases in activation driven by local blood oxygenation. While changes in oxygenation correlate with changes in postsynaptic neural activity, researchers still have not determined how the two are physiologically linked.[29] Dopamine neurons have been consistently implicated in reward processing, and project from the ventral tegmental area (VTA) of the midbrain to both the NAcc and MPFC. Electrophysiological studies show that these dopamine neurons fire at an average rate of approximately 5 impulses per second,[30] and this firing temporarily increases when animals anticipate rewards or receive unexpected rewards, but temporarily decreases when animals fail to receive expected rewards.[10] Further, monkey research suggests that injection of dopamine-releasing agents, such as amphetamine, can increase FMRI activation in the NAcc, while concurrent dopamine depletion abolishes these amphetamine-induced increases.[31] Thus, dopaminergic modulation of postsynaptic targets may contribute to changes in activation visualized with FMRI.[32] However, the independence of signal changes in the NAcc and MPFC suggest that activation in these regions is not modulated solely by dopamine release, raising the possibility that NAcc and MPFC may exert differential control on VTA dopamine neurons via descending projections.[33] While not discussed here in detail (due to potential artifactual warping of midbrain regions), VTA activation more closely correlated with AD GP than GPE (see TABLE 1), a pattern that has been replicated in other FMRI studies.[34,35] Future studies with enhanced temporal resolution may better elucidate the influence of NAcc and MPFC activation on VTA activity.

The AD split of the TD error term offers both practical and theoretical benefits as a tool for predicting changes in FMRI activation. From a practical standpoint, models that enhance sensitivity for detecting activation promise to save investigators both time and money. These models may also help investigators to decompose and better understand symptoms related to psychiatric disorders. To have clinical relevance, a computational model must not only predict brain activity, but the predicted brain activity must then correlate with behavioral phenomena (e.g., affect, behavior, cognition) of psychiatric importance. In stable incentive processing tasks, NAcc activation has been correlated with the experience of positive but not negative aroused affect in healthy individuals undergoing both FMRI[24] and PET studies.[36] Deficits in positive arousal have been documented in psychiatric disorders ranging from affective disorders to addiction to schizophrenia. In the context of the MID task, we have recently observed that schizophrenics (both never-medicated and treated with traditional neuroleptics) show specific deficits in NAcc activation during GP. Further, the extent of this blunting correlates with a chronic absence of positive arousal (called "negative symptoms" in the psychiatric literature).[20,37] Thus, some schizophrenics may suffer from a deficit in GP, which in turn manifests as negative symptoms. A model that conflates GP and GPE would not have elucidated this specific deficit. Thus, this set of findings exemplifies

TABLE 3. AD versus TD regressor direct comparison foci ($P<.01$, uncorrected; cluster $= 3$ voxels; gain regressors only)

	Region	Z-score	R	A	S
AD GP vs. TD GPE**					
AD > TD	R cingulate	3.07	7	21	42
		2.86	18	3	53
	L medial frontal gyrus	3.74	−3	2	54
	R anterior insula	3.52	32	17	7
	L anterior insula	2.83	−31	15	6
	R NAcc/putamen	4.80	12	12	0
	L putamen	4.28	−18	14	−5
		3.35	−25	5	5
	R thalamus	3.29	15	−22	10
	L thalamus	3.35	−18	−15	2
	R SNc/midbrain	2.95	4	−17	−3
	L SNc/midbrain	3.26	−2	−15	−2
	R precentral gyrus/BA 6	3.95	27	−11	50
	L precentral gyrus/BA 6	3.05	−29	−16	60
		3.51	−33	−23	46
TD > AD	R MPFC	−3.88	3	56	8
		−3.44	3	49	0
		−3.84	6	48	25
	L superior frontal gyrus	−4.30	−15	36	39
	R subgenual cingulate	−2.85	2	18	−10
	R posterior cingulate	−4.15	10	−53	15
	L posterior cingulate	−3.86	−3	−50	33
	R parahippocampal gyrus	−3.23	12	−7	−15
	L parahippocampal gyrus	−3.89	−18	−15	−15
AD GPE vs. TD GPE					
AD > TD	MPFC	3.00	0	53	2
	R posterior cingulate	3.36	7	−53	30
		2.89	7	−53	16
	R insula	3.53	41	−12	15
TD > AD	R cingulate	−3.14	7	15	37
		−3.98	5	10	52
	L cingulate	−3.28	−4	4	45
		−4.22	−4	−16	49
	R anterior insula	−4.37	24	19	−1
	L anterior insula	−4.64	−30	19	−3
	L caudate/ putamen	−4.76	−16	9	4
	R caudate	−4.09	11	17	7
		−3.14	14	5	15
	L caudate	−4.60	−16	11	15
		−3.44	−12	3	15
	R putamen	−4.14	16	12	−1
		−3.99	22	1	3
	L putamen	−3.79	−20	11	−1
		−3.36	−22	−7	8
	R thalamus	−4.17	5	−16	6
	L thalamus	−4.35	−6	−16	5

Continued.

TABLE 3. (Continued)

Region	Z-score	R	A	S
AD GP vs. TD GPE**				
PAG/midbrain	−3.30	−4	−31	−11
R SNc/midbrain	−3.19	3	−18	−7
L SNc/midbrain	−3.80	−7	−16	−4
R precentral gyrus/BA 6	−3.34	15	−18	64
L precentral gyrus/BA 6	−3.59	−33	−19	58
R superior parietal lobule	−3.42	25	−49	54

**3 neighboring voxels at $P < 0.01$.

how properly specified computational models may help investigators to better probe covert phenomena (e.g., affective experience) that nonetheless can have an overt impact (e.g., on feelings, thoughts, or behavior) on psychiatric health.

From a theoretical standpoint, computational models must not only correlate with brain activation, but should also approximate the operation of neural mechanisms. The selection of models that fit neural constraints can thus be framed as a continuing journey of closer approximations. The AD split augments the TD error term by suggesting that the brain distinguishes between uncertain anticipation and certain outcomes, as well as between gain and loss.[38] Of course, such distinctions raise new questions about which neural circuits support the computation of different error terms, and how information from these circuits then combines to coherently inform learning and behavior. Eventual answers to such questions may help investigators to better isolate distinct neuropsychological components that contribute to complex disorders.

Distinct spatial correlates of AD GP and GPE imply that output from lower and higher centers may combine, either cooperatively or competitively, to produce a given behavior. For instance, if the NAcc computes GP while the MPFC computes GPE, MPFC lesions might severely impair relearning of reward associations, but paradoxically spare or even accentuate preexisting reward associations.[5] How and where these terms combine to channel behavior remains to be discovered. But as Jackson prophesied long ago, optimal function may require dynamic coordination of brain circuits both high and low.

ACKNOWLEDGMENTS

We thank Jamil Bhanji for technical assistance, as well as Jeffrey C. Cooper, Matthew T. Kaufman, Paul King, and an anonymous reviewer for comments on previous drafts of this manuscript. During manuscript preparation, BK was supported by a National Alliance for Research on Schizophrenia and Depression Young Investigator Award and NIDA Grant DA020615-01.

REFERENCES

1. TAYLOR, J. 1958. Selected Writings of John Hughlings Jackson. Basic Books. New York.
2. MACLEAN, P.D. 1990. The Triune Brain in Evolution. Plenum Press. New York.
3. LEDOUX, J.E. 1993. Emotional memory: in search of systems and synapses. *In* Brain Mechanisms: Papers in Memory of Robert Thompson, Vol. 702. F.M. Crinella & J. Yu, Eds.: 149–157. Ann. N.Y. Acad. Sci. New York.
4. PANKSEPP, J. 1998. Affective Neuroscience: The Foundations of Human and Animal Emotions. Oxford University Press. New York.
5. KNUTSON, B. & J.C. COOPER. 2005. Functional magnetic resonance imaging of reward prediction. Curr. Opin. Neurol. **18:** 411–417.
6. CRAIG, W. 1918. Appetites and aversions as constituents of instincts. Biol. Bull. **34:** 91–107.
7. KNUTSON, B. *et al.* 2003. A region of mesial prefrontal cortex tracks monetarily rewarding outcomes: characterization with rapid event-related FMRI. Neuroimage **18:** 263–272.
8. SUTTON, R.S. & A.G. BARTO. 1990. Time-derivative models of Pavlovian reinforcement. *In* Learning and Computational Neuroscience: Foundations of Adaptive Networks. M. Gabriel & J. Moore, Eds.: 497–537. MIT Press. Boston.
9. MONTAGUE, P.R., P. DAYAN & T.J. SEJNOWSKI. 1996. A framework for mesencephalic dopamine systems based on predictive Hebbian learning. J. Neurosci. **16:** 1936–1947.
10. SCHULTZ, W., P. DAYAN & P.R. MONTAGUE. 1997. A neural substrate of prediction and reward. Science **275:** 1593–1599.
11. MIRENOWICZ, J. & W. SCHULTZ. 1996. Preferential activation of midbrain dopamine neurons by appetitive rather than aversive stimuli. Nature **379:** 449–451.
12. FIORILLO, C.D., P.N. TOBLER & W. SCHULTZ. 2003. Discrete coding of reward probability and uncertainty by dopamine neurons. Science **299:** 1898–1902.
13. KAHNEMAN, D. & A. TVERSKY. 1979. Prospect theory: an analysis of decision under risk. Econometrica **47:** 263–291.
14. MCCLURE, S.M., G.S. BERNS & P.R. MONTAGUE. 2003. Temporal prediction errors in a passive learning task activate human striatum. Neuron **38:** 338–346.
15. O'DOHERTY, J.P. *et al.* 2003. Temporal difference models and reward-related learning in the human brain. Neuron **38:** 329–337.
16. KIM, H., S. SHIMOJO & J.P. O'DOHERTY. 2006. Is avoiding an aversive outcome rewarding? Neural substrates of avoidance learning in the human brain. Public Library of Science Biology **4:** e233.
17. DAW, N.D. & K. DOYA. 2006. The computational neurobiology of learning and reward. Curr. Opin. Neurobiol. **16:** 199–204.
18. KNUTSON, B. *et al.* 2001. Dissociation of reward anticipation and outcome with event-related FMRI. Neuroreport **12:** 3683–3687.
19. HARUNO, M. *et al.* 2004. A neural correlate of reward-based behavioral learning in caudate nucleus: a functional magnetic resonance imaging study of a stochastic decision task. J. Neurosci. **24:** 1660–1665.
20. JUCKEL, G. *et al.* 2006. Dysfunction of ventral striatal reward prediction in schizophrenia. Neuroimage **29:** 409–416.
21. SCHERES, A. *et al.* 2007. Ventral striatal hyporesponsiveness during reward anticipation in Attention Deficit/Hyperactivity Disorder. Biol. Psychiatry **61:** 720–724.

22. PAVLOV, I.P. 1927. Conditioned Reflexes: An Investigation of the Physiological Activity of the Cerebral Cortex. Oxford University Press. London.
23. KNUTSON, B. *et al.* 2000. FMRI visualization of brain activity during a monetary incentive delay task. Neuroimage **12:** 20–27.
24. KNUTSON, B. *et al.* 2001. Anticipation of increasing monetary reward selectively recruits nucleus accumbens. J. Neurosci. **21:** RC159.
25. COHEN, M.S. 1997. Parametric analysis of fMRI data using linear systems methods. Neuroimage **6:** 93–103.
26. COX, R.W. 1996. AFNI: software for analysis and visualization of functional magnetic resonance images. Comp. Biomed. Res. **29:** 162–173.
27. GOTTFRIED, J.A., J. O'DOHERTY & R.J. DOLAN. 2003. Appetitive and aversive olfactory learning in humans studied using event-related functional magnetic resonance imaging. J. Neurosci. **22:** 10829–10837.
28. O'DOHERTY, J. *et al.* 2004. Dissociable roles of ventral and dorsal striatum in instrumental conditioning. Science **304:** 452–454.
29. LOGOTHETIS, N.K. & B.A. WANDELL. 2004. Interpreting the BOLD signal. Ann. Rev. Physiol. **66:** 735–769.
30. WIGHTMAN, R.M. & D.L. ROBINSON. 2002. Transient changes in mesolimbic dopamine and their association with 'reward.' J. Neurochem. **82:** 721–735.
31. JENKINS, B.G. *et al.* 2004. Mapping dopamine function in primates using pharmacological magnetic resonance imaging. J. Neurosci. **24:** 9553–9560.
32. KNUTSON, B. & S.E.B. GIBBS. 2007. Linking nucleus accumbens dopamine and blood oxygenation. Psychopharmacology **191:** 813–822.
33. SHIZGAL, P. 1997. Neural basis of utility estimation. Curr. Opin. Neurobiol. **7:** 198–208.
34. KNUTSON, B. *et al.* 2005. Distributed neural representation of expected value. J. Neurosci **25:** 4806–4812.
35. ADCOCK, R.A. *et al.* 2006. Reward-motivated learning: mesolimbic activation precedes memory formation. Neuron **50:** 507–517.
36. DREVETS, W.C. *et al.* 2001. Amphetamine-induced dopamine release in human ventral striatum correlates with euphoria. Biol. Psychiatry **49:** 81–96.
37. JUCKEL, G. *et al.* 2006. Dysfunction of ventral striatal reward prediction in schizophrenic patients treated with typical but not atypical neuroleptics. Psychopharmacology **187:** 222–228.
38. YACUBIAN, J. *et al.* 2006. Dissociable systems for gain-and loss-related value predictions and errors of prediction in the human brain. J. Neurosci. **26:** 9530–9537.

Reward-Related Responses in the Human Striatum

MAURICIO R. DELGADO

Department of Psychology, Rutgers University, Newark, New Jersey, USA

ABSTRACT: Much of our knowledge of how reward information is processed in the brain comes from a rich animal literature. Recently, the advancement of neuroimaging techniques has allowed researchers to extend such investigations to the human brain. A common finding across species and methodologies is the involvement of the striatum, the input structure of the basal ganglia, in a circuit responsible for mediating goal-directed behavior. Central to this idea is the role of the striatum in the processing of affective stimuli, such as rewards and punishments. The goal of this article is to probe the human reward circuit, specifically the striatum and its subdivisions, with an emphasis on how the affective properties of outcomes or feedback influence the underlying neural activity and subsequent decision making. Discussion will first focus on anatomical and functional considerations regarding the striatum that have emerged from animal models. The rest of the article will center on how human neuroimaging studies map to findings from the animal literature, and how more recently, this research can be extended into the social and economic domains.

KEYWORDS: reward; punishment; striatum; incentive; valence; neuroeconomics; instrumental conditioning; goal-directed behavior; caudate nucleus; nucleus accumbens; dopamine

INTRODUCTION

Rewards can broadly be defined as desirable outcomes that serve to influence behavior. Information conveyed by rewards is important for learning about and deciding between different courses of action. As our behavior is motivated by the outcomes of our actions, day-to-day activities, such as going to work, are performed routinely to either achieve a reward (e.g., receiving a paycheck) or to avoid a punishment (e.g., losing your job). Thus, rewards have been posited to serve various functions, such as inducing subjective feelings of pleasure, eliciting exploratory or approach behavior, and increasing the frequency and

Address for correspondence: Mauricio Delgado, Department of Psychology, Rutgers University, 101 Warren Street, 340 Smith Hall, Newark, NJ 07102. Voice: +973-353-5440, ext: 241; fax: +973-353-1171.

delgado@psychology.rutgers.edu

Ann. N.Y. Acad. Sci. 1104: 70–88 (2007). © 2007 New York Academy of Sciences.

doi: 10.1196/annals.1390.002

intensity of behaviors that lead to rewards.[1] Central to understanding how rewards impact goal-directed behavior is an appreciation of how rewards affect typical neural processes, which in turn lead to changes in behavior. Therefore, a necessary step in understanding behavior is to understand how knowledge of rewards and punishments is represented in our brain, and how such knowledge leads to learning of new associations that serve to guide goal-directed behavior (e.g., going to work leads to monetary income).

The goal of this article is to survey the current literature on the neural correlates of reward-related processing, with an emphasis on how the affective properties of outcomes or feedback influence choice behavior and, consequently, goal-directed behavior. One brain structure in particular, the striatum, has been thought to be involved in reward-related processes. The striatum is the recipient of cortical and dopaminergic projections, being centrally positioned in functional loops that exert an influence over motor and cognitive aspects of behavior.[2–4] Thus, the focus of the discussion will be on the contributions of the human striatum to reward-related processing.

The first section of this article involves a brief review of basal ganglia anatomy and function, including support from studies in rodents and non-human primates, which highlight anatomical and functional divisions within the striatum. The second section outlines early efforts aimed at characterizing the response of the human striatum to simple behaviors and reward outcomes. Based on current neuroimaging data, the third section then considers the potential role of the human striatum and its specific subdivisions in reward-related processing. The fourth section looks at how the striatum signal can be modulated by properties of reward (e.g., probability of consumption), followed by a discussion in the fifth section of how such signals in the striatum are influenced by more complex social interactions that occur during day-to-day behavior (e.g., trust). Finally, the last section briefly describes some outstanding questions in the field and possible extensions of this work.

THE STRIATUM: ANATOMICAL AND FUNCTIONAL CONSIDERATIONS

The basal ganglia complex is formed by a group of structures that exert various functions primarily related to motor and learning aspects of behavior. The main structures that form the basal ganglia are the striatum, the globus pallidus, the subthalamic nucleus, and the substantia nigra. The globus pallidus and substantia nigra can further be subdivided into smaller components, and most of these subsections, working in conjunction with the glutamatergic projections from the subthalamic nucleus, serve as output structures of the basal ganglia.[5] A specific portion of the substantia nigra (the pars compacta division) produces dopamine (DA), a neurotransmitter found to innervate areas of the basal ganglia complex, such as the striatum, that is involved in motor and reward processes.[6]

The main input unit of the basal ganglia is the striatum, which receives synaptic input from cortical and subcortical afferents, such as motor cortical input and dopaminergic projections from substantia nigra (but also other midbrain nuclei, such as the ventral tegmental area[2,4,5,7–9]). The striatum can be further subdivided into dorsal and ventral components. The dorsal striatum primarily consists of the caudate nucleus, an extensive structure that lies medially in the brain (adjacent to the lateral ventricles) and the putamen, which expands ventrally and laterally to the caudate nucleus. The dorsal striatum receives extensive projections from dorsolateral prefrontal cortex, as well as other surrounding frontal regions (e.g., premotor cortex, frontal eye fields[5,8]). The ventral striatum consists primarily of the nucleus accumbens (although portions of the putamen and ventral caudate are also considered part of the ventral striatum) and receives extensive projections from ventral frontal regions (orbitofrontal, ventromedial, and ventrolateral cortex[10,11]). As previously mentioned, both the dorsal and ventral striatum also receive dopaminergic input from substantia nigra and ventral tegmental area, respectively. In addition, the striatum (especially ventral portions) has connections with limbic areas implicated in emotional processing, such as the amygdala.[10]

Historically, the basal ganglia complex has been considered a collection of structures involved in motor functions. This is predominantly due to observations of motor deficits in patients afflicted with Parkinson's disease, a neurodegenerative disorder that affects the microcircuitry of the basal ganglia. More recently, however, research has shown that the multiple corticostriatal loops that connect the basal ganglia with the rest of the brain may serve different functions, ranging from the control of eye movements[12,13] to more motivational behaviors.[14–17] The striatum, in particular, has been linked to various aspects of learning (for review see Ref. 18), such as habit formation,[19] skill learning,[20] and reward-related learning.[21,22] The multifaceted striatum, therefore, has been posited to integrate information regarding cognition, motor control, and motivation. For example, Kawagoe and colleagues[13] used a memory-guided saccade task with an asymmetric reward schedule to show that the nonhuman primate caudate nucleus is important in connecting both action (i.e., eye movements) and motivation (i.e., reward expectation) information. In this experiment, neuronal activity in the nonhuman primate caudate nucleus representing visual and memory responses was sensitive to changes in reward contingency, resulting in earlier and faster saccades during trials that led to rewards.

Within the striatum, further subdivisions (e.g., dorsal and ventral) exist that are considered to be functionally distinct. Studies in rodents have suggested that the ventral striatum, in particular the nucleus accumbens, is involved in affective and motivational processing. For instance, lesions in the rat ventral striatum lead to deficits in approach behavior.[23] In contrast, lesions in the rat dorsal striatum lead to consummatory deficits (i.e., failure to consume a reward) or shortfalls in stimulus response learning,[24] suggesting a role for the dorsal striatum in more cognitive and sensorimotor functions.[2,18,25] Although much of the existing animal research supports a functional division in the striatum between dorsal and ventral components, there are also theories that posit that a gradient of information shifts from more ventromedial (i.e., caudate and nucleus accumbens) to dorsolateral (i.e., putamen) structures during affective learning.[26] Irrespective of how information flows through the basal ganglia, due to its heterogeneity in terms of connectivity and functionality, the striatum finds itself in a prime position to integrate affective, motor, and cognitive information and influence goal-directed behavior.

A vast array of research implicates the striatum in reward-related processing. Neurons in the nonhuman primate striatum, for example, have been shown to respond to the anticipation[13,27] and delivery[12,28] of rewards. These striatal neurons have also been found to fire more vigorously for preferred rewards,[29] also showing modulation of activity based on different magnitudes of reward.[30] Further, significant increases in DA release in both dorsal and ventral striatum have been observed during cocaine self-administration in rats.[31,32] Thus, the existing animal models suggest that presentation of affective, unpredictable outcomes contingent on a learned behavior involve the reward systems of the brain, specifically the striatum.

THE HUMAN STRIATUM AND REWARD

Recently, advances in neuroimaging methodology have allowed investigators to confirm and extend the findings of a rich animal literature. The initial efforts aimed at understanding the link between rewards and human neural responses were performed using positron emission tomography (PET). An example is a report of increased DA release in both dorsal and ventral striatum (measured by displacement of raclopride binding to D2 receptors by endogenous DA release) when participants played a video game for incentives.[33] Similar results were obtained with food rewards, as DA release has been reported to rise in the dorsal striatum of hungry participants when stimulated with food items.[34] Breiter and colleagues[35] used functional magnetic resonance imaging (fMRI) to study the human brain's reward circuit and addiction. By giving cocaine addicts injections of cocaine while in the scanner, the authors were able to show that activity in the human ventral striatum correlated to feelings of craving, while activity in the dorsal striatum correlated to feelings of the rush felt after receiving the drug. Thus, early neuroimaging experiments built on the existing animal literature by demonstrating that the human striatum is involved in reward processing, particularly when primary rewards were available for consumption (e.g., food, cocaine).

A drawback of the early PET and fMRI studies was the limitations in the experimental designs. For instance, although activation of the striatum was observed in a video game where rewards were present,[33] it is difficult to assess what the signal is due to (i.e., anticipation, delivery, or even magnitude of rewards). Thus, it was imperative for new designs to isolate specific components of the reward response. One specific paradigm was developed with the goal of probing neural responses to the delivery of monetary rewards and punishments.[36] In the form of a card-guessing game, participants were asked to guess if the value of a "card" was higher (values 6–9) or lower (values 1–4) than the number 5 (FIG. 1A). Following the choice, participants received the outcome of the card (the actual value) and a feedback symbol that indicated if the participant was correct (reward, monetary gain of $1.00) or incorrect (punishment, monetary loss of $0.50). The difference in magnitude between a positive and negative outcome is attributed to prospect theory and the idea that the impact of losses loom larger than gains.[37] Each trial in this event-related paradigm corresponded to one guess and one outcome, although unbeknownst to the participants, the outcomes were predetermined to ensure that each participant played the same reinforcement schedule and received the same feedback. Thus, the goal of the simple card game was to recruit the striatum with repeated presentations of unpredictable delivery of rewards and punishments.

Increases in blood oxygenation level-dependent (BOLD) responses in both dorsal and ventral striatum were observed using this paradigm, showing differential responses according to the feedback received (FIG. 1B[36]). Specifically, at the onset of a trial, an increase in the BOLD response was observed as participants were faced with a choice and made a prediction and subsequent guess. This hemodynamic response was sustained and slowly returned to baseline when reward feedback was given, but decreased more rapidly before returning to baseline when punishment feedback was given. Replications of this paradigm were also carried out in different populations, such as adolescents[38] and nicotine addicts.[39]

This was the first observation in humans of differential hemodynamic responses in the striatum to monetary outcomes of different valence (reward and punishment). Further, the findings could be mapped to the existing animal literature, supporting a role for the striatum in reward processing, and were concurrent with reports from early human imaging studies involving either rewards[33,35,40] or delivery of positive or negative feedback.[41–44] Other studies soon followed building on new imaging techniques and investigating different facets of the reward response. Many of these involved elegant paradigms that implicated the striatum during anticipation of both primary and secondary rewards.[45–49]

The next step in understanding how the human striatum responded to rewards was to modulate the characterized BOLD response. An alteration to the paradigm allowed for investigation of potential changes in the striatum

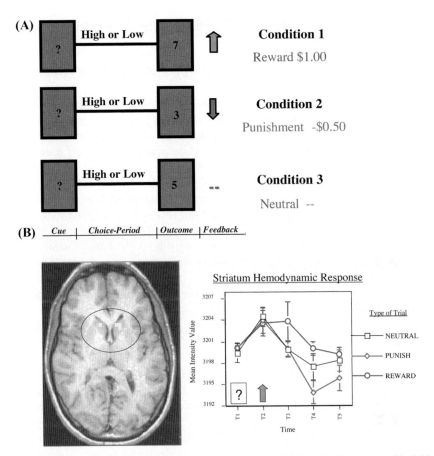

FIGURE 1. (**A**) The card-guessing task—a random "high or low" guess could yield one of three affective outcomes (reward, punishment, neutral). (**B**) Activation of both dorsal (pictured) and ventral striatum showing a differential response to reward and punishment outcomes. Time depicted as repetition time (TR) or scans, each time epoch lasting 3 sec. *Arrow* represents time of reward delivery—hemodynamic response occurs 6–9 sec after stimulus presentation. (Figure adapted from Delgado et al.[36] used with permission.)

according to variations in feedback properties, such as magnitude. The basic card-guessing game was modified to include delivery of four potential feedbacks of unpredictable valence (reward and punishment) and magnitude (large and small). While activity was once again observed in both dorsal and ventral striatum showing differential responses according to valence, magnitude differences were mostly in the dorsal striatum, where a parametric ranking of the BOLD signal according to both valence and magnitude was observed.[50] More recently, magnitude differences have been observed in the striatum with different paradigms[46,51,52] although the response may be context-dependent.[53,54]

INTERPRETATION OF THE STRIATUM REWARD RESPONSE

Using event-related fMRI designs allowed researchers to characterize the response of the human striatum through different phases of reward processing. Two interesting questions surfaced, however, with respect to the striatum signal during affective outcomes. First, while activation in both dorsal and ventral striatum was observed during delivery of rewards and punishments in the card-guessing game, the intensity of the fMRI signal was higher in the dorsal striatum, predominantly the head of the caudate nucleus.[36] This was a slightly surprising finding—in contrast with research in animals, which often highlights the role of the ventral striatum (chiefly the nucleus accumbens) in reward processes—raising a question about the role of the human dorsal striatum. Second, what exactly did dorsal striatum activation in this paradigm mean? Were the increases in BOLD signal in the caudate nucleus due to the delivery of rewards? Or were there other features of the card-guessing paradigm (such as making a choice) that recruited the striatum?

To test these questions, a study was designed to mimic elements of the card-guessing game, while attempting to isolate the response to reward.[55] In an oddball-like paradigm, participants were presented with a series of purple squares in succession. At random intervals, an "oddball" would be presented and the participant was to press a button to indicate recognition. There were three oddballs: a reward (green upward arrow worth $1.50), a punishment (red downward arrow worth –$0.75), and a neutral oddball (a blue dash worth no monetary value). If the dorsal striatum is involved in the detection of rewards, then increases in BOLD signal in the striatum should be observed during the delivery of reward oddballs. A region of interest (ROI) analysis in the caudate nucleus defined previously, however[36], revealed no significant increases in BOLD response to any of the affective oddballs presented. An additional analysis found little activation in the ventral striatum as well, although the primary analysis was in the dorsal striatum and the fMRI signal in the nucleus accumbens was not optimized, thus deeming the ventral striatum results exploratory. It is worthy to note that in an oddball paradigm, the intensity of the affective stimulus is a potential issue, and a secondary reward, such as money, may not be as intense as a primary reward, such as juice, that led to activation in the ventral striatum when delivered in an unpredictable manner.[56] The goal of the study, however, was to determine what reward-related process was responsible for robustly recruiting the dorsal striatum. This experiment suggested that the caudate nucleus is not activated by the mere delivery of rewards and punishments.[55]

If the caudate nucleus response in the card-guessing game is not attributed to the reward itself, then perhaps there was something about the way the reward was attained. That is, participants had a feeling of agency, as they believed their choice or guess directly led to the reward, suggesting that perhaps learning mechanisms may be involved in this task. This hypothesis was tested in a separate study that built on the oddball paradigm.[55] Participants were once again shown a series of purple squares and were told that intermittent presentations of affective outcomes (rewards and punishments) would occur throughout. However, they were also told that they would see two anticipatory cues that predicted delivery of either a reward or a punishment. If a yellow circle was presented, participants were instructed to press one of two buttons to identify recognition of oddball. They were aware that an affective outcome followed the circle, but that the button press had no effect on the valence of the outcome. In contrast, if a blue circle was presented, participants were instructed to press one of two buttons with the perception that said button press could influence the outcome. Thus, the blue circle was much like the question mark in the card-guessing game, which elicited a button press and a prediction. In both experiments, participants felt that the reward was contingent on their behavior.

The dissociation between rewards and punishments observed previously was only replicated during presentation of the blue circle, when rewards were contingent on behavior (FIG. 2A). When participants were making a noncontingent button press (i.e., yellow circle), no differential activity was observed (FIG. 2B[55]). The combined findings of both experiments in the Tricomi *et al.*[55] study suggest that the dorsal striatum, specifically the caudate nucleus, responds to the reinforcement of an action, rather than the reward per se.

Evidence from some animal studies supports a potential role for the dorsal striatum in reinforcement-based and instrumental learning.[57,58] In a series of microdialysis studies,[32,59] Ito and colleagues showed that dopamine levels in the dorsal striatum are elevated when rats are presented with a conditioned stimulus in which cocaine delivery is contingent upon a behavior (i.e., drug-seeking behavior), but not when a noncontingent-conditioned stimulus is presented (which leads to increases in DA release in the ventral striatum). Corroboration of these findings was also found in other neuroimaging studies using different paradigms.[60,61] One distinct study used a conditioning paradigm incorporating fMRI and reinforcement learning models.[62] Building on the ideas put forth by the actor–critic model,[63] O'Doherty and colleagues[62] found that the ventral striatum was activated during both an instrumental and classical conditioning session, thus behaving like a "critic"; that is, it is involved in predicting potential rewards. In contrast, the dorsal striatum was activated solely during instrumental conditioning, leading the authors to posit that the dorsal striatum may be the "actor," maintaining the reward outcome of actions to optimize future choices that will lead to reward. Hence, this collection of results strongly suggest that the human dorsal striatum is involved in reward processing, specifically learning and updating actions that lead to reward, rather than representing and identifying rewards, a function postulated to occur in the frontal cortex.[21,64,65]

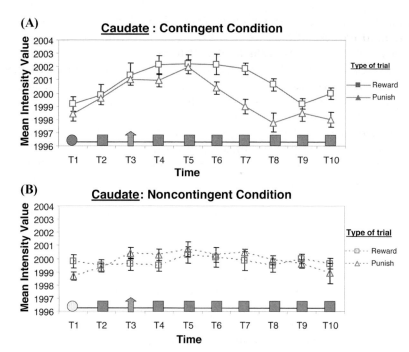

FIGURE 2. Affective oddball paradigm results. Time epochs are 1.5 sec each. Anticipatory circle, feedback arrows, and squares are depicted in both graphs. (**A**) Differential blood oxygenation level-dependent (BOLD) responses in the caudate nucleus to reward and punishment feedback when participants perceive a contingency between action and outcome. (**B**) BOLD response in the caudate nucleus during noncontingent delivery of feedback. There is no difference between the two affective conditions (reward and punishment). (Figure adapted from Tricomi et al.[55]—reprinted from *Neuron,* with permission from Elsevier.)

THE STRIATUM AND REWARD-RELATED LEARNING

Research suggests that the human striatum is part of a circuit important during learning, being involved in evaluating current rewards to guide future actions.[66,67] Research in patients with Parkinson's disease further suggests a link between the striatum and trial and error learning. For instance, compared with control subjects, Parkinson's patients are slower during initial learning of associative paradigms,[68] showing deficits during a feedback-based learning task, as opposed to intact learning during a nonfeedback version of the same paradigm.[69] Neuroimaging of similar cognitive learning paradigms involving feedback has resulted in activation of the striatum (mostly the dorsal striatum) differentiating between positive and negative feedback,[43,44,70] substantiating the neuropsychological data.

The deficit in feedback-based learning in Parkinson's patients can be attributed to the low levels of DA in the striatum, as restoration of DA via agonist medication (e.g., L-dopa) also restores sensitivity to positive feedback.[71] Dopamine neurons, a strong source of input into the striatum, have been associated with reward-related learning processes by neurophysiological recordings in nonhuman primates.[72] In such studies, dopamine neurons fire first to the delivery of unpredictable rewards (e.g., juice), and second, after learning, to the earliest predictor of rewards (e.g., a light). Further, these neurons show a depression in neuronal firing when an expected reward fails to occur, suggesting a role for dopamine neurons in coding for prediction errors during affective learning. In humans, prediction errors have elicited activation in both dorsal and ventral striatum, particularly in the putamen.[73–75]

There are parallels between the neurophysiological role of DA in learning and the increases in BOLD response observed in the striatum that were previously described in the card-guessing and affective oddball paradigms. In both experiments there was no explicit learning as participants believed the outcomes were random. Perception of control is important, however, and participants may have made predictions when a choice determined the reward. In such a scenario, the question mark or anticipatory cue may have served as the earliest predictor of a potential reward, eliciting activation at the onset of the trial. This BOLD response was then sustained if a reward was the outcome, but decreased if the choice resulted in punishment, indicating a withdrawal of an expected reward and a signal to adjust predictions or choices. The same dissociation between positive and negative feedback was observed in the caudate nucleus of participants performing a perceptual learning task.[76] Interestingly, the same participants also performed the card-guessing game in the same session, showing analogous responses between feedback received during the game and the learning task, suggesting a critical role for the caudate nucleus as a moderator of the influence of feedback on learning.

Although these feedback-based paradigms suggest a role for the dorsal striatum in reinforcement learning, the lack of explicit measures of learning does not allow such conclusion. Instead, it is necessary to measure how the reward feedback serves to shape future behavior by increasing the frequency of the optimal choice. Given the previous findings in the caudate nucleus, it can be hypothesized that during learning, caudate activation should be modulated by the perceived "value" or information provided by the reward feedback, leading to better choices for continued rewards. A modified version of the card-guessing task attempted to address this question by introducing a probabilistic cue prior to participants' guesses.[22] The cue indicated the probability that the card was of high or low value, and once learned would lead to maximization of available rewards. There were three main types of cues: easy to learn (e.g., cue predicted a high card 100% of the time), harder to learn (e.g., cue predicted a high card 67% of the time), and random (e.g., cue predicted a high card 50% of the time). Overall, participants were quicker and more accurate at learning

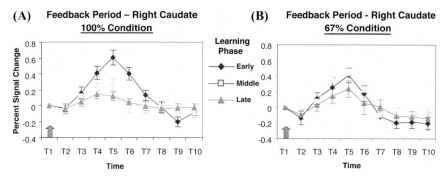

FIGURE 3. Probabilistic card-guessing task results—functional magnetic resonance imaging (fMRI) signal in caudate nucleus during delivery of positive feedback. (**A**) Activation during deterministic (100%) condition. The BOLD response to positive feedback is larger during the early trials of learning (when feedback is informative) compared to late trials (when feedback is predictable). (**B**) Activation during probabilistic condition (67%). The BOLD response to positive feedback is similar across all stages of learning, since the feedback is still informative. Each time epoch is 1.5 sec and *arrow* represents time of feedback delivery. (Figure adapted from Delgado *et al.*,[22] reprinted from *NeuroImage,* with permission from Elsevier.)

the easy associations (100%). Brain imaging data suggested that the caudate nucleus was integral for performance, being activated during the initial early stages of learning (i.e., the first few trials) when feedback was most valuable. As learning progressed, however, the response in the caudate nucleus to a "reward" decreased, as the 100% cue became predictable (FIG. 3A). In contrast, more probabilistic cues (e.g., 67%) did not differ between early and late stages of learning (FIG. 3B). That is, where the reward feedback was still valuable in educating predictions, caudate activation was still observed. These data and others[77] suggest that the human caudate nucleus is an integral component of a circuit involved in learning and updating current rewards with the purpose of guiding action that will maximize reward consumption.

THE SOCIAL BRAIN: THE STRIATUM
AND SOCIAL BEHAVIOR

Over the past few years, human neuroimaging studies have been successful in replicating and extending ideas put forth by animal models of reward processes. Central to these models is the role of the striatum in processing reward-related information and mediating goal-directed behavior. Yet, much of our knowledge comes from studies investigating simple rewards and behaviors. The challenge for researchers in the upcoming years is to broaden this knowledge by drawing parallels between these simple behaviors to more complex behaviors that occur in everyday human life. For instance, learning to trust someone new is a trial and error procedure in which social interaction is crucial for developing the feeling of trust. In a thought-provoking experiment, King-Casas et al.[78] investigated neural responses of individuals involved in multiple rounds of economic exchange (i.e., the trust game). Participants learned through trial and error if a partner was trustworthy or not, developing trust through reciprocity. Neural signals in the dorsal striatum (the head of the caudate nucleus) mirrored the behavioral results, as the response in the caudate nucleus shifted to the earliest sign of potential "trust"—a pattern analogous to responses observed in DA models of reinforcement learning, suggesting the development of reputation.

The decision to trust can be developed through trial and error experience, but can also be influenced by previous biases or information learned through different methods. Indeed, moral beliefs can influence economic behavior and the choices we make. An individual may be willing to work for a lower salary, for example, if they believe their employer's mission is morally praiseworthy.[79] The observation that not all learning occurs through trial and error motivated a social learning experiment involving a variation of the trust game.[80] In this interactive game, participants were instructed that they would have a choice between keeping $1.00 or sharing the money with a partner (resulting in a $3.00 investment). In a subsequent feedback phase, the participant then received either positive or negative feedback regarding their share decision by finding out if the partner was trustworthy (split the investment resulting in a $1.50 reward) or untrustworthy (kept the money), respectively. Participants were also instructed they would be playing multiple rounds of this game with three fictional characters, depicted by individual bios to be of good, bad, or neutral moral character. Despite the assigned personalities, however, each partner played exactly the same. That is, each partner's reinforcement rate was exactly 50%, suggesting that participants should learn over time to overcome their initial bias and improve their decision making.

Participants showed a degree of explicitly learning that the partners played similarly, illustrated by ratings of trustworthiness acquired during pre- and post-experimental sessions. Yet behaviorally, participants chose to share more frequently with the "good" partner, and keep more often with the "bad" partner, while no differences were observed with the "neutral" partner (FIG. 4A). This preference in choice behavior was observed both in early and late trials of the experiment, suggesting that although participants showed some explicit learning, implicitly they were still influenced by their previous bias.

As expected, the caudate nucleus was activated during the feedback phase showing differential responses to positive compared to negative feedback (FIG. 4B). This was especially true for the "neutral" partner trials, where little or no information was available to bias decisions. The "neutral" partner trials results are the most analogous to the original card-guessing task[36] and formation of reputation previously observed in the trust game,[78] as participants learn through trial and error how to ameliorate their decisions. When prior

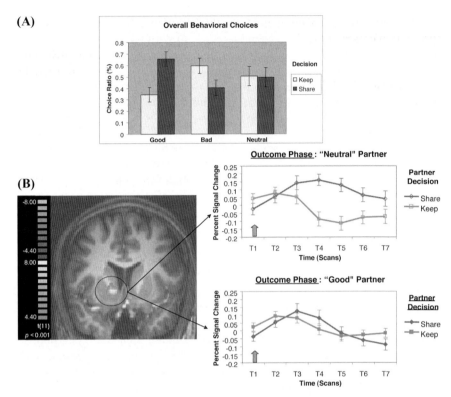

FIGURE 4. Trust game results. (**A**) Average choices by participants performing a multitrial trust game with three partners depicted to be of a certain moral character. (**B**) Activation of the ventral caudate nucleus during the outcome phase showing differential activation between positive and negative feedback during trials involving the neutral partner (where little information was available to influence decisions), but not the good partner (where previous information biased decision making). *Time epoch* represents 2 sec each and *arrow* represents time of feedback delivery. (Figure adapted from Delgado et al.[80])

information, such as perceived moral character, was available, however, activation in the caudate nucleus did not differentiate between positive and negative feedback.[80] There are two potential interpretations of the caudate activation. First, the strength of prior beliefs can lead participants to bypass current negative feedback thus delaying adjustments in the decision-making process, resulting in a lack of differential responses to feedback. Alternatively, the surprise that occurs when a "good" partner does not share, for example, could elicit activation in the striatum due to the saliency of the omission[61] or perhaps other emotions confounded together (e.g., disbelief, regret, disappointment).

Other experiments have also investigated the role of the striatum during social and economic interactions. Many of these studies have found that the striatum is involved during interactions that result in cooperation,[81] or in response to pictures of previous cooperators,[82] and even when exerting revenge on previous defectors.[83] From such experiments and the two trust experiments previously described, it is clear that the dorsal striatum, particularly the caudate nucleus, is involved in social learning. It is also apparent that prior beliefs can influence the neural mechanisms of trial and error learning, perhaps leading participants to not optimize their choice behavior.

FINAL THOUGHTS: OUTSTANDING ISSUES AND FUTURE RESEARCH

Much progress has occurred over the last few years regarding our understanding of human reward circuits. Yet, some issues and questions still remain to be resolved. First and foremost is investigating how classes of affective reinforcers differentially affect the human reward circuit. For example, are there differences in how rewards and punishments influence learning and how they are processed in the brain? Some studies suggest that while DA is involved in learning from positive feedback, it is not necessary for inferring information from negative feedback[71] (e.g., punishment), which could be mediated by serotonin.[84] An extension of this question is a specific investigation of the role of the striatum in aversive processing. While this is still an open question, recent studies suggest that the human striatum is not only involved in appetitive or reward processing, but also in aversive processing,[85,86] and even fear conditioning.[87] One possibility is that the striatum is coding for the saliency of a stimulus.[61] Another possibility, however, is that the striatum is involved in affective learning, and both appetitive and aversive stimuli can motivate behavior and lead to learning.

What about differences between primary and secondary reinforcers? Due to social evolution, is money (a secondary reinforcer) as potent as a primary reinforcer (e.g., juice)? The overlap in findings between studies that use either primary or secondary reinforcers is evident, such as activation of the ventral striatum during anticipation of either type of reward.[46,88] Nevertheless, discrepancies are also apparent, as some studies where juice rewards are used typically find activation in the ventral putamen (perhaps due to gustatory stimulation or motor properties of the tongue), while in experiments with monetary rewards, the locus of activation tends to be around the nucleus accumbens and ventral caudate. Some studies have started to directly compare primary and secondary reinforcers within subjects,[89] and although early results point to similarities in how both reinforcers affect learning, it is difficult to titrate primary and secondary reinforcers with respect to intensity, subjective value, and time of delivery (i.e., immediate vs. delayed).

A second question with respect to the human striatum's role in reward processing is its functional and anatomical hierarchy. Often in neuroimaging studies researchers refer to activation in the striatum, without specifically parceling out the subdivisions. Even within dorsal and ventral striatum, sometimes it is unclear if the locus of activity is in the nucleus accumbens or ventral head of the caudate nucleus (see Ref. 90, for example, of human ventral striatum boundaries and DA receptor distribution). Thus, future work would enhance our knowledge about the gradient of information flow in the human striatum and how functional and anatomical subdivisions are established.

Finally, a more recent trend is to understand the social and affective human brain; that is, to investigate how social rewards modulate brain circuits involved in goal-directed behavior. There are some promising initial studies that extend findings from studies involving simple processes (e.g., reward learning) to more complex social behaviors (e.g., developing trust). In the future, investigators will also probe interactions between the human reward system in social situations and memory processes. Are there memory processes mediated by medial temporal lobe structures or prefrontal areas that, by encoding and reactivating memory traces, inhibit the striatum during trial and error learning?[80] How are prediction errors calculated during social interactions and how do they influence decision processes?

Due to anatomical considerations, the striatum is thought to be involved in possibly integrating information of reward-related information in the brain, receiving input from cortical and limbic regions that may be further modulated or shaped by mesencephalic dopaminergic projections.[91] Both dorsal and ventral striatum are involved in reward processing, specifically aspects of affective learning. Although a variety of studies ranging across species suggest a link between action–outcome learning and the dorsal striatum, additional research is needed to enhance our knowledge about information flow in the striatum and potential contributions of its subdivisions to goal-directed behavior. Future investigations will also aim to extend our understanding of the human reward circuit by probing the interaction between natural rewards and punishments and social situations that occur in day-to-day interactions.

ACKNOWLEDGMENTS

The author would like to acknowledge Susan Ravizza and Dominic Fareri for helpful comments and discussion, Elizabeth Tricomi for assistance, and John O'Doherty for constructive feedback and suggestions.

REFERENCES

1. SCHULTZ, W. 2000. Multiple reward signals in the brain. Nat. Rev. Neurosci. **1:** 199–207.
2. GRAYBIEL, A.M. *et al.* 1994. The basal ganglia and adaptive motor control. Science **265:** 1826–1831.

3. MIDDLETON, F.A. & P.L. STRICK. 2000. Basal ganglia output and cognition: evidence from anatomical, behavioral, and clinical studies. Brain Cogn. **42:** 183–200.

4. MIDDLETON, F.A. & P.L. STRICK. 1997. New concepts about the organization of basal ganglia output. Adv. Neurol. **74:** 57–68.

5. ALEXANDER, G.E. & M.D. CRUTCHER. 1990. Functional architecture of basal ganglia circuits: neural substrates of parallel processing. Trends Neurosci. **13:** 266–271.

6. SCHULTZ, W. 1998. The phasic reward signal of primate dopamine neurons. Adv. Pharmacol. **42:** 686–690.

7. KIMURA, M. & A.M. GRAYBIEL. 1995. Role of basal ganglia in sensory motor association learning. In Functions of the Cortico-Basal Ganglia Loop. M. Kimura & A. M. Graybiel, Eds.: 2–17. Springer-Verlag. Tokyo.

8. MIDDLETON, F.A. & P.L. STRICK. 2000. Basal ganglia and cerebellar loops: motor and cognitive circuits. Brain Res. Brain Res. Rev. **31:** 236–250.

9. GRAYBIEL, A.M. 2000. The basal ganglia. Curr. Biol. **10:** R509–R511.

10. GROENEWEGEN, H.J. et al. 1999. Convergence and segregation of ventral striatal inputs and outputs. Ann. N. Y. Acad. Sci. **877:** 49–63.

11. LYND-BALTA, E. & S.N. HABER. 1994. The organization of midbrain projections to the ventral striatum in the primate. Neuroscience **59:** 609–623.

12. HIKOSAKA, O., M. SAKAMOTO & S. USUI. 1989. Functional properties of monkey caudate neurons. III. Activities related to expectation of target and reward. J. Neurophysiol. **61:** 814–832.

13. KAWAGOE, R., Y. TAKIKAWA & O. HIKOSAKA. 1998. Expectation of reward modulates cognitive signals in the basal ganglia. Nat. Neurosci. **1:** 411–416.

14. ROBBINS, T.W. & B.J. EVERITT. 1996. Neurobehavioural mechanisms of reward and motivation. Curr. Opin. Neurobiol. **6:** 228–236.

15. IKEMOTO, S. & J. PANKSEPP. 1999. The role of nucleus accumbens dopamine in motivated behavior: a unifying interpretation with special reference to reward-seeking. Brain Res. Rev. **31:** 6–41.

16. JENTSCH, J.D. & J.R. TAYLOR. 1999. Impulsivity resulting from frontostriatal dysfunction in drug abuse: implications for the control of behavior by reward-related stimuli. Psychopharmacology (Berl). **146:** 373–390.

17. BALLEINE, B.W. 2005. Neural bases of food-seeking: affect, arousal and reward in corticostriatolimbic circuits. Physiol. Behav. **86:** 717–730.

18. PACKARD, M.G. & B.J. KNOWLTON. 2002. Learning and memory functions of the basal ganglia. Annu. Rev. Neurosci. **25:** 563–593.

19. JOG, M.S. et al. 1999. Building neural representations of habits. Science **286:** 1745–1749.

20. POLDRACK, R.A. et al. 1999. Striatal activation during acquisition of a cognitive skill. Neuropsychology **13:** 564–574.

21. O'DOHERTY, J.P. 2004. Reward representations and reward-related learning in the human brain: insights from neuroimaging. Curr. Opin. Neurobiol. **14:** 769–776.

22. DELGADO, M.R. et al. 2005. An fMRI study of reward-related probability learning. Neuroimage **24:** 862–873.

23. ROBBINS, T.W. & B.J. EVERITT. 1992. Functions of dopamine in the dorsal and ventral striatum. Sem. Neurosci. **4:** 119–127.

24. ROBBINS, T.W. et al. 1989. Limbic-striatal interactions in reward-related processes. Neurosci. Biobehav. Rev. **13:** 155–162.

25. WHITE, N.M. & R.J. MCDONALD. 2002. Multiple parallel memory systems in the brain of the rat. Neurobiol. Learn. Mem. **77:** 125–184.
26. VOORN, P. *et al.* 2004. Putting a spin on the dorsal-ventral divide of the striatum. Trends Neurosci. **27:** 468–474.
27. APICELLA, P. *et al.* 1992. Neuronal activity in monkey striatum related to the expectation of predictable environmental events. J. Neurophysiol. **68:** 945–960.
28. APICELLA, P. *et al.* 1991. Responses to reward in monkey dorsal and ventral striatum. Exp. Brain Res. **85:** 491–500.
29. HASSANI, O.K., H.C. CROMWELL & W. SCHULTZ. 2001. Influence of expectation of different rewards on behavior-related neuronal activity in the striatum. J. Neurophysiol. **85:** 2477–2489.
30. CROMWELL, H.C. & W. SCHULTZ. 2003. Effects of expectations for different reward magnitudes on neuronal activity in primate striatum. J. Neurophysiol. **89:** 2823–2838.
31. Di CHIARA, G. & A. IMPERATO. 1988. Drugs abused by humans preferentially increase synaptic dopamine concentrations in the mesolimbic system of freely moving rats. Proc. Natl. Acad. Sci. USA **85:** 5274–5278.
32. ITO, R. *et al.* 2002. Dopamine release in the dorsal striatum during cocaine-seeking behavior under the control of a drug-associated cue. J. Neurosci. **22:** 6247–6253.
33. KOEPP, M.J. *et al.* 1998. Evidence for striatal dopamine release during a video game. Nature. **393:** 266–268.
34. VOLKOW, N.D. *et al.* 2002. "Nonhedonic" food motivation in humans involves dopamine in the dorsal striatum and methylphenidate amplifies this effect. Synapse **44:** 175–180.
35. BREITER, H.C., R. L. Gollub, R. M. Weisskoff, *et al.* 1997. Acute effects of cocaine on human brain activity and emotion. Neuron **19:** 591–611.
36. DELGADO, M.R. *et al.* 2000. Tracking the hemodynamic responses to reward and punishment in the striatum. J. Neurophysiol. **84:** 3072–3077.
37. KAHNEMAN, D. & A. TVERSKY. 1979. Prospect theory: an analysis of decision under risk. Econometrica **47:** 263–291.
38. MAY, J.C. *et al.* 2004. Event-related functional magnetic resonance imaging of reward-related brain circuitry in children and adolescents. Biol. Psychiatry **55:** 359–366.
39. WILSON, S.J. *et al.* 2005. Instructed smoking expectancy modulates cue-elicited neural activity: a preliminary study. Nicotine Tob. Res. **7:** 637–645.
40. THUT, G. *et al.* 1997. Activation of the human brain by monetary reward. Neuroreport **8:** 1225–1228.
41. ELLIOTT, R., C.D. FRITH & R.J. DOLAN. 1998. Differential neural response to positive and negative feedback in planning and guessing tasks. Neuropsychologia **35:** 1395–1404.
42. ELLIOTT, R. *et al.* 1998. Abnormal neural response to feedback on planning and guessing tasks in patients with unipolar depression. Psychol. Med. **28:** 559–571.
43. POLDRACK, R.A. *et al.* 2001. Interactive memory systems in the human brain. Nature **414:** 546–550.
44. SEGER, C.A. & C.M. CINCOTTA. 2005. The roles of the caudate nucleus in human classification learning. J. Neurosci. **25:** 2941–2951.
45. KNUTSON, B. & J.C. COOPER. 2005. Functional magnetic resonance imaging of reward prediction. Curr. Opin. Neurol. **18:** 411–417.
46. KNUTSON, B. *et al.* 2001. Anticipation of increasing monetary reward selectively recruits nucleus accumbens. J. Neurosci. **21:** RC159.

47. O'DOHERTY, J.P. *et al.* 2002. Neural responses during anticipation of a primary taste reward. Neuron **33:** 815–826.
48. BREITER, H.C. *et al.* 2001. Functional imaging of neural responses to expectancy and experience of monetary gains and losses. Neuron **30:** 619–639.
49. KIRSCH, P. *et al.* 2003. Anticipation of reward in a nonaversive differential conditioning paradigm and the brain reward system: an event-related fMRI study. Neuroimage **20:** 1086–1095.
50. DELGADO, M.R. *et al.* 2003. Dorsal striatum responses to reward and punishment: effects of valence and magnitude manipulations. Cogn. Affect. Behav. Neurosci. **3:** 27–38.
51. KNUTSON, B. *et al.* 2005. Distributed neural representation of expected value. J. Neurosci. **25:** 4806–4812.
52. GALVAN, A. *et al.* 2005. The role of ventral frontostriatal circuitry in reward-based learning in humans. J. Neurosci. **25:** 8650–8656.
53. NIEUWENHUIS, S. *et al.* 2005. Activity in human reward-sensitive brain areas is strongly context dependent. Neuroimage **25:** 1302–1309.
54. DELGADO, M.R., V.A. STENGER & J.A. FIEZ. 2004. Motivation-dependent responses in the human caudate nucleus. Cereb. Cortex **14:** 1022–1030.
55. TRICOMI, E.M., M.R. DELGADO & J.A. FIEZ. 2004. Modulation of caudate activity by action contingency. Neuron **41:** 281–292.
56. BERNS, G.S. *et al.* 2001. Predictability modulates human brain response to reward. J. Neurosci. **21:** 2793–2798.
57. SAMEJIMA, K. *et al.* 2005. Representation of action-specific reward values in the striatum. Science **310:** 1337–1340.
58. BALLEINE, B.W. & A. DICKINSON. 1998. Goal-directed instrumental action: contingency and incentive learning and their cortical substrates. Neuropharmacology **37:** 407–419.
59. ITO, R. *et al.* 2000. Dissociation in conditioned dopamine release in the nucleus accumbens core and shell in response to cocaine cues and during cocaine-seeking behavior in rats. J. Neurosci. **20:** 7489–7495.
60. ELLIOTT, R. *et al.* 2004. Instrumental responding for rewards is associated with enhanced neuronal response in subcortical reward systems. Neuroimage **21:** 984–990.
61. ZINK, C.F. *et al.* 2004. Human striatal responses to monetary reward depend on saliency. Neuron **42:** 509–517.
62. O'DOHERTY, J. *et al.* 2004. Dissociable roles of ventral and dorsal striatum in instrumental conditioning. Science **304:** 452–454.
63. BARTO, A.G. 1995. Adaptive critics and the basal ganglia. *In* Models of Information Processing in the Basal Ganglia. J.C. Houk, J.L. Davis & D. G. Beiser Eds. MIT Press. Cambridge, MA.
64. KNUTSON, B. *et al.* 2003. A region of mesial prefrontal cortex tracks monetarily rewarding outcomes: characterization with rapid event-related fMRI. Neuroimage **18:** 263–272.
65. KRINGELBACH, M.L. 2005. The human orbitofrontal cortex: linking reward to hedonic experience. Nat. Rev. Neurosci. **6:** 691–702.
66. MONTAGUE, P.R., B. KING-CASAS & J.D. COHEN. 2006. Imaging valuation models in human choice. Annu. Rev. Neurosci. **29:** 417–448.
67. MONTAGUE, P. & G. BERNS. 2002. Neural economics and the biological substrates of valuation. Neuron **36:** 265.
68. MYERS, C.E. *et al.* 2003. Dissociating hippocampal versus basal ganglia contributions to learning and transfer. J. Cogn. Neurosci. **15:** 185–193.

69. SHOHAMY, D. *et al*. 2004. Cortico-striatal contributions to feedback-based learning: converging data from neuroimaging and neuropsychology. Brain **127:** 851–859.
70. FILOTEO, J.V. *et al*. 2005. Cortical and subcortical brain regions involved in rule-based category learning. Neuroreport **16:** 111–115.
71. FRANK, M.J., L.C. SEEBERGER & C. O'REILLY R. 2004. By carrot or by stick: cognitive reinforcement learning in parkinsonism. Science **306:** 1940–1943.
72. SCHULTZ, W., P. DAYAN, P.R. MONTAGUE. 1997. A neural substrate of prediction and reward. Science **275:** 1593–1599.
73. MCCLURE, S.M., G.S. BERNS & P.R. MONTAGUE. 2003. Temporal prediction errors in a passive learning task activate human striatum. Neuron **38:** 339–346.
74. O'DOHERTY, J.P. *et al*. 2003. Temporal difference models and reward-related learning in the human brain. Neuron **38:** 329–337.
75. PAGNONI, G. *et al*. 2002. Activity in human ventral striatum locked to errors of reward prediction. Nat. Neurosci. **5:** 97–98.
76. TRICOMI, E. *et al*. 2006. Performance feedback drives caudate activation in a phonological learning task. J. Cogn. Neurosci. **18:** 1029–1043.
77. HARUNO, M. *et al*. 2004. A neural correlate of reward-based behavioral learning in caudate nucleus: a functional magnetic resonance imaging study of a stochastic decision task. J. Neurosci. **24:** 1660–1665.
78. KING-CASAS, B. *et al*. 2005. Getting to know you: reputation and trust in a two-person economic exchange. Science **308:** 78–83.
79. FRANK, R.H. 2004. What Price the Moral High Ground? Princeton University Press. Princeton, NJ.
80. DELGADO, M.R., R.H. FRANK & E.A. PHELPS. 2005. Perceptions of moral character modulate the neural systems of reward during the trust game. Nat. Neurosci. **8:** 1611–1618.
81. RILLING, J. *et al*. 2002. A neural basis for social cooperation. Neuron **35:** 395–405.
82. SINGER, T. *et al*. 2004. Brain responses to the acquired moral status of faces. Neuron **41:** 653–662.
83. DE QUERVAIN, D.J. *et al*. 2004. The neural basis of altruistic punishment. Science **305:** 1254–1258.
84. DAW, N.D., S. KAKADE & P. DAYAN. 2002. Opponent interactions between serotonin and dopamine. Neural Netw. **15:** 603–616.
85. JENSEN, J. *et al*. 2003. Direct activation of the ventral striatum in anticipation of aversive stimuli. Neuron **40:** 1251–1257.
86. SEYMOUR, B. *et al*. 2004. Temporal difference models describe higher-order learning in humans. Nature **429:** 664–667.
87. PHELPS, E.A. *et al*. 2004. Extinction learning in humans: role of the amygdala and vmPFC. Neuron **43:** 897–905.
88. O'DOHERTY, J. *et al*. 2001. Representation of pleasant and aversive taste in the human brain. J. Neurophysiol. **85:** 1315–1321.
89. DELGADO, M.R., C.D. LABOULIERE & E.A. PHELPS. 2006. Fear of losing money? Aversive conditioning with secondary reinforcers. Soc. Cogn. Affect. Neurosci. **1:** 250–259.
90. MAWLAWI, O. *et al*. 2001. Imaging human mesolimbic dopamine transmission with positron emission tomography: I. Accuracy and precision of D(2) receptor parameter measurements in ventral striatum. J. Cereb. Blood Flow Metab. **21:** 1034–1057.
91. SCHULTZ, W., L. TREMBLAY & J.R. HOLLERMAN. 1998. Reward prediction in primate basal ganglia and frontal cortex. Neuropharmacology **37:** 421–429.

Integration of Cognitive and Motivational Information in the Primate Lateral Prefrontal Cortex

MASAMICHI SAKAGAMI[a] AND MASATAKA WATANABE[b]

[a]Brain Science Research Center, Tamagawa University Brain Science Institute, Machida, Tokyo, Japan

[b]Department of Psychology, Tokyo Metropolitan Institute for Neuroscience, Fuchu, Tokyo, Japan

ABSTRACT: The prefrontal cortex (PFC), particularly the lateral prefrontal cortex (LPFC), has an important role in cognitive information processing. The area receives projections from sensory association cortices and sends outputs to motor-related areas. Neurons in LPFC code the behavioral significance of stimuli, which can be abstract precursors for complex motor commands and are structured hierarchically. Loss of these neurons leads to a lack of flexibility in decision making, such as seen in stereotyped behaviors. However, to make more appropriate decisions the code for behavioral significance has to reflect the subject's own desires and demands. Indeed, LPFC has connections with reward-related areas, such as the orbitofrontal cortex (OFC), basal ganglia, and medial prefrontal cortex. Recently, many studies have reported reward modulation of neural codes of behavioral significance. Using an asymmetric reward paradigm, we can investigate the functional specificity of LPFC neurons that code both cognitive information and motivational information. In this review, we will discuss details of neuronal properties of LPFC neurons from the viewpoints of cognitive information processing and motivational information processing, and the question of how these two pieces of information are integrated. Abstract coding and contextual representations in the cognitive information processing are functional characteristics of LPFC. Such functional specificity in LPFC cognitive processes is supported by a long-term scale of reward history in the motivational information processing. The integration enables us to make an elaborate decision with respect to goal-directed behavior in complex circumstances.

Address for correspondence: Masamichi Sakagami, Brain Science Research Center, Tamagawa University Brain Science Institute, Tamagawa-gakuen, 6-1-1, Machida, Tokyo 194-8610, Japan. Voice: +81-42-739-8679; fax: +81-42-739-8663.

sakagami@lab.tamagawa.ac.jp

Ann. N.Y. Acad. Sci. 1104: 89–107 (2007). © 2007 New York Academy of Sciences.
doi: 10.1196/annals.1390.010

KEYWORDS: behavioral significance; caudate nucleus; decision making; goal-directed behavior; punishment; reward

INTRODUCTION

The ability to make decisions is one of the most important higher-order brain functions and allows humans and other animals to select an appropriate action in given circumstances. To make a better decision, we have to not only convert sensory information into associated action or behavioral information but also take into account the outcome of the behavior. The prefrontal cortex (PFC), particularly the lateral prefrontal cortex (LPFC), is a good candidate for the computational stage of integration of cognitive and motivational information. Two lines of evidence support this idea, namely, anatomical and clinical.

As the anatomical basis for cognitive processing, LPFC receives projections from sensory areas and sends outputs to the higher-order motor cortices. Anatomical data reveal that the inferotemporal cortex (IT) has strong neuronal projections to PFC, especially the vetrolateral prefrontal cortex (VLPFC).[1–4] The parietal cortex, in turn, has strong neuronal projections to the dorsolateral prefrontal cortex (DLPFC).[5–8] Being the basis for motivational processing as well, LPFC receives reward-related information through the dense interconnection within PFC, particularly the orbitofrontal cortex (OFC)[7,9] and from the striatum, particularly the caudate nucleus (CD), and A8-A10 complex, including the ventral tegmental area (VTA) and substantia nigra pars compacta (SNc).[10,11] The anatomical data also show that LPFC projects directly to the premotor (PM) and supplementary motor areas (SMA).[4,12–17]

Patients with prefrontal pathology, including LPFC lesions, have difficulty in performing tasks that require contextual decisions and monitoring response–outcome relations. In experimental situations, such as the antisaccade eye movement task,[18] the Stroop task,[19] the go/no-go task,[20] and the Wisconsin Card Sorting Test,[21] prefrontal patients are often unable to suppress prepotent responses evoked by currently irrelevant stimuli. This inability to inhibit stereotyped responding is especially apparent in environments that require PFC patients to evaluate and make decisions based on rapidly changing circumstances.[22] The Iowa Gambling task is another decision-making task sensitive to PFC lesions.[23,24] In the task with four card decks, patients are required to maximize the total amount of money, choosing the safe deck (smaller gain but minimal loss) and avoiding the risky deck (larger gain but even larger loss). PFC patients often fail to avoid the risky deck. Experimental lesion studies with nonhuman primates have more clearly suggested that LPFC, particularly VLPFC, is responsible for the integration of cognitive and motivational information.[25–30]

Here we will discuss results of single-unit recording experiments in primate LPFC involved in cognitive information processing and reward (motivational)

information processing, as well as how these pieces of information are integrated to understand the underlying neuronal mechanisms for the decision-making process in LPFC.

BEHAVIORAL SIGNIFICANCE IN LPFC

Single unit studies in LPFC have been conducted with nonhuman primates using various discrimination tasks. These studies have suggested that neurons in primate LPFC code the representation of the response, which should be made to the cue stimulus to obtain reward, given that the activity of these neurons during the cue period predicts the monkey's impending action.[31–43] The activity pattern correlates with the behavior to be executed later but does not have a direct causal relationship with it at the time of response execution. The representation is characterized best as the behavioral meaning or significance of the cue stimulus.[35,38,44] For example, in Sakagami and Niki,[38] monkeys were trained to make a go or no-go response depending on the physical features of the cue stimulus; for example, a green or purple color indicated that the monkey should make a go response to obtain reward, and a red or yellow color indicated that the monkey should make a no-go response (FIG. 1A). Many neurons in LPFC showed differential visual responses to the cue stimulus, not depending on the physical features, but on whether the feature of the cue stimulus indicated a go or no-go response (FIG. 1B). The differential activity of these neurons was not simply related to the motor execution of a go or no-go response as there was a delay period between the cue presentation and the response.

Sakagami and Tsutsui[43] reported the hierarchical organization of meaning detection in LPFC. They used moving colored random dots as cues in a selective attention task with go/no-go response (FIG. 2). Monkey subjects had to attend to one of two visual dimensions, color or motion direction, indicated by the color of fixation spot, to distinguish between a go or no-go response. In the color condition, go was signaled by a green color, and no-go by red, regardless of the motion direction, whereas in the motion condition, go was signaled by upward motion and no-go by downward motion (FIG. 2A). In VLPFC, the majority of neurons showed go/no-go differential activity in the color condition only (color go/no-go neurons [C neurons] shown in the upper panels of FIG. 2B) and only a few showed differential activity in the motion condition. In contrast, in DLPFC and the arcuate area, particularly the dorsal part, including PM region, the neurons tended to show go/no-go differential activity in both the color and motion conditions (color/motion go/no-go neurons [CM neurons] shown in the lower panels of FIG. 2B) and a minority of neurons coded go/no-go differential activity only in the motion condition (motion go/no-go neurons [M neurons] shown in the middle panels of FIG. 2B). This asymmetric distribution of go/no-go neurons sensitive to color versus motion cues suggests that VLPFC and DLPFC have different functions and that these neurons code the behavioral significance of external sensory stimuli in a hierarchical fashion.

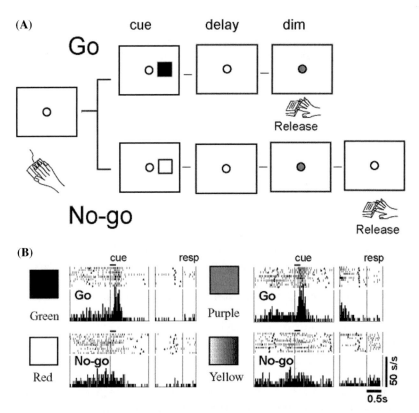

FIGURE 1. Schematic illustration of the sequence of a trial, and an example of a lateral prefrontal neuron coding the behavioral significance of colors. (**A**) The trial began when the monkey pressed the lever. The monkey was required to focus his gaze on the fixation spot. A cue (200 msec) was then presented at one of the positions; right, left, upper, and lower, followed by a delay period (1–2 sec) until the fixation spot dimmed. A green or purple color indicated that the monkey should make a go response to obtain reward, and a red or yellow color indicated that the monkey should make a no-go response. If the cue indicated a go response, the monkey had to release the lever immediately after the fixation spot dimmed (<800 msec). If the cue indicated a no-go response, the monkey had to continue pressing the lever throughout the dim period (1.2 sec) and release the lever after the fixation spot was reilluminated. (**B**) Example of a go-type neuron (all rasters and histograms indicate the response from the same neuron). The rasters and histograms are split in two; the left, aligned on target onset, and the right, aligned on lever release. One line in the rasters indicates one trial and a small vertical line segment represents the timing of neuronal activity. The figure appears in color online.

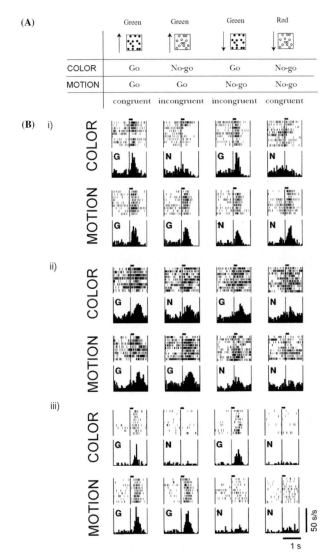

FIGURE 2. Examples of cues in a selective attention task with go/no-go response, and activity patterns of three types of go/no-go neurons. (**A**) The monkey had to attend to one of two visual dimensions of a moving random dot pattern, either color or motion direction, to decide on a go or no-go response. In the color condition, a go response was instructed by a green color, and no-go by red, regardless of the motion direction. In the motion condition, a go response was instructed by upward motion and no-go by downward motion. A black dot represents green; a white dot represents red. (**B**) Examples of neuronal activities of C neuron (i), M neuron (ii), and CM neuron (iii); each raster and histogram illustrated the neuronal response to the cue for the period from 1 sec before until 1 sec after the onset. The upper panel represents the neuronal activity in the color condition, and the lower panel, that in the motion condition. Adapted from Sakagami & Tsutsui[43] (with permission from Elsevier, copyright 1999). The figure appears in color online.

BEHAVIORAL SIGNIFICANCE IN LPFC
AND TASK PERFORMANCE

Many animal lesion studies, however, have suggested that monkeys without LPFC could learn discrimination tasks as long as they were not complex.[45,46] Niki et al.[36] investigated the change in activity of go/no-go neurons during learning processes with a learning set paradigm. In the experiment, after finding a go/no-go neuron in LPFC with a well-learned pair of stimuli, Niki et al. introduced a new pair, one element of which indicated a go response while the other element of the pair indicated no-go. A typical example of LPFC neuronal activity in this paradigm is shown in FIGURE 3. After introducing a new pair of stimuli, the monkey's performance gradually increased through trial and error. However, even in the stage of 100% correct performance, the go/no-go neuron could not show well-differentiated activity for go and no-go stimuli. In the course of overlearning, the neuron improved its go/no-go differential activity. From the activity pattern, it seems that go/no-go neurons do not contribute to learning and so cannot contribute to the executive guide for behavior, at least in the early stages of learning.

What kind of functional roles do the go/no-go neurons play for selection of behavior? Kobayashi et al.[47] analyzed the relation between the monkey's performance and the activity of go/no-go neurons. In this analysis, they concentrated on C neurons that showed go/no-go differential activity only in the color condition because the physiological properties of C neurons are different from those of M and CM neurons, which represent less distinctive populations both in terms of response latency and anatomical location.[43] The cue to which a given neuron showed increased activity is referred to as preferred cue (go or no-go). Sixty-four percent of C neurons preferred the go cue, whereas the remaining 36% preferred the no-go cue.[48] The activity distributions of C neurons showed a clear difference between the preferred and nonpreferred cues in correct trials (FIG. 4). In this task, the cues can be classified into two types, congruent and incongruent (FIG. 2A). In congruent trials, both the color and motion direction features indicate the same type of response (e.g., a go response in both the color and the motion condition). In incongruent trials, the cues require different responses in the two conditions (e.g., a go response in the color condition, but a no-go response in the motion condition). We have previously shown that incongruent cues generate more errors and longer reaction times.[49] However, we could not find a congruency effect in the go/no-go differential activity of C neurons in correct trials of color attending condition.[50] When we observed the activity in error trials separately for congruent and incongruent trials, we found that the activity in incongruent trials could predict whether the monkey would make a correct response or an error. To understand the general tendency of C neurons in LPFC, we made a population analysis with Bayesian estimation (FIG. 4). With this we could estimate the monkey's upcoming response after cue presentation based on the

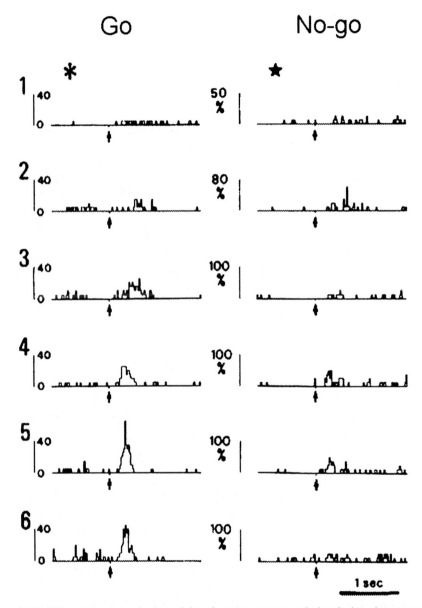

FIGURE 3. The change in the activity of a go/no-go neuron during the learning process in LPFC. These histograms show the activity change of one neuron in the time course 1 through 6 (from as soon as a new pair of stimuli was introduced until the stage of over-learning). The left panel represents the response to the go cue; the right panel represents the response to the no-go cue. These histograms indicate the average pulses in every 10 correct trials. The number between go and no-go panels represents the correct rate (percentile) in each block.

population activity of C neurons in the LPFC. The estimated probabilities of the upcoming response similarly increased after cue presentation up to 300 msec in congruent (solid line in FIG. 4A) and incongruent correct trials (solid line in FIG. 4B). However, in incongruent trials, the estimated probability was significantly lower in error trials than in correct trials (FIG. 4B). In congruent

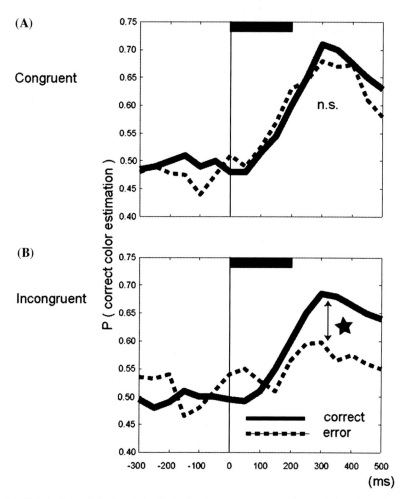

FIGURE 4. Probability of discrimination between go and no-go trials in color go/no-go neurons. A Bayesian approach was applied to population data of 123 C neurons to estimate the probabilities at each sampled point. The horizontal axis indicates time from the onset of the cue presentation indicated by the vertical bars. Solid lines and dotted lines represent the probabilities in the correct trials and those in the error trials, respectively. There was no significant difference in activity between the correct and error trials in congruent trials (panel **A**), whereas there was a significant difference in the incongruent trials (panel **B**, $P < 0.05$).

trials, on the other hand, the probability looked similar for error and correct trials (FIG. 4A). These results suggest that the go/no-go differential activity of C neurons in LPFC is critical for the subjects' behavior only in incongruent trials.

To confirm this tendency behaviorally we injected muscimol in VLPFC in one hemisphere of one monkey and analyzed the behavioral changes. When the monkey performed the task in the motion condition, the muscimol injection did not generate any behavioral change (FIG. 5C, D). In the color condition, if the incongruent cue was presented in the hemifield contralateral to the injection hemisphere, there was a dramatic behavioral change; the performance significantly decreased compared to the saline control condition (right panel in FIG. 5B). Even in the color condition, the muscimol injection was not effective

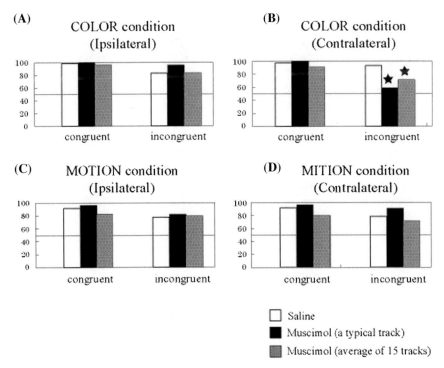

FIGURE 5. Correct response rates of go trials after muscimol injection in VLPFC. White bars indicate the correct rates after saline injection. Black bars indicate the worst correct rate after muscimol injection across tracks. Gray bars indicate the average correct rates of 15 tracks after muscimol injection in the VLPFC. The upper two panels (A and B) and lower two panels (C and D) represent correct rates in the color and motion condition, respectively. Panels A and C indicate correct rates in case the cue was presented in the side ipsilateral to the injection hemisphere. Panels B and D indicate correct rates in case the cue was presented in the contralateral side. An asterisk indicates a significant difference in correct rate compared to that following saline injection.

in case of ipsilateral cue presentation (FIG. 5A) or in case of a congruent cue in the contralateral hemifield (left panel in FIG. 5B). The behavioral changes were observed only in go trials, not in no-go trials in any condition. The muscimol injection is thought to suppress the output of the local injection area. The change of output from VLPFC, where C neurons are densely distributed, affected the behavior in incongruent color trials but not in congruent trials. These results are highly consistent with those from the error activity analysis.

In their series of studies using the Stroop task, Cohen and colleagues argued that the conflict is detected in the anterior cingulate cortex and the signal is sent to LPFC, which then controls task-related processes to overcome noise.[51,52] However, in our study, activities of go/no-go neurons in LPFC were similar in both congruent and incongruent correct trials in color condition,[50] although we could have expected differential activity patterns for congruent and incongruent stimuli based on the hypothesis of Cohen's group.

In general, to solve a problem, the nervous system could use several possible pathways. Some may mostly depend on cortical pathways and others may depend on subcortical pathways. Even in cortical circuits, to reach motor output regions some may go through LPFC but others may not. The results of our studies suggest that, even in the same task, the contribution of LPFC to the behavioral decision can be changed depending on the complexity of the experimental situation. Pasupathy and Miller[53] compared the activities of CD neurons with those of LPFC neurons in the course of learning with a learning set paradigm. They found that in the earlier stages of learning, CD neurons contribute to task performance, whereas in later stages LPFC neurons take charge. The brain thus depends on the rough but quickly learned signal from CD nucleus in the earlier stages and, later, on the accurate but slowly learned coding in LPFC.

BEHAVIORAL SIGNIFICANCE MODULATED BY REWARD PREDICTION IN THE LPFC

In the previous section we overviewed the behavioral significance in LPFC, which represents an instructional code associated with external stimulus. However, our decision making depends on the motivational state as well as the cognitive knowledge. Watanabe[54] reported LPFC neurons that showed reward-related activity during the delay period in a modified version of the spatial delayed response task (FIG. 6). In the experiment, after the indication of target position (left or right), the monkey had to remember the position during a 5-sec delay, and after the delay period, a go signal instructed the monkey to respond manually to the indicated position. A correct response was rewarded with different kinds of food or liquid depending on the block. In this block design, then, the monkey could predict what kind of food or liquid would be delivered as reward. Many neurons showed differential activity during the delay for

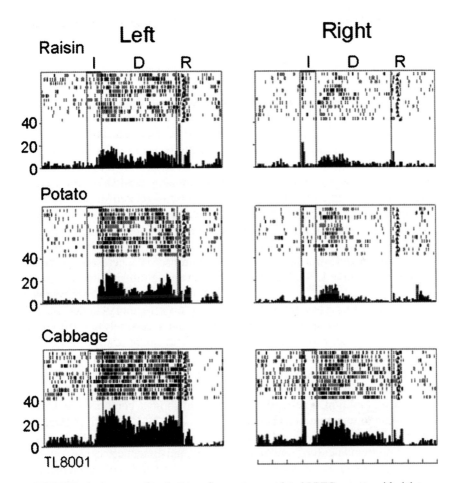

FIGURE 6. An example of a reward-expectancy-related LPFC neuron with delay period activity. The neuronal activity is shown separately for each reward block in raster and histogram displays, with left-sided displays for the left trials and right-sided displays for the right trials. For each display, the first two vertical lines from the left indicate the instruction onset and offset, and the third line indicates the end of the delay period. The horizontal and the vertical axes indicate the impulses/sec and the timescale. Only correctly performed trials are displayed. Adapted from Watanabe[54] (with permission from the Nature Publishing Group, copyright 1996).

left- and right-target trials. Furthermore, the spatial information of some delay neurons was also modulated by reward type. In the example shown in FIG-URE 6, the delay activity of the neuron increased more in left-target trials than right-target trials and the amount of enhancement was dependent on the kind of reward predicted. In the example, cabbage evoked the largest enhancement, and raisin the smallest, in accordance with the monkey's reward preference.

Leon and Shadlen[55] reported that a similar difference in activity of LPFC delay neurons could be found in relation to reward magnitude. Accordingly, the spatial and reward information might be integrated in LPFC. Kobayashi et al.[56] suggested that the method of integration is not a simple summation of the components but an increase of the amount of transmitted information with respect to target position. In other words, the reward information is used for increasing the discrimination of target positions, which leads to enhancement of performance.[57]

Many LPFC neurons showed predictive activity for nonreward (or smaller reward).[56,58,59] Those neurons increased their activity for the target cue period and/or following delay period when the monkey predicted nonreward or smaller reward. Some of them showed the activity related to prediction regarding which reward would be absent; when more preferred reward would be absent, they showed more activity than when less preferred reward would be absent.[58] This predictive code for the absence of reward could be integrated with spatial information in LPFC neurons.[56,59] These results lead to the question of how LPFC neurons process aversive information. Kobayashi et al.[60] compared the effects of reward and punishment in the cognitive neural code in LPFC. They trained the monkey on a memory-guided saccade task in three conditions (reward, punishment, and sound only). In the reward condition, a correct response was reinforced with liquid reward; in the punishment condition, an erroneous response was followed by an air puff; and in the sound only condition, a correct response was followed only by a feedback signal in the form of a sound. The monkey's performance was best in the reward condition, intermediate in the punishment condition, and worst in the sound only condition. The majority of task-related neurons showed cue and/or delay activity modulated only by reward prediction (FIG. 7A). Kobayashi et al.[60] also found neurons modulated only by punishment effect (FIG. 7B), although the number was much smaller than that of reward-modulated neurons. Interestingly, it turned out not so many neurons modulated by both positive and negative reinforcers, which may reflect general reinforcement or attentional processes, although both reinforcers significantly enhanced behavioral performance. The results indicate that information on reward and punishment is processed differentially in LPFC and it can contribute to careful planning in goal-directed behavior. However, the results also suggest that the integration is not yet completed in LPFC and further study is necessary to understand how the brain processes aversive information, and how information about reinforcers is integrated with cognitive information.

INTEGRATION OF COGNITIVE AND MOTIVATIONAL INFORMATION IN THE LPFC AND THE CD

The basal ganglia are another stage for the integration of cognitive and motivational information.[61] Particularly, CD receives cognitive information from

(A)

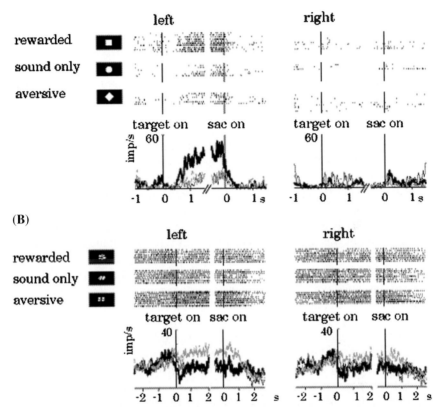

FIGURE 7. Example of neurons with outcome-dependent activity. (**A**) Reward type neuron. Rastergrams are shown separately for each target position and each reinforcement type. Vertical lines in the rastergram indicate the time of target onset (left) and saccade onset (right). Histograms at bottom compare mean discharge rates (thick black line, rewarded trials; thin black line, sound-only trials; gray line, aversive trials). Correct performance rates were 86.4%, 80.0%, and 64.7% in rewarded, aversive, and sound-only trials, respectively. Only correctly performed trials are displayed. (**B**) Aversive type neuron. Correct performance rates were 100%, 61.1%, and 43.8% in rewarded, aversive, and sound-only trials, respectively. Adapted from Kobayashi et al.[60] (with permission from Springer Science, copyright 2006).

various cortical areas including LPFC, PM, and parietal cortices and motivational information from SNc. Many neurophysiological studies also support the possibility. For example, Kawagoe et al.[62] found that reward-modulated CD neurons tended to show an overriding reward effect so that spatial characteristics observed in a symmetric reward schedule often disappeared in an asymmetric schedule.

Kobayashi *et al.*[57] compared the reward-modulated neuronal activity between LPFC and CD in an asymmetrically rewarded memory-guided saccade task, in which a correct saccade to one spatial position was followed by a liquid reward and a correct saccade to the other position was not followed by reward. Within a block of 40–60 trials, the rewarded position was fixed, so the monkey could predict whether he could receive reward in a given trial after the target cue presentation. Kobayashi *et al.* observed many neurons that reflect the spatial information processes and/or the reward information processes in both LPFC and CD. There were two major differences in neuronal characteristics between these two areas. First, neurons coding pure spatial information were more common in LPFC. Second, neurons with reward-anticipating precue activity were more prevailing in CD. Neurons reflecting both spatial and reward information were found in both areas (FIG. 8). The activity of these neurons was highest in trials in which the target cue was presented in the receptive field and the position was associated with liquid reward. However, a substantial difference was found in trials in which the target cue was presented in the receptive field but the position was associated with nonreward. In the population histogram of LPFC neurons, the cue response was still present although somewhat degraded relative to the most preferable cueing condition (FIG. 8A). In contrast, CD neurons could not increase their activity to the presentation of a cue in the receptive field in nonrewarded trials (FIG. 8B). There may be many possible interpretations for the activity difference. Among these, the most interesting is concerned with short-term versus long-term reward effects. The monkeys performed the task well even in nonrewarded trials because a correction method was used in which the monkeys had to make a correct response in a nonrewarded trial to proceed to the next trial. In other words, the monkeys must have some motivation, although smaller, for the performance in nonrewarded trials in terms of future reward. LPFC neurons showed a similar activity pattern to the actual behavior. So LPFC integration neurons may use the long-term scale of reward history for their learning process, whereas the scale for CD integration neurons may be smaller. Recently, Ding and Hikosaka[63] found a similar contrast between frontal eye field neurons and CD neurons. Consistent data were also reported in fMRI studies.[64,65]

The brain has several circuits to make a behavioral decision that is appropriate in given circumstances. Some are based on a relatively simple input–output relation, while others depend on the integration of various pieces of complex information. The ways of integration for different kinds of information are also distinct. These circuits for decision making seem to work relatively independently in the brain.[61] Among the circuits, one including LPFC can generate the most elaborate decision. Barraclough *et al.*[66] recorded LPFC neurons in an experimental situation in which the monkey had to make an appropriate decision considering an enemy (computer)'s changeable strategy, and found that LPFC neurons could provide information reflecting the history of responses and their outcome. The characteristics of LPFC include, in terms

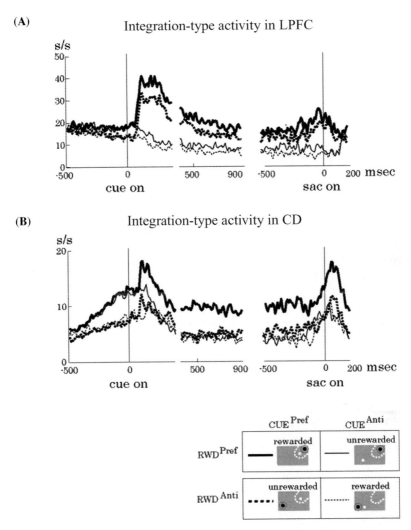

FIGURE 8. Population histograms of integration-type neurons in the LPFC (**A**) and in the caudate nucleus (CD) (**B**). The cue was presented either in the preferred direction (CUEPref, white circle inside the white dashed line in the inset; thick line in the histograms) or antipreferred direction (CUEAnti, white circle out of the white dashed line in the inset; thin line in the histograms). Immediate reward was associated either with the preferred direction (RWDPref, bull's eye mark inside the white dashed line in the inset; solid line in the histograms) or nonpreferred direction (RWDAnti, bull's eye mark out of the white dashed line; dashed line line in the histograms). Immediate reward was available in the CUEPref–RWDPref condition and the CUEAnti-RWDAnti-condition, whereas immediate reward was not available in the CUEPref-RWDAnti condition and the CUEAnti–RWDPref condition. The left vertical line indicates the time of cue onset and the right vertical line indicates the time of saccade onset. Adapted from Kobayashi et al.[57] (with permission from Elsevier, copyright 2007). The figure appears in color online.

of cognitive information processing, the ability to code abstract and contextual representations, and, in terms of motivational information processing, the ability to reflect reward history on a long-term scale. Particularly, learning on long-term reward history is a basis for long-term prediction of reward, which is prerequisite for systematic planning. The integration could enable elaborate decisions with respect to goal-directed behavior in complex circumstances.

ACKNOWLEDGMENTS

We thank J. Lauwereyns for critical comments and A. Noritake and M. Yamamoto for technical assistance. Supported by Human Frontier Science Program (MS), Precursory Research for Embryonic Science and Technology (MS), Grant-in-Aid for Target-Oriented Research and Development in Brain Science (MW) from Japan Science and Technology Corporation, and Grant-in-Aid for Scientific Research on Priority Areas (MW and MS) and Tamagawa University COE (MS) from Japanese Ministry of Education, Science, Sports and Culture.

REFERENCES

1. UNGERLEIDER, L.G., D. GAFFAN & V.S. PELAK. 1989. Projections from inferior temporal cortex to prefrontal cortex via the uncinate fascicle in rhesus monkeys. Exp. Brain Res. **76:** 473–484.
2. WEBSTER, M.J., J. BACHEVALIER & L.G. UNGERLEIDER. 1994. Connections of inferior temporal areas TEO and TE with parietal and frontal cortex in macaque monkeys. Cereb. Cortex. **4:** 470–483.
3. CARMICHAEL, S.T. & J.L. PRICE. 1995. Sensory and premotor connections of the orbital and medial prefrontal cortex of macaque monkeys. J. Comp. Neurol. **363:** 642–664.
4. PETRIDES, M. & D.N. PANDYA. 2002. Comparative cytoarchitectonic analysis of the human and the macaque ventrolateral prefrontal cortex and corticocortical connection patterns in the monkey. Eur. J. Neurosci. **16:** 291–310.
5. PANDYA, D.N. & B. SELTZER. 1982. Intrinsic connections and architectonics of posterior parietal cortex in the rhesus monkey. J. Comp. Neurol. **204:** 196–210.
6. PETRIDES, M. & D.N. PANDYA. 1984. Projections to the frontal cortex from the posterior parietal region in the rhesus monkey. J. Comp. Neurol. **228:** 105–116.
7. BARBAS, H. & M.M. MESULAM. 1985. Cortical afferent input to the principalis region of the rhesus monkey. Neuroscience **15:** 619–637.
8. MORRIS, R., D.N. PANDYA & M. PETRIDES. 1999. Fiber system linking the mid-dorsolateral frontal cortex with the retrosplenial/presubicular region in the rhesus monkey. J. Comp. Neurol. **407:** 183–192.
9. BARBAS, H. & D.N. PANDYA. 1989. Architecture and intrinsic connections of the prefrontal cortex in the rhesus monkey. J. Comp. Neurol. **286:** 353–375.
10. ALEXANDER, G.E., M.R. DELONG & P.L. STRICK. 1986. Parallel organization of functionally segregated circuits linking basal ganglia and cortex. Annu. Rev. Neurosci. **9:** 357–381.

11. GASPAR, P., I. STEPNIEWSKA & J.H. KAAS. 1992. Topography and collateralization of the dopaminergic projections to motor and lateral prefrontal cortex in owl monkeys. J. Comp. Neurol. **325:** 1–21.
12. GOLDMAN, P.S. & W.J. NAUTA. 1976. Autoradiographic demonstration of a projection from prefrontal association cortex to the superior colliculus in the rhesus monkey. Brain Res. **116:** 145–149.
13. BARBAS, H. & D.N. PANDYA. 1987. Architecture and frontal cortical connections of the premotor cortex (area 6) in the rhesus monkey. J. Comp. Neurol. **256:** 211–228.
14. BATES, J.F. & P.S. GOLDMAN-RAKIC. 1993. Prefrontal connections of medial motor areas in the rhesus monkey. J. Comp. Neurol. **336:** 211–228.
15. SCHMAHMANN, J.D. & D.N. PANDYA. 1997. Anatomic organization of the basilar pontine projections from prefrontal cortices in rhesus monkey. J. Neurosci. **17:** 438–458.
16. PETRIDES, M. & D.N. PANDYA. 1999. Dorsolateral prefrontal cortex: comparative cytoarchitectonic analysis in the human and the macaque brain and corticocortical connection patterns. Eur. J. Neurosci. **11:** 1011–1036.
17. MIYACHI, S., X. LU, S. INOUE, *et al.* 2005. Organization of multisynaptic inputs from prefrontal cortex to primary motor cortex as revealed by retrograde transneuronal transport of rabies virus. J. Neurosci. **25:** 2547–2556.
18. GUITTON, D., H.A. BUCHTEL & R.M. DOUGLAS. 1985. Frontal lobe lesions in man cause difficulties in suppressing reflexive glances and in generating goal-directed saccades. Exp. Brain Res. **58:** 455–472.
19. PERRET, E. 1974. The left frontal lobe of man and the suppression of habitual responses in verbal categorical behaviour. Neuropsychologia **12:** 323–330.
20. DREWE, E.A. 1975. Go-Nogo learning after frontal lobe lesions in humans. Cortex **11:** 8–16.
21. MILNER, B. 1964. Some effects of frontal lobectomy in man: the frontal granular cortex and behavior. *In* J.M. Warren & K. Akert, Eds.: 313–334. McGraw-Hill. New York.
22. LHERMITTE, F., B. PILLON & M. SERDARU. 1986. Human autonomy and the frontal lobes. Part I: imitation and utilization behavior: a neuropsychological study of 75 patients. Ann. Neurol. **19:** 326–334.
23. BECHARA, A., A.R. DAMASIO, H. DAMASIO, *et al.* 1994. Insensitivity to future consequences following damage to human prefrontal cortex. Cognition **50:** 7–15.
24. MANES, F., B. SAHAKIAN, L. CLARK, *et al.* 2002. Decision-making processes following damage to the prefrontal cortex. Brain **125:** 624–639.
25. BUTTERS, N. & D. PANDYA. 1969. Retention of delayed-alternation: effect of selective lesions of sulcus principalis. Science **165:** 1271–1273.
26. IVERSEN, S.D. & M. MISHKIN. 1970. Perseverative interference in monkeys following selective lesions of the inferior prefrontal convexity. Exp. Brain Res. **11:** 376–386.
27. BUTTERS, N., C. BUTTER, J. ROSEN, *et al.* 1973. Behavioral effects of sequential and one-stage ablations of orbital prefrontal cortex in the monkey. Exp. Neurol. **39:** 204–214.
28. PASSINGHAM, R. 1975. Delayed matching after selective prefrontal lesions in monkeys (*Macaca mulatta*). Brain Res. **92:** 89–102.
29. MISHKIN, M. & F.J. MANNING. 1978. Non-spatial memory after selective prefrontal lesions in monkeys. Brain Res. **143:** 313–323.

30. DIAS, R., T.W. ROBBINS & A.C. ROBERTS. 1996. Dissociation in prefrontal cortex of affective and attentional shifts. Nature **380:** 69–72.
31. NIKI, H. 1974. Prefrontal unit activity during delayed alternation in the monkey. II. Relation to absolute versus relative direction of response. Brain Res. **68:** 197–204.
32. NIKI, H. 1974. Differential activity of prefrontal units during right and left delayed response trials. Brain Res. **70:** 346–349.
33. KOMATSU, H. 1982. Prefrontal unit activity during a color discrimination task with go and no-go responses in the monkey. Brain Res. **244:** 269–277.
34. WATANABE, M. 1986. Prefrontal unit activity during delayed conditional go/no-go discrimination in the monkey. I. Relation to go and no-go responses. Brain Res. **382:** 1–14.
35. YAJEYA, J., J. QUINTANA & J.M. FUSTER. 1988. Prefrontal representation of stimulus attributes during delay tasks. II. The role of behavioral significance. Brain Res. **474:** 222–230.
36. NIKI, H., S. SUGITA & M. WATANABE. 1990. Modification of the activity of primate frontal neurons during learning of a go/no-go discrimination and its reversal: a progress report. Vision, Memory and the Temporal Lobe. *In* E. Iwai & M. Mishkin, Eds.: 295–304. Elsevier. New York.
37. YAMATANI, K., T. ONO, H. NISHIJO, *et al.* 1990. Activity and distribution of learning-related neurons in monkey (*Macaca fuscata*) prefrontal cortex. Behav. Neurosci. **104:** 503–531.
38. SAKAGAMI, M. & H. NIKI. 1994. Encoding of behavioral significance of visual stimuli by primate prefrontal neurons: relation to relevant task conditions. Exp. Brain Res. **97:** 423–436.
39. SAKAGAMI, M. & H. NIKI. 1994. Spatial selectivity of go/no-go neurons in monkey prefrontal cortex. Exp. Brain Res. **100:** 165–169.
40. BICHOT, N.P., J.D. SCHALL & K.G. THOMPSON. 1996. Visual feature selectivity in frontal eye fields induced by experience in mature macaques. Nature **381:** 697–699.
41. MILLER, E.K., C.A. ERICKSON & R. DESIMONE. 1996. Neural mechanisms of visual working memory in prefrontal cortex of the macaque. J. Neurosci. **16:** 5154–5167.
42. ASSAD, J.A., G. RAINER & E.K. MILLER. 1998. Neural activity in the primate prefrontal cortex during associative learning. Neuron **21:** 1399–1407.
43. SAKAGAMI, M. & K. TSUTSUI. 1999. The hierarchical organization of decision making in the primate prefrontal cortex. Neurosci. Res. **34:** 79–89.
44. WANATABE, M. 1986. Prefrontal unit activity during delayed conditional go/no-go discrimination in the monkey. II. Relation to go and no-go responses. Brain Res. **382:** 15–27.
45. PASSINGHAM, R.E. 1993. The Frontal Lobes and Voluntary Action. Oxford University Press. New York.
46. FUSTER, J.M. 1997. The Prefrontal Cortex: Anatomy, Physiology, and Neuropsychology of the Frontal Lobe. Lippincott-Reven. Philadelphia.
47. KOBAYASHI, S., J. LAUWEREYNS, M. KOIZUMI, *et al.* 2001. Generation of behaviorally relevant coding in macaque prefrontal cortex. Soc. Neurosci. Abstr. 574. 10.
48. SAKAGAMI, M., K. TSUTSUI, J. LAUWEREYNS, *et al.* 2001. A code for behavioral inhibition on the basis of color, but not motion, in ventrolateral prefrontal cortex of macaque monkey. J. Neurosci. **21:** 4801–4808.

49. LAUWEREYNS, J., M. KOIZUMI, M. SAKAGAMI, *et al.* 2000. Interference from irrelevant features on visual discrimination by Macaques (*Macaca fuscata*): a
behavioral analogue of the human stroop effect. J. Exp. Psychol. Anim. Behav.
Processes **26:** 352–357.

50. LAUWEREYNS, J., M. SAKAGAMI, K. TSUTSUI, *et al.* 2001. Responses to task-
irrelevant visual features by primate prefrontal neurons. J. Neurophysiol. **86:**
2001–2010.

51. MACDONALD, A.W., III, J.D. COHEN, V.A. STENGER, *et al.* 2000. Dissociating the
role of the dorsolateral prefrontal and anterior cingulate cortex in cognitive control. Science **288:** 1835–1838.

52. KERNS, J.G., J.D. COHEN, A.W. MACDONALD III, *et al.* 2004. Anterior cingulate
conflict monitoring and adjustments in control. Science **303:** 1023–1026.

53. PASUPATHY, A. & E.K. MILLER. 2005. Different time courses of learning-related
activity in the prefrontal cortex and striatum. Nature **433:** 873–876.

54. WATANABE, M. 1996. Reward expectancy in primate prefrontal neurons. Nature
382: 629–632.

55. LEON, M.I. & M.N. SHADLEN. 1999. Effect of expected reward magnitude on the
response of neurons in the dorsolateral prefrontal cortex of the macaque. Neuron
24: 415–425.

56. KOBAYASHI, S., J. LAUWEREYNS, M. KOIZUMI, *et al.* 2002. Influence of reward
expectation on visuospatial processing in macaque lateral prefrontal cortex. J.
Neurophysiol. **87:** 1488–1498.

57. KOBAYASHI, S., R. KAWAGOE, Y. TAKIKAWA, *et al.* 2007. Functional differences
between macaque prefrontal cortex and caudate nucleus during eye movements
with and without reward. Exp. Brain Res. **176:** 341–355.

58. WATANABE, M., K. HIKOSAKA, M. SAKAGAMI, *et al.* 2002. Coding and monitoring
of motivational context in the primate prefrontal cortex. J. Neurosci. **22:** 2391–
2400.

59. WATANABE, M., K. HIKOSAKA, M. SAKAGAMI, *et al.* 2005. Functional significance
of delay-period activity of primate prefrontal neurons in relation to spatial working memory and reward/omission-of-reward expectancy. Exp. Brain Res. **166:**
263–276.

60. KOBAYASHI, S., K. NOMOTO, M. WATANABE, *et al.* 2006. Influences of rewarding
and aversive outcomes on activity in macaque lateral prefrontal cortex. Neuron
51: 861–870.

61. HIKOSAKA, O., K. SAKAI, H. NAKAHARA, *et al.* 2000. Role of the basal ganglia in
the control of purposive saccadic eye movements. Physiol. Rev. **80:** 953–978.

62. KAWAGOE, R., Y. TAKIKAWA & O. HIKOSAKA. 1998. Expectation of reward modulates cognitive signals in the basal ganglia. Nat. Neurosci. **1:** 411–416.

63. DING, L. & O. HIKOSAKA. 2006. Comparison of reward modulation in the frontal
eye field and caudate of the macaque. J. Neurosci. **26:** 6695–6703.

64. MCCLURE, S.M., D.I. LAIBSON, G. LOEWENSTEIN, & J.D. COHEN. 2004. Separate
neural systems value immediate and delayed monetary rewards. Science **306:**
503–507.

65. TANAKA, S.C., K. DOYA, G. OKADA, *et al.* 2004. Prediction of immediate and
future rewards differentially recruits cortico-basal ganglia loops. Nat. Neurosci.
7: 887–893.

66. BARRACLOUGH, D.J., M.L. CONROY & D. LEE. 2004. Prefrontal cortex and decision
making in a mixed-strategy game. Nat. Neurosci. **7:** 404–410.

Mechanisms of Reinforcement Learning and Decision Making in the Primate Dorsolateral Prefrontal Cortex

DAEYEOL LEE AND HYOJUNG SEO

Department of Neurobiology, Yale University School of Medicine, New Haven, Connecticut, USA

ABSTRACT: To a first approximation, decision making is a process of optimization in which the decision maker tries to maximize the desirability of the outcomes resulting from chosen actions. Estimates of desirability are referred to as utilities or value functions, and they must be continually revised through experience according to the discrepancies between the predicted and obtained rewards. Reinforcement learning theory prescribes various algorithms for updating value functions and can parsimoniously account for the results of numerous behavioral, neurophysiological, and imaging studies in humans and other primates. In this article, we first discuss relative merits of various decision-making tasks used in neurophysiological studies of decision making in nonhuman primates. We then focus on how reinforcement learning theory can shed new light on the function of the primate dorsolateral prefrontal cortex. Similar to the findings from other brain areas, such as cingulate cortex and basal ganglia, activity in the dorsolateral prefrontal cortex often signals the value of expected reward and actual outcome. Thus, the dorsolateral prefrontal cortex is likely to be a part of the broader network involved in adaptive decision making. In addition, reward-related activity in the dorsolateral prefrontal cortex is influenced by the animal's choices and other contextual information, and therefore may provide a neural substrate by which the animals can flexibly modify their decision-making strategies according to the demands of specific tasks.

KEYWORDS: cognitive control; game theory; neuroeconomics; reward; utility theory; working memory

INTRODUCTION

During the process of decision making, possible outcomes from each alternative action are evaluated, and an action that is expected to optimize the

Address for correspondence: Daeyeol Lee, Ph.D., Department of Neurobiology, Yale University School of Medicine, 333 Cedar Street, SHM C303, New Haven, CT 06510. Voice: 203-785-3527; fax: 203-785-5263.

daeyeol.lee@yale.edu

Ann. N.Y. Acad. Sci. 1104: 108–122 (2007). © 2007 New York Academy of Sciences.
doi: 10.1196/annals.1390.007

outcomes according to some subjective criteria is selected. The complexity of such decision-making process varies greatly. In simple cases, a stereotyped behavioral response, such as blinking, is triggered, often deterministically, by specific sensory stimuli, such as an air puff to an eye. Such reflex-like behaviors can be supported by hardwired neural circuitry laid down by a set of genetic and epigenetic instructions and tuned by the evolutionary process. The process of decision making increases its complexity as it becomes more difficult for animals to predict the outcomes of their actions. The ability to predict accurately the outcome of a particular action often requires an extensive knowledge of the animal's environment, and in a dynamic environment, this can be accomplished only through experience. In this article, we describe how this dynamic process of decision making can be formally described by reinforcement learning theory, and compare several different types of behavioral tasks that can be used to test the predictions of such theory in human and animal choice behaviors. We also discuss the recent findings from neurobiological studies of lateral prefrontal cortex that focused on reward-based decision making.

In reinforcement learning theory,[1] it is assumed that the amount of reward received by an animal at each time step is probabilistically determined by the animal's action and the state of its environment. In addition, the objective of learning is to maximize a weighted sum of all future rewards, referred to as return, rather than only an immediate reward. The return is commonly defined as a temporally discounted sum of future rewards, and cannot be known completely because of the animal's limited knowledge of its environment. Instead, the animal estimates the expected return, which is referred to as value function. In reinforcement learning theory, two different types of value functions can be distinguished. The estimate for the return from a particular state of the environment is referred to as state value function, whereas the estimate for the return expected from a particular action chosen in a particular state of the environment is referred to as action value function. These two quantities are closely related, since the state value function for a given state is the sum of action value functions weighted by the probability of taking each action from the same state. Since value functions are based on estimates for all future rewards, the change in the value functions at two different time steps corresponds to the reward expected during the transition between them. If the value functions correctly predict the future rewards, the expected reward estimated from the value functions and the actual reward received by the animal will be the same. Otherwise, the difference between the expected and actual rewards, referred to as a reward prediction error, can be used to update the value functions.

Formal analyses based on optimality principles, such as utility theory and reinforcement learning, are not always successful in predicting the actual choice behaviors of humans and animals.[2,3] Nevertheless, they still provide useful insights into the neural basis of decision making under various circumstances.[4]

For example, dopamine neurons in the primate midbrain increase their activity phasically when the animal receives an unexpected reward or when a reward-predicting stimulus is delivered, indicating that they can signal the reward prediction error postulated in the reinforcement learning theory.[5] Subsequently, single-neuron recording and metabolic imaging studies have identified signals related to reward prediction error in the main targets of the projections from the dopamine neurons, such as the striatum and the orbitofrontal cortex.[6–8] In addition, signals related to expected rewards that resemble expected utilities or value functions have been identified in a large number of cortical regions as well as in the striatum.[2,4] In a majority of these studies, however, the animals were rewarded deterministically upon producing actions instructed by specific sensory stimuli. The results from these previous studies have, therefore, provided somewhat limited insights into how signals related to reward prediction errors and expected rewards are processed and how such signals in a particular brain area contribute to the animal's decisions. To obtain such insights, behavioral tasks must render the outcomes of the animal's decisions unpredictable. In this article, we first discuss several behavioral tasks that have been adopted in neurophysiological studies of decision making in nonhuman primates. We then focus on the recent findings that provided new insights into the functions of the primate dorsolateral prefrontal cortex in relation to reinforcement learning and decision making. We conclude with a possible reinterpretation of the functions traditionally attributed to the dorsolateral prefrontal cortex, such as working memory and cognitive control, within the framework of reinforcement learning theory.

BEHAVIORAL STUDIES OF DECISION MAKING IN PRIMATES

Many behavioral paradigms have been used to investigate different aspects of the neural processes involved in decision making. In classical or Pavlovian conditioning paradigms, for example, a neutral sensory stimulus is repeatedly paired with an appetitive or aversive stimulus, referred to as an unconditioned stimulus, and eventually gains the ability to elicit a so-called conditioned response that resembles the original, unconditioned response. In the terminology of reinforcement learning theory, a stimulus or a state of the environment inherits the value of the primary reinforcer weighted according to the probability that the stimulus predicts the reinforcer correctly. As a result of such conditioning, actions that allow the animal to obtain conditioned stimuli with high value functions would increase the overall reward. Therefore, despite its simplicity, the paradigms of classical conditioning can be used to characterize neural activity related to the utilities and value functions associated with conditioned stimuli.

In most physiological studies on sensorimotor processes in primates, animals are trained in operant or instrumental conditioning paradigms and rewarded with a fixed amount of reward. Recently, many of the same brain areas that have been implicated in sensorimotor transformations have been studied using behavioral tasks in which the magnitude or timing of reward was systematically manipulated. These studies have found that the animal's performance, such as reaction times and error rates, is systematically affected by subjective values or utilities of expected reward,[9–12] indicating that some signals related to the values and utilities are maintained in the brain even during simple operant conditioning tasks.

Whereas classical and operant conditioning paradigms focus on simple stimulus–action or action–reward associations, principles governing the process of choosing from multiple behavioral alternatives have been most frequently studied using concurrent schedules of reinforcements. For example, in a concurrent variable–interval (VI) schedule, each of two alternative actions is rewarded according to its own VI schedule. Namely, each action can be rewarded again only if a minimum delay passes after the delivery of reward for the same action, and this delay is chosen randomly from a particular probability distribution associated with each action. In most tasks based on concurrent schedules of reinforcements, the animal's task is to operate one of the two different devices, such as levers, or to shift its gaze toward one of the two visual targets. In addition, once a particular device or target is baited with a reward, the reward remains available until the animal selects the required action and harvests the reward, and therefore probability of obtaining a reward from an unchosen target increases with time. Accordingly, as long as both options are rewarded with nonzero probabilities, choosing exclusively only one option is not an optimal strategy to maximize the overall reward.[13] Indeed, choice behaviors of animals trained under such concurrent schedules of reinforcement often conform to the so-called matching law, according to which the fraction of a particular choice is equated with the fraction of reward harvested from that choice. Recent studies in nonhuman primates have shown that this matching law might result from the process akin to reinforcement learning.[14,15] However, it should be noted that the behavior predicted by the matching law is different from the so-called probability matching, which states that the probability of choosing a particular action is matched to the probability that the same action would be rewarded. Unlike the behavior predicted by the matching law, probability matching is not optimal.

An optimal strategy under certain concurrent schedules of reinforcements might be to choose a deterministic sequence of actions. For example, denoting the two alternative targets as A and B, an optimal strategy under certain concurrent VI schedules might be to repeat $AAAB$.[13] This leaves open the possibility that the animal may discontinue the process of evaluating its choice outcomes as the process of generating a particular sequence of movements becomes automatic.[16] Similar tendencies were found when the two alternative choices in an oculomotor free-choice task were rewarded randomly with equal probabilities.[17] To investigate the brain mechanisms responsible for adjusting and optimizing the animal's decision-making strategy, it is therefore advantageous to adopt a behavioral paradigm that requires the animal to produce a stochastic sequence of actions.

STOCHASTIC DECISION MAKING IN COMPETITIVE GAMES

In a stationary environment, the probability of reward for a given action does not change over time. In this case, the optimal decision-making strategy is simply to choose exclusively the action that maximizes the expected value of reward. In contrast, stochastic choices may become an optimal strategy in a social context, especially during competitive social interactions. In economics, decision making in a social setting is analyzed by game theory,[18] which seeks to find mathematically an optimal strategy for a group of socially interacting decision makers. In game theory, a game can be specified by a set of decision makers or players, a set of alternative actions available to each player, and a rule that specifies the outcome for each player according to the combination of actions chosen by all players. It is also assumed that each player tries to maximize his or her payoff, and therefore, optimal solutions need to be specified for all players simultaneously. Nash equilibrium provides one such solution. A set of strategies is defined as Nash equilibrium when no players can increase their payoffs by changing their strategies individually. A game might have one or more Nash equilibriums. Consider, for example, the game known as prisoner's dilemma, illustrated by the payoff matrix shown in FIGURE 1A. The rows and columns of this payoff matrix correspond to the alternative options for each of the two players. Each player can choose to cooperate or defect, and the two numbers shown for each combination of choices indicate the payoffs given to the two players, respectively. For example, if both players choose to cooperate, each player will receive the payoff of 3. This is, however, not a Nash equilibrium, because each player can increase his or her payoff by defecting if the other player continues to cooperate. When both players decide to defect, the payoffs for both players will be 1. Although this is worse for both players than the outcome of mutual cooperation, this is the Nash equilibrium, because switching to cooperate while the other player continues to defect decreases the payoff of the cooperative player from 1 to 0.

In a given game, a player might choose one of the available actions exclusively, and this is referred to as a pure strategy, whereas choosing multiple options with positive probabilities is referred to as a mixed strategy. Nash equilibriums often consist of pure strategies, as in the prisoner's dilemma. In other words, an optimal strategy in such games corresponds to choosing one of the options exclusively. In contrast, in some competitive games, such as a rock-paper-scissors game, an optimal strategy is to choose multiple actions randomly with some nonzero probabilities, and these games are referred to as mixed-strategy games. In a rock-paper-scissors game, for example, choosing

(A) Prisoner's dilemma game

Player II

		Cooperate	Defect
	Cooperate	(3, 3)	(0, 5)
Player I			
	Defect	(5, 0)	(1, 1)

(B) Matching pennies game

Computer

		Left	Right
	Left	(1, −1)	(0, 0)
Monkey			
	Right	(0, 0)	(1, −1)

FIGURE 1. Payoff matrix for a prisoner's dilemma game (**A**) and a matching pennies game (**B**). For each game, the two alternative choices available to each player are indicated at the top or to the left of the matrix. Two numbers within parentheses correspond to the payoffs of the row player (A, player I; B, monkey) and the column player (A, player II; B, computer), respectively.

exclusively one of the options, such as rock or paper, can be easily exploited by the opponent, and therefore is suboptimal.

Previous studies in our laboratories investigated the extent to which monkeys can make stochastic choices required in simple competitive zero-sum games.[17,19,20] In one study, monkeys were required to make binary choices against a computer opponent in a oculomotor free-choice task that simulated the matching pennies game (FIG. 1B). In this task, the animal began each trial by fixating a small square presented at the center of a computer screen. After a brief fore-period, two peripheral targets were presented along the horizontal meridian, and the animal was required to shift its gaze toward one of these targets when the central target was extinguished. The animal was rewarded only when it chose the same target as the computer opponent, and the behavior of the computer opponent was determined by one of three different algorithms. In the first algorithm (algorithm 0), the computer chose the two targets randomly with equal probabilities, which corresponds to the Nash equilibrium for the matching pennies game. In zero-sum games, such as the matching pennies, the strategy of a player has no effect on his or her payoff, if the other players choose to play according to a Nash equilibrium. Therefore, in algorithm 0, the expected payoff or reward for the animal was fixed regardless of its strategy. Indeed, the strategy of the animals for algorithm 0 was always significantly different from the equilibrium strategy and changed idiosyncratically across different animals.[17,19]

In the second algorithm (algorithm 1), the computer opponent saved the entire sequence of the choices made by the animal in a given session. From this, the computer estimated a series of conditional probabilities that the animal would choose a particular target in a given trial. If any of these conditional probabilities did not differ significantly from 0.5, then the computer selected the two targets randomly with equal probabilities as in algorithm 0. Otherwise, the computer biased its choice against the target that the animal was more likely to choose. Therefore, the optimal strategy for the animal in algorithm 1 was to choose the two targets randomly with equal probabilities. It should be noted that many possible deterministic strategies, if adopted by the animal, were quickly detected and punished. For example, if the animal tended to alternate between the two targets, the estimate for the first-order conditional probability that the animal would choose the right ward target given that it has chosen the left ward target would quickly increase and this would be exploited by the computer opponent. After the onset of algorithm 1, the animal's choice behavior became quite close to the Nash equilibrium. However, they still displayed a strong tendency to make their choices according to the so-called win-stay-lose-switch strategies.[19] For example, if the animal was rewarded after choosing the right ward target, it was more likely to choose the same target again, whereas if it was not rewarded after choosing the right ward target, it was more likely to choose the left ward target in the next trial. This win-stay-lose-switch strategy was not penalized in algorithm 1, because the computer opponent did not use the information about the animal's reward history.

The animal's choice behavior in algorithm 1 was modeled using a relatively simple reinforcement learning algorithm to quantify how the animal's decision-making strategy was affected by the animal's choices and rewards received in previous trials. In this model, it is assumed that the animal's choice is governed by the difference between the value functions associated with the two targets. In addition, value functions were adjusted according to the animal's choice and its outcome after each trial. For example, the value function for the right ward target increased after the animal chose the right ward target and was rewarded, and decreased if the animal was not rewarded after making the same choice. Thus, this simple reinforcement learning formally implements the win-stay-lose-switch strategy. The changes in the value functions in rewarded and unrewarded trials and the decay factor for the value functions were estimated using a maximum likelihood method, and the statistical significance of these model parameters was evaluated using the method of a profile likelihood interval.[19] This model provided a good account of the animal's behavior during algorithm 1. Results from other studies suggest that the matching behavior of monkeys in concurrent schedules of reinforcement might also result from similar reinforcement learning algorithms.[14,15]

To determine whether monkeys can adjust their natural strategies, such as the win-stay-lose-switch strategy, according to the rewards resulting from their previous choices, the computer opponent was reprogrammed to analyze the reward history of the animal in combination with its choice history. Now, a frequent use of the win-stay-lose-switch strategy was reflected in a relatively large conditional probability that the animal would choose the right ward target, after the animal was rewarded for choosing the right ward target, or after the animal was not rewarded for choosing the left ward target. Hence, such biases would be exploited by the computer. In response to the introduction of this new algorithm (algorithm 2), the probability of using the win-stay-lose-switch was indeed significantly reduced in all animals, indicating that the dependency of the animal's choice on the outcomes of its previous choices was weakened. However, such dependency was not completely removed. Accordingly, the animal's choice behavior was still consistent with the same reinforcement learning algorithm described above (FIG. 2A), although changes made in the value functions after each trial became significantly smaller.[17,19] These results indicate that the parameters of the reinforcement learning algorithms can be adjusted through experience, and therefore provide an example of so-called meta-learning.[21] It has been demonstrated that such meta-learning can be implemented in a

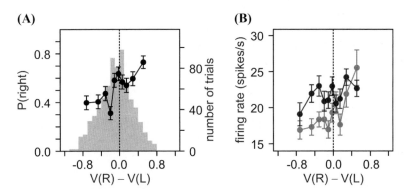

FIGURE 2. The effect of the value functions on the animal's choice (**A**) and the activity of an example neuron in the dorsolateral prefrontal cortex that was modulated by the difference in the value functions associated with the two alternative targets (**B**). (**A**) Histogram shows the distribution of trials as a function of the difference in the value functions, and the black dots indicate the probability that the animal would choose the right-hand target for each decile of the value function difference. (**B**) Average firing rate during the delay period is plotted for the same deciles in the value function difference, separately for the trials in which the animal chose the right ward (*black*) or left ward (*gray*) target. Error bars, SEM.

network model of spiking neurons equipped with plastic synapses.[22] Thus, simple reinforcement learning models, especially when combined with meta-learning algorithms, provide a good account of stochastic choice behaviors. Nevertheless, it is likely that reinforcement learning algorithms might be too simplistic for decision making in more complex situations. For example, the predictions for the choice behavior of monkeys can be improved if the models incorporate serial correlations in the animal's choices.[14,19] In addition, there is evidence that human subjects can adjust their decision-making strategies more efficiently than predicted by a simple reinforcement learning model.[3]

PREFRONTAL SIGNALS RELATED TO VALUE FUNCTIONS

Damages to prefrontal cortex in humans result in a large range of behavioral deficits, and they can be often characterized as impairments in decision making. For example, during the Wisconsin card-sorting task (WCST), the subjects need to maintain information about the relevant dimension of the sensory stimuli to make appropriate choices and update this information based on behavioral feedback. Pathology in the lateral prefrontal cortex produces deficits in the WCST performance.[23,24] In addition, numerous neurophysiological studies have examined the role of the dorsolateral prefrontal cortex, and the results from these studies have often been interpreted in the context of working memory[25] and cognitive control.[26] For example, during a delayed oculomotor response task or delayed match-to-sample task, activity of individual neurons in the dorsolateral prefrontal cortex is often modulated during the delay period according to the information necessary to perform the task correctly.[27] In addition, the type and magnitude of reward expected at the end of a successful trial in working memory tasks often influence the activity of neurons in this cortical area, although information about the expected reward was not necessary for the successful task performance.[28–32] Imaging studies in human subjects have also shown that the activation in the dorsolateral prefrontal cortex is modulated by the amount of monetary reward.[33,34] These results raise the possibility that the activity in the dorsolateral prefrontal cortex may encode the value functions associated with behaviorally relevant stimuli presented during the task or the animal's motor responses.[35,36]

In most previous studies mentioned above, the animals were required to perform their tasks based on a one-to-one stimulus–response mapping. For example, during a choice reaction time task or a memory saccade task, the animal's movement in a given trial would be rewarded only if it is directed to a particular target location. This makes it difficult to determine whether neural activity related to expected reward encodes state or action value functions, which are more closely related to the stimulus and chosen action, respectively. The results from a recent study by Amemori and Sawaguchi[37] suggest that the activity in the dorsolateral prefrontal cortex might be more closely related to state value functions. In this study, the animal was first informed about the magnitude of reward available in a given trial, and then required to remember the location of a visual cue. After a short delay, the animal was given a cue indicating whether it was required to shift its gaze toward (prosaccade) or away from (antisaccade) the previously cued location. This was followed by another delay period, at the end of which the animal produced the required eye movement. Accordingly, the animal could finalize its motor output only during the second delay period. Interestingly, the expected reward influenced the activity in the dorsolateral prefrontal cortex during the first delay period, but not after a specific motor response was chosen during the second delay period. Therefore, it is possible that action value functions and the process of selecting an optimal action may be localized in brain areas more directly involved in controlling specific motor responses, such as the striatum[38] or the posterior parietal cortex,[39-41] rather than the dorsolateral prefrontal cortex. The exact nature of signals in the prefrontal cortex related to expected reward or value functions can potentially provide important insights as to how such signals are used during the process of action selection, and needs to be examined more carefully in further studies. For example, it is possible that more complex value-related signals might emerge in the dorsolateral prefrontal cortex when the animal is required to adjust its decision-making strategy more dynamically based on the animal's previous reward history.

As described above, the process of decision making becomes more complex when the environment changes dynamically, as occurs during the concurrent schedules of reinforcement or competitive games. If the reward-related signals in the dorsolateral prefrontal cortex contribute to the animal's decision-making process, such signals must be systematically related to the value functions that are adjusted dynamically according to the outcomes of the animal's previous choices. To test this prediction, neural activity was recorded from the dorsolateral prefrontal cortex of monkeys making binary decisions in a matching pennies task. The results showed that activity of some neurons in this cortical area indeed encoded the difference in the value functions associated with the two alternative choice targets (FIG. 2B).[17]

PREFRONTAL SIGNALS RELATED TO DECISION OUTCOMES

In addition to the signals related to expected reward or value functions, many neurons in the primate dorsolateral prefrontal cortex encode signals related to the rewards resulting from the animal's decisions.[17,35,42] Signals related to rewards or reward prediction errors were also found in the human dorsolateral prefrontal cortex.[43,44] Similar to the activity of midbrain dopamine neurons, this outcome-related activity in the prefrontal cortex might play an important role in enabling the neural processes involved in updating the value functions and thereby improving the animal's decision-making strategies. Consistent with the broad cortical and subcortical projection of dopamine neurons, activity related to reward and error has been found in many different brain regions, including supplementary eye field,[45,46] anterior[42,47] and posterior[48] cingulate cortex, orbitofrontal cortex,[49] and striatum.[50] Nevertheless, the possibility that such outcome-related activity might subserve different types of computations across these brain areas has not been fully tested and will be an important topic for future studies.

Recent single-neuron recording studies have found that the outcome-related activity in the primate dorsolateral prefrontal cortex often carries additional information about the animal's behavioral responses and other contextual information. For example, when the animal's eye movements in a memory saccade task were rewarded unpredictably after short or long delays, prefrontal neurons often modulated their activity according to the conjunction of saccade direction and reward timing.[51] Similarly, when monkeys made binary choices between the two alternative targets during a competitive game, activity related to action–outcome conjunctions was often maintained across intertrial intervals, suggesting that this outcome-related activity may be temporally integrated to update action value functions, as predicted by reinforcement learning theory (FIG. 3).[17,52] Although the precise mechanism for such temporal integration is still poorly understood, prefrontal activity related to action–outcome associations can be specific to a particular behavioral context. For example, outcome-related activity in the prefrontal cortex can vary depending on whether the animal is performing a memory saccade task or a visually guided saccade task,[53] and whether the animal is performing a free-choice task or a visual search task.[17,52] Outcome-related activity specific to a particular task can be used to update the value functions selectively for the same task without affecting unnecessarily the value functions for similar behaviors during other tasks. Therefore, context-dependent outcome-related activity in the primate prefrontal cortex may play an important role in allowing the animal to adjust its decision-making strategy flexibly according to the demands of specific tasks.

CONCLUSIONS

Previously, the functions of the dorsolateral prefrontal cortex have been largely described by theories that focus on specific cognitive processes, such as working memory and inhibitory control. Each of these theories successfully accounts for a broad spectrum of empirical findings obtained from a variety of methods, including lesions and neurophysiological recordings. More recently, however, neurophysiological studies in nonhuman primates as well as imaging studies in human subjects have characterized various motivational processes in the prefrontal cortex in increasing details. Due to their axiomatic nature, theories developed in economics and machine learning, such as game theory

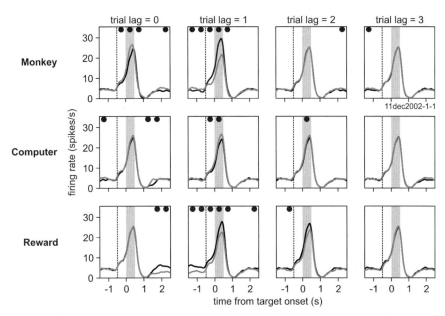

FIGURE 3. An example neuron in the dorsolateral prefrontal cortex that modulated its activity according to the animal's choice, computer's choice, and reward in the previous trial (trial lag = 1). This is the same neuron shown in FIGURE 2B. Each panel shows a pair of spike density functions for two groups of trials sorted by each of the three behavioral events (monkey's choice, computer's choice, and reward) in the current trial (trial lag = 0) or previous trials (trial lag = 1 to 3). In the top two rows, the gray and black lines correspond to the right ward and left ward choices, whereas in the bottom row, they correspond to the rewarded and unrewarded trials, respectively. The dotted vertical line in each panel corresponds to the time when the animal fixated the central target, and black dots shown at the top of each panel indicate that a given behavioral variable has a significant effect on spike counts in the corresponding 0.5-sec bin according to a multiple linear regression model.[52]

and reinforcement learning theory, provide powerful tools that can sharpen questions and hypotheses as to how different types of signals found in the prefrontal cortex can be integrated for the purpose of improving the animal's decision-making strategies. For example, decision-making tasks inspired by game theory, such as a matching pennies task, often include well-defined optimal decision-making strategies that are stochastic, and this makes it possible to investigate the neural mechanisms involved in integrating information about the outcomes of previous choices and updating the animal's decision-making strategies in a dynamic environment. Reinforcement learning theory formally analyzes such adaptive processes of decision making, and has begun to impact the way of theorizing the functions of the prefrontal cortex and other brain areas. Delay activity, although traditionally viewed as a neural substrate of working memory, is often modulated by the motivational significance of stored information, and therefore may be better viewed as coding a certain form of value functions. Outcome-related activity in the prefrontal cortex is heterogeneous and often context dependent, and this may reflect the need for highly flexible cognitive control. Important findings will emerge from future studies directed to understand how various prefrontal signals related to expected and actual outcomes eventually lead to the animal's chosen action.

ACKNOWLEDGMENTS

This study was supported by the grants from the National Institute of Health (MH059216, MH073246, NS044270, and NS048328).

REFERENCES

1. SUTTON, R.S. & A.G. BARTO. 1998. Reinforcement Learning: An Introduction. MIT Press. Cambridge, MA.
2. LEE, D. 2006. Neural basis of quasi-rational decision making. Curr. Opin. Neurobiol. **16:** 191–198.
3. HAMPTON, A.N., P. BOSSAERTS & J. P. O'DOHERTY. 2006. The role of the ventromedial prefrontal cortex in abstract state-based inference during decision making in humans. J. Neurosci. **26:** 8360–8367.
4. DAW, N.D. & K. DOYA. 2006. The computational neurobiology of learning and reward. Curr. Opin. Neurobiol. **16:** 199–204.
5. SCHULTZ, W. 1998. Predictive reward signal of dopamine neurons. J. Neurophysiol. **80:** 1–27.
6. SCHULTZ, W., L. TREMBLAY & J.R. HOLLERMAN. 2000. Reward processing in primate orbitofrontal cortex and basal ganglia. Cereb. Cortex **10:** 272–283.
7. O'DOHERTY, J.P., P. DAYAN, K. FRISTON, et al. 2003. Temporal difference models and reward-related learning in the human brain. Neuron **38:** 329–337.
8. MCCLURE, S.M., G.S. BERNS & P.R. MONTAGUE. 2003. Temporal prediction errors in a passive learning task activate human striatum. Neuron **38:** 339–346.
9. ZEAMAN, D. 1949. Response latency as a function of the amount of reinforcement. J. Exp. Psychol. **39:** 466–483.
10. WATANABE, M., H.C. CROMWELL, L. TREMBLAY, et al. 2001. Behavioral reactions reflecting different reward expectations in monkeys. Exp. Brain Res. **140:** 511–518.
11. TAKIKAWA, Y., R. KAWAGOE, H. ITO, et al. 2002. Modulation of saccadic eye movements by predicted reward outcome. Exp. Brain Res. **142:** 284–291.
12. SOHN, J.-W. & D. LEE. 2006. Effects of reward expectancy on sequential eye movements in monkeys. Neural Netw. **19:** 1181–1191.
13. STADDON, J.E.R., J.M. HINSON & R. KRAM. 1981. Optimal choice. J. Exp Anal. Behav. **35:** 397–412.
14. LAU, B. & P.W. GLIMCHER. 2005. Dynamic response-by-response models of matching behavior in rhesus monkeys. J. Exp. Anal. Behav. **84:** 555–579.
15. CORRADO, G.S., L.P. SUGRUE, H.S. SEUNG, et al. 2005. Linear-nonlinear-poisson models of primate choice dynamics. J. Exp. Anal. Behav. **84:** 581–617.

16. HIKOSAKA, O., H. NAKAHARA, M.K. RAND, *et al.* 1999. Parallel neural networks for learning sequential procedures. Trends Neurosci. **22:** 464–471.

17. BARRACLOUGH, D.J., M.L. CONROY & D. LEE. 2004. Prefrontal cortex and decision making in a mixed-strategy game. Nat. Neurosci. **7:** 404–410.

18. VON NEUMANN, J. & O. MORGENSTERN. 1944. Theory of games and economic behavior. Princeton University Press. Princeton, NJ.

19. LEE, D., M.L. CONROY, B.P. MCGREEVY, *et al.* 2004. Reinforcement learning and decision making in monkeys during a competitive game. Cogn. Brain Res. **22:** 45–58.

20. LEE, D., B.P. MCGREEVY & D.J. BARRACLOUGH. 2005. Learning and decision making in monkeys during a rock-paper-scissors game. Cogn. Brain Res. **25:** 416–430.

21. DOYA, K. 2002. Metalearning and neuromodulation. Neural Netw. **15:** 495–506.

22. SOLTANI, A., D. LEE & X.-J. WANG. 2006. Neural mechanism for stochastic behavior during a competitive game. Neural Netw. **19:** 1075–1090.

23. MILNER, B. 1963. The effects of different brain lesions on card sorting. Arch. Neurol. **9:** 90–100.

24. STUSS, D.T., B. LEVINE, M.P. ALEXANDER, *et al.* 2000. Wisconsin card sorting test performance in patients with focal frontal and posterior brain damage: effects of lesion location and test structure on separable cognitive processes. Neuropsychologia **38:** 388–402.

25. LEVY, R. & P.S. GOLDMAN-RAKIC. 2000. Segregation of working memory functions within the dorsolateral prefrontal cortex. Exp. Brain Res. **133:** 23–32.

26. MILLER, E.K. & J.D. COHEN. 2001. An integrative theory of prefrontal cortex function. Ann. Rev. Neurosci. **24:** 167–202.

27. FUNAHASHI, S., C.J. BRUCE & P.S. GOLDMAN-RAKIC. 1989. Mnemonic coding of visual space in the monkey's dorsolateral prefrontal cortex. J. Neurophysiol. **61:** 331–349.

28. WATANABE, M. 1996. Reward expectancy in primate prefrontal neurons. Nature **382:** 629–632.

29. LEON, M.I. & M.N. SHADLEN. 1999. Effect of expected reward magnitude on the response of neurons in the dorsolateral prefrontal cortex of the macaque. Neuron **24:** 415–425.

30. KOBAYASHI, S., J. LAUWEREYNS, M. KOIZUMI, *et al.* 2002. Influence of reward expectation on visuospatial processing in macaque lateral prefrontal cortex. J. Neurophysiol. **87:** 1488–1498.

31. WATANABE, M., K. HIKOSAKA, M. SAKAGAMI, *et al.* 2005. Functional significance of delay-period activity of primate prefrontal neurons in relation to spatial working memory and reward/omission-of-reward expectancy. Exp. Brain Res. **166:** 263–276.

32. ICHIHARA-TAKEDA, S. & S. FUNAHASHI. 2006. Reward-period activity in primate dorsolateral prefrontal and orbitofrontal neurons is affected by reward schedules. Cereb. Cortex **18:** 212–226.

33. POCHON, J.B., R. LEVY, P. FOSSATI, *et al.* 2002. The neural system that bridges reward and cognition in humans: an fMRI study. Proc. Natl. Acad. Sci. USA **99:** 5669–5674.

34. TAYLOR, S.F., R.C. WELSH, T.D. WAGER, *et al.* 2004. A functional neuroimaging study of motivation and executive function. Neuroimage **21:** 1045–1054.

35. WATANABE, M. 1990. Prefrontal unit activity during associative learning in the monkey. Exp. Brain Res. **80:** 296–309.

36. WATANABE, M. 1992. Frontal units of the monkey coding the associative significance of visual and auditory stimuli. Exp. Brain Res. **89:** 233–247.
37. AMEMORI, K. & T. SAWAGUCHI. 2006. Contrasting effects of reward expectation on sensory and motor memories in primate prefrontal neurons. Cereb. Cortex **16:** 1002–1015.
38. SAMEJIMA, K., Y. UEDA, K. DOYA, et al. 2005. Representation of action-specific reward values in the striatum. Science **310:** 1337–1340.
39. PLATT, M.L. & P.W. GLIMCHER. 1999. Neural correlates of decision variables in parietal cortex. Nature **400:** 233–238.
40. SUGRUE, L.P., G.S. CORRADO & W.T. NEWSOME. 2004. Matching behavior and the representation of value in the parietal cortex. Science **304:** 1782–1787.
41. DORRIS, M.C. & P.W. GLIMCHER. 2004. Activity in posterior parietal cortex is correlated with the relative subjective desirability of action. Neuron **44:** 365–378.
42. NIKI, H. & M. WATANABE. 1979. Prefrontal and cingulate unit activity during timing behavior in the monkey. Brain Res. **171:** 213–224.
43. DREHER, J-D., P. KOHN & K.F. BERMAN. 2006. Neural coding of distinct statistical properties of reward information in humans. Cereb. Cortex **16:** 561–573.
44. PAULUS, M.P., J.S. FEINSTEIN, S.F. TAPERT, et al. 2004. Trend detection via temporal difference model predicts inferior prefrontal cortex activation during acquisition of advantageous action selection. Neuroimage **21:** 733–743.
45. AMADOR, N., M. SCHLAG-REY & J. SCHLAG. 2000. Reward-predicting and reward-detecting neuronal activity in the primate supplementary eye field. J. Neurophysiol. **84:** 2166–2170.
46. STUPHORN, V., T.L. TAYLOR & J.D. SCHALL. 2000. Performance monitoring by the supplementary eye field. Nature **408:** 857–860.
47. ITO, S., V. STUPHORN, J.W. BROWN, et al. 2003. Performance monitoring by the anterior cingulate cortex during saccade countermanding. Science **302:** 120–122.
48. MCCOY, A.N., J.C. CROWLEY, G. HAGHIGHIAN, et al. 2003. Saccade reward signals in posterior cingulate cortex. Neuron **40:** 1031–1040.
49. TREMBLAY, L. & W. SCHULTZ. 2000. Reward-related neuronal activity during go-nogo task performance in primate orbitofrontal cortex. J. Neurophysiol. **83:** 1864–1876.
50. HOLLERMAN, J.R., L. TREMBLAY & W. SCHULTZ. 1998. Influence of reward expectation on behavior-related neuronal activity in primate striatum. J. Neurophysiol. **80:** 947–963.
51. TSUJIMOTO, S. & T. SAWAGUCHI. 2004. Neuronal representation of response-outcome in the primate prefrontal cortex. Cereb. Cortex **14:** 47–55.
52. SEO, H., D.J. BARRACLOUGH & D. LEE. 2007. Dynamic signals related to choices and outcomes in the dorsolateral prefrontal cortex. Cereb. Cortex. In press.
53. TSUJIMOTO, S. & T. SAWAGUCHI. 2005. Context-dependent representation of response-outcome in monkey prefrontal neurons. Cereb. Cortex **15:** 888–898.

Resisting the Power of Temptations

The Right Prefrontal Cortex and Self-Control

DARIA KNOCH[a,b,c] AND ERNST FEHR[a,c]

[a]Institute for Empirical Research in Economics, University of Zurich, Blümlisalpstrasse, Zurich, Switzerland

[b]Department of Neurology, University Hospital Zurich, Switzerland

[c]Collegium Helveticum, Schmelzbergstrasse, Zurich, Switzerland

ABSTRACT: Imagine you are overweight and you spot your favorite pastry in the storefront of a bakery. How do you manage to resist this temptation? Or to give other examples, how do you manage to restrain yourself from overspending or succumbing to sexual temptations? The present article summarizes two recent studies stressing the fundamental importance of inhibition in the process of decision making. Based on the results of these studies, we dare to claim that the capacity to resist temptation depends on the activity level of the right prefrontal cortex (PFC).

KEYWORDS: decision making; prefrontal cortex; self-control; transcranial magnetic stimulation; laterality

INTRODUCTION

The siren call of our impulses, desires, and urges often tempts us, and many of our decisions involve a conflict between our deliberate and our pleasure-seeking sides. From the standpoint of adaptive self-regulation, an appropriate response to temptations involves exercising self-control.[1–3] This conscious control of thought, action, and emotions may be considered as a distinctive feature of human cognition. Moreover, the ability to override immediate urges is not only relevant for adaptive *individual* decision making but also contributes to harmonious *social* interactions. For example, suppressing a desire to retaliate may be necessary to prevent the escalation of interpersonal conflict. Thus, our capacity to suppress the unlimited pursuit of immediate self-interest has been suggested to be a hallmark of civilized life.[4]

Address for correspondence: D. Knoch, Department of Neurology, University Hospital Zurich, CH–8091 Zürich. Voice: +41-44-634-4809; fax +41-44-634-4907.
dknoch@iew.unizh.ch

Ann. N.Y. Acad. Sci. 1104: 123–134 (2007). © 2007 New York Academy of Sciences.
doi: 10.1196/annals.1390.004

A considerable amount of research has shown that the resistance to immediate self-interests is often greatly diminished in people with injuries to the prefrontal cortex (PFC).[5–7] This seems particularly true for patients with *right-sided* lesions.[8–10] However, patient studies are sometimes difficult to conduct due to limited opportunities for experimental manipulations. In addition, confounding variables, that is, the possibility of functional reorganization after brain lesions, may hamper the interpretation of the results, and studies often have a low number of patients (see Ref. 11 for limitations of the lesion method). Some functional imaging studies suggest that the *right* PFC may be particularly critical for self-regulation and self-control[12–15] or behavioral adjustments.[16] These studies, however, only passively measure brain activity correlated with a specific task, but do not reveal a causal relationship between changes in brain activity and their respective behavioral consequences. A direct investigation of such a causal brain–behavior relationship would require a controlled manipulation of brain activity where the impact on behavior or cognition can be quantified.[17] The technique of transcranial repetitive magnetic stimulation (TMS) allows for such a manipulation by inducing brief electric currents within discrete brain areas via pulsed magnetic fields on the corresponding scalp location. The "virtual lesion" technique in particular, that is, low-frequency repetitive transcranial magnetic stimulation (rTMS) over the course of several minutes, allows a transient disruption of cortical functions.[18] We applied this technique to examine whether self-control can be modified in healthy individuals in the context of both individual and social decision making.

Individual Decision Making: Diminished Self-Control Leads to Increased Risk-Taking Behavior

Adolescents generally exhibit riskier behavior than do adults. Their decision-making behavior is also thought to be a manifestation of an immature PFC,[19] and patients with traumatic brain injuries or other pathologies affecting the PFC show a tendency for riskier, "out-of-character" decision making, and an apparent disregard for negative consequences of their actions.[20,21] Clinical impressions tell us that this is particularly true for patients with right-sided lesions.[9,10] We designed a "virtual lesion" study[22] with healthy volunteers to investigate hemispheric asymmetries in risk-taking behavior directly. We used low-frequency rTMS to disrupt left or right dorsolateral prefrontal cortex (DLPFC) function transiently before applying a well-known gambling paradigm that provides a measure of risk taking ("risk task").[23] In this task, subjects have to decide between a relatively safe choice, which provides a low reward with a high probability, and a risky choice, which provides a substantially higher reward with a relatively low probability. Subjects were presented with binary choices between a safe and a risky choice. If subjects perceive the high reward to be salient, there may be a temptation to decide in favor of the risky choice. However, this choice has the drawback that the reward is only available with a low probability.

In the task, a series of six boxes on the screen indicated the probabilities of the different outcomes. The boxes can be either pink or blue, and the proportion of blue and pink boxes changed from one trial to another (5:1, 4:2, or 3:3; FIG. 1). The subject was told that the computer had arbitrarily hidden a "winning token" inside one of the blue or pink boxes. The subject had to decide whether the token was hidden in a box by pressing a button of the corresponding color.

Findings were straightforward. Participants stimulated over the right DLPFC ($N = 9$) were more likely to choose the high-risk prospect than those stimulated over the left DLPFC ($N = 9$) or those who received sham stimulation ($N = 9$) (FIG. 2A). We thus demonstrated that individuals display a significantly

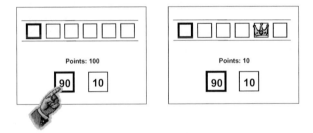

FIGURE 1. This figure shows one exemplary trial as displayed on the computer screen. Subjects were presented with six boxes colored pink or blue (in this black and white version of the figure pink boxes are replaced by white boxes with thick borders; blue boxes are replaced by boxes with thin borders). The number of pink and blue boxes varied from trial to trial according to a fixed pseudorandom sequence (sample shows "level of risk": 5:1). Subjects were asked to find the winning token. They did not have to pick the individual box hiding the winning token, but simply had to select the color of the box it was hidden in (illustrated in the left panel by a schematic hand pointing to the pink box in this example). Subjects were told that each box, regardless of color, was equally likely to hide the winning token. Thus, the likelihood of finding the winning token was directly related to the ratio of blue to pink boxes. For a trial showing 5 blue boxes and 1 pink box, there would be a probability of 5/6 that the winning token was hidden in a blue box, but only a 1/6 chance that it was hidden in the single pink box. Importantly, subjects are rewarded or penalized depending on whether they pick the correct color box or not. There is a fixed reward associated with either choice of boxes' color ("balance of reward": 10 vs. 90, 20 vs. 80, 30 vs. 70, and 40 vs. 60). The larger reward (and penalty) is always associated with choice of the high-risk prospect (i.e., the color with the fewer number of boxes), whereas the smallest reward (and penalty) is associated with choice of the low-risk prospect. A correct choice results in the addition of the number of points associated with that particular scenario, while an incorrect choice results in the subtraction of the same amount (sample in the right panel shows an incorrect choice that results in a subtraction of 90 points). Adapted from Knoch *et al.* 2006[22] (copyright 2006 by the Society for Neuroscience). This figure appears in color online.

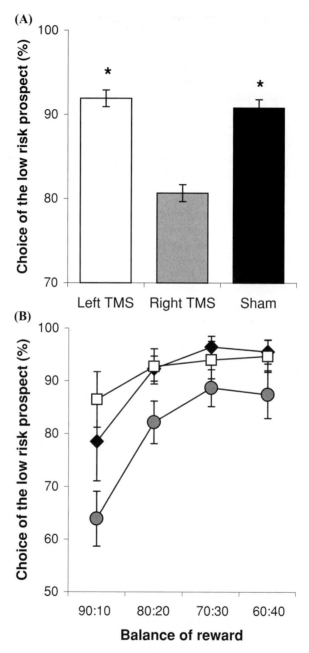

FIGURE 2. Percentage choice of the low-risk prospect (means ± SEM) (**A**) for all three groups, $P < 0.05$ and (**B**) as a function of the balance of reward (□ = Left TMS; ● = Right TMS; ◆ = Sham). Adapted from Knoch *et al.* [22] (copyright 2006 by the Society for Neuroscience).

stronger preference for the risky prospect, choosing the larger potential reward even at the risk of greater penalty, following disruption of the right, but not the left, DLPFC. We therefore suggest that the "risk task" primarily requires suppression of an option that appears seductive because of its large reward. That is, subjects are initially attracted to the high-reward/high-risk options by virtue of their higher payoffs, a tendency normally suppressed over the course of the task by top-down control mechanisms.

At this point we will discuss two alternative possible explanations of the rTMS effect, which are not considered in the original paper. For a discussion of additional alternative explanations, we ask the reader to refer to Ref. 22. The task requires subjects to measure the ratio of pink and blue boxes, and to take the various amounts of money associated with the different choices into account to calculate the expected value of the two options. Moreover, it requires subjects to integrate information about the rewarding and punishing consequences of an action. It may thus be that subjects receiving right prefrontal rTMS are impaired at calculating the riskiness of the choices or at integrating information about the choices' consequences. However, these interpretations are less convincing since repeated measures' analysis of variance (ANOVA) of group (left TMS, right TMS, sham) × level of risk (5:1, 4:2) × balance of reward (90:10, 80:20, 70:30, 60:40) revealed a main effect of balance of reward ($P < 0.001$) and, importantly, no interaction between group and balance of reward ($P = 0.414$; see FIG. 2B). In other words, if subjects were impaired in calculating or integrating the consequences of different choices, we should observe that subjects who received right rTMS are unable to discriminate between the different balances of reward. In addition, if subjects who received rTMS over the right PFC were impaired at calculating the riskiness of the choice, their deliberation time would probably be longer than that of subjects who received rTMS over left PFC or subjects who received sham stimulation. This was not the case. ANOVA of group × level of risk × balance of reward for the decision times revealed no main effect for group ($P = 0.737$). Therefore, we favor the hypothesis that the right PFC plays a crucial role in the suppression of superficially seductive options. We further speculate that the substantial differences among individuals in risk proneness in real-life scenarios may correspond to different levels of activity in the right PFC. The higher this level is, the lower one's "appetite for risk." If this turns out to be true, high-frequency rTMS (which increases cortical excitability) could be used to increase activity of the right PFC in a therapeutic framework to enhance cognitive control and adaptive decision making.

The results of this study led us to speculate whether increasing rather than decreasing the level of activity in the right PFC would diminish, rather than raise, subjects' "appetite for risk." Indeed, preliminary results show that participants with increased activity in the right DLPFC chose the safe prospect more often than did the sham group.[24]

Social Decision Making: Diminished Self-Control
Leads to Selfish Behavior

The human species is unique to the extent that social norms that constrain the unrestricted pursuit of self-interest govern behavior. Overcoming the self's natural, impulsive nature requires self-control. The ultimatum game (UG) provides a useful tool for studying the neural mechanisms of self-control in the context of social decision making, as it illustrates the tension between economic self-interest on the one hand and fairness goals on the other. In this bargaining game, two anonymous individuals, a "proposer" and a "responder," have to agree on the division of a given amount of money, say $10, according to the following rules: The proposer can make exactly one suggestion on how the $10 should be allocated among the two by making an integer offer X to the responder. If the responder accepts, each player keeps the amount the proposer allocates. If the responder rejects the offer, neither player receives any money. If economic self-interest alone motivates the responder, he will accept even a very low offer, say $1, because $1 is better than $0. However, if concerns for reciprocity[25] and equity[26] drive him, he may reject low offers, because he views them as insultingly unfair and inequitable. Thus, the responder faces a conflict in case of low offers between his economic self-interest, which drives him toward accepting the offer, and his fairness goals, which encourage him to reject it. Strong evidence[27,28] suggests that many people reject low offers in the game even if the stake level is as high as 3 months' income.[29] Rejection rates up to 80% have been observed[30] for offers below 25% of the available money, and a neuroimaging study[31] showed that both the anterior insula—a brain area involved in the evaluation and representation of negative emotional states[32]— and the DLPFC are activated when responders decide whether to accept or reject an unfair offer. For our purposes, it is particularly interesting that both the right and left DLPFC are more strongly activated when subjects face unfair offers compared to when they face fair offers. These areas are widely thought to be involved in executive control, goal maintenance, and the inhibition of prepotent responses.[33] All these functions are relevant for the responder in UG because there are likely to be several competing goals—fairness goals and self-interest—and the question is which of them should be maintained, that is, given priority, and which motivational impulse should be restrained.

The fact that the DLPFC is more strongly active when subjects are confronted with an unfair offer compared with a fair offer[31] cannot provide conclusive evidence that DLPFC activity is crucial for the responders' decisions. In principle, it is even possible that this area is not causally involved in the decision to accept or reject unfair offers. To address this question, we applied prefrontal low-frequency rTMS to 52 subjects[34] (left DLPFC, $N = 17$; right DLPFC, $N = 19$; sham, $N = 16$) who were in the role of the responder in an anonymous UG with a stake size of CHF 20 (CHF $1 \approx \$ 0.80$). To generate enough observations on the responders' side, we limited the proposer's strategy space, meaning that only offers of CHF 10, 8, 6, or 4 were possible. Obviously, CHF 10 is the fairest offer, because it splits the stake size equally, while CHF 4 is the most unfair offer.

If the DLPFC is involved in overriding selfish impulses that drive a subject toward acceptance of unfair offers, low-frequency rTMS of this brain region should *increase* the acceptance rate for unfair offers relative to the sham stimulation condition, as this kind of stimulation leads to a disruption of neuronal firing in the stimulated brain region.[18] In other words, if we disrupt activity in a brain region that is hypothesized to override selfish impulses, we should functionally weaken the inhibitory control and, selfish impulses should thus have a stronger impact on decision making; as a consequence, the acceptance rate of unfair offers should increase. We focus on acceptance behavior with regard to the lowest offer in this case because the tension between fairness and self-interest is greatest here. FIGURE 3A indicates that the right DLPFC group has a significantly higher acceptance rate than the left DLPFC and the sham rTMS group. Importantly, these differences across conditions cannot be attributed to different fairness judgments across groups. Immediately after the ultimatum game experiment, we elicited subjects' fairness judgments with regard to different offers on a 7-point scale. Subjects in all three treatment groups judged the lowest offer of 4 as rather unfair and there are no differences in fairness judgments across groups (FIG. 3B). Thus, despite the fact that subjects in all three groups judge low offers as very unfair, subjects whose right DLPFC has been disrupted exhibit much higher acceptance rates. Similar results were found for the other unfair offer of 6 CHF.

These results suggest that subjects who received right prefrontal TMS are less able to resist the economic temptation to accept unfair offers. Our findings are also interesting in the light of evidence suggesting that patients with right prefrontal lesions are characterized by the inability to behave in normatively appropriate ways despite the fact that they possess the social knowledge that is necessary for normative behavior.[35] Our findings are also congruent with the observation of empathy deficits in patients with predominantly right frontal lesions,[36,37] as an inhibitory component is required to regulate and tone down the prepotent self-perspective to allow the perception and evaluation of others' perspectives.[38] Further support comes from findings in patients with frontotemporal dementia (FTD), showing that symptoms may be influenced by the relative involvement of the right versus the left hemisphere, with left-sided FTD manifesting language changes and right-sided FTD presenting with aggressive, antisocial, and other socially undesirable behaviors.[39]

Note that if we suggest that right DLPFC is involved in overriding self-interest motives, we do not necessarily imply that this brain region directly suppresses other brain areas that represent self-interest. Instead, we believe that right DLPFC is involved in top-down control (or executive control), the overall effect of which is a reduction in the weight of self-interested impulses on an individual's action. Thus, rather than directly suppressing neural activities that

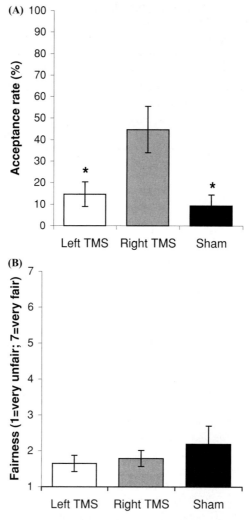

FIGURE 3. Behavioral responses and fairness judgments (means ± SEM) related to the most unfair offer of CHF 4. (**A**) Acceptance rates across treatment groups. Subjects whose right DLPFC is disrupted exhibit a much higher acceptance rate than subjects in the other two treatments (Mann–Whitney U tests, two-tailed, $P < 0.05$). (**B**) Perceived unfairness across treatments (1 = very unfair; 7 = very fair). Subjects in all three treatment groups perceive an offer of 4 as very unfair, and there are no significant differences across groups. Responders were shown a list of all feasible offers in the human- and the computer-offer condition and asked to report on a 7-point scale to what extent they perceived an offer as fair or unfair (1 = very unfair; 7 = very fair). As the ultimatum game experiment lasted only a short time, these fairness assessments took place roughly 4–5 min after the 15-min offline stimulation with rTMS. Thus, when subjects assessed the fairness of different offers, their DLPFC was still disrupted if they had received real rTMS stimulation. Adapted from Knoch et al.[34].

represent self-interested impulses, the DLPFC may be part of a network that modulates the relative impact of fairness motives and self-interest goals on decision making, and the final outcome of this modulation may then be a weakening of the impact of self-interest motives on decision making.

Evidence from human lesion studies has implicated the right hemisphere in the processing of emotional and social information.[40,41] Can a reduced general emotional responsiveness explain the observed rTMS effect after right prefrontal rTMS stimulation? This seems to be unlikely for the following reasons: rTMS of the right DLPFC reduces fair behaviors but not fairness judgments. As there is no reason to believe that emotions are less involved in fairness judgments than in fair behaviors, a general reduction in emotional responsiveness should have also affected fairness judgments. Moreover, we had an additional treatment condition that we did not yet mention in this article. In our experiment, a responder not only played the usual ultimatum game for 10 rounds where the human partner proposes a division of the available money (human-offer condition), but every subject also played 10 rounds with human partners who could not make proposals themselves. A computer randomly generated the offers in these trials (computer-offer condition). The motive to punish the human partner for an unfair offer cannot play a role in the computer-offer condition, because the partner is not responsible for it. According to the theory of inequity aversion,[26] many subjects find accepting low offers per se aversive, even if made by a computer. Indeed, a considerable share (33%) of low offers was rejected in the computer-offer condition. A general reduction in emotional responsiveness by rTMS of the right DLPFC also should have affected behavior in the computer-offer condition significantly because there is no reason to believe that an aversion against inequality is associated with less emotional involvement. We could not find such an effect, however; low-frequency rTMS of the right DLPFC did not increase the acceptance rate in the computer-offer condition ($P = 0.306$). Finally, if rTMS simply reduces subjects' emotional responsiveness, we should also expect an impact on subjects' mood. We measured mood before and after rTMS using visual analogue scales and could not find any effect of rTMS on mood (see FIG. 4). We therefore favor the hypothesis that the right DLPFC is causally involved in a neural network that controls or regulates the impact of the economic temptation on the acceptance decision. This interpretation can explain why rTMS affects fair behavior but not fairness judgments because no economic temptation is involved in fairness *judgments*, whereas the rejection of an unfair but positive offer requires foregoing economic gains.

As the right frontal lobe seems to be relevant for the integration of information[42,43] one might be tempted to suggest alternatively that right prefrontal rTMS causes an impairment in the ability to integrate different decision values—the positive monetary value of the gain and the negative value of unfairness. This hypothesis, however, cannot easily account for the fact that we observed treatment differences regarding the response times in accepting

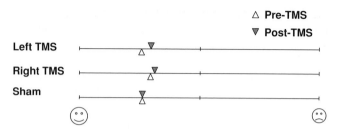

FIGURE 4. Subjects' moods as indicated on a visual analogue scale. Subjects' moods are predominantly positive in all treatments. Neither before the application of rTMS (Kruskall–Wallis Test, $P = 0.659$) nor after the application (Kruskall–Wallis Test, $P = 0.773$) did we observe mood differences across treatments. There are also no treatment differences in the change in mood (Kruskall–Wallis Test, $P = 0.970$).

unfair offers, while there are no differences in response times when subjects face a fair offer. If integrating different decision values is more difficult after right rTMS of the DLPFC, subjects should exhibit longer response times in all situations. However, rTMS of the right DLPFC did not affect response times when subjects faced fair offers, and the response time *decreases* strongly for subjects whose right PFC was disrupted when they faced unfair offers of 4 in the human-offer condition.

In conclusion, the findings in these studies suggest that cortical stimulation can modulate the fundamental human capacity of self-control, and the results thus confirm the asymmetric role of the PFC in decision making. They also indicate that the capacity for restraint depends on the activity level of the right PFC. In response override, one must stop a prepotent response to a stimulus, either because the response needs to be withheld or because a less prepotent response is more appropriate. Without this capacity, we would be slaves of our emotional impulses, temptations, and desires and thus unable to behave socially adequately. Our findings also illustrate the importance of rTMS for further progress in understanding the neural basis of decision making. Although neuroimaging data indicate that responders' left DLPFC is activated when they face unfair offers,[32] disruption of the left DLPFC does not change their behavior, suggesting that neuroimaging studies need to be complemented with techniques, such as rTMS, to reach more firm conclusions. Since our data suggest that the right lateral PFC is causally involved in overriding self-interest, it would be interesting to investigate whether decreasing the activity level of right PFC results in immoral decision making, as moral decisions often require the inhibition of self-interest.

Even though the application of TMS enables us to claim a causal role of lateral PFC activity for self-control processes in the context of individual and social decision making, the exact neuronal mechanisms underlying the TMS-induced virtual lesion will remain unknown until a simultaneous examination of underlying brain activity during task performance can be performed. Therefore, the combination of TMS-induced virtual lesions and fMRI during a given task promises important additional insights for studying the neural mechanisms of self-control.

REFERENCES

1. BAUMEISTER, R.F., T.F. HEATHERTON & D.M. TICE. 1993. When ego threats lead to self-regulation failure: negative consequences of high self-esteem. J. Pers. Soc. Psychol. **64:** 141–156.
2. MURAVEN, M. & R.F. BAUMEISTER. 2000. Self-regulation and depletion of limited resources: does self-control resemble a muscle? Psychol. Bull. **126:** 247–259.
3. METCALFE, J. & W. MISCHEL. 1999. A hot/cool-system analysis of delay of gratification: dynamics of willpower. Psychol. Rev. **106:** 3–19.
4. FREUD, S. 1930/1961. Civilization and Its Discontents. Norton. New York.
5. DAMASIO, A.R. 1996. The somatic marker hypothesis and the possible functions of the prefrontal cortex. Philos. Trans. R. Soc. Lond. B. Biol. Sci. **351:** 1413–1420.
6. SHALLICE, T. & P.W. BURGESS. 1991. Deficits in strategy application following frontal lobe damage in man. Brain **114:** 727–741.
7. STUSS, D.T. & D.F. BENSON. 1986. The Frontal Lobes. Raven. New York.
8. STARKSTEIN, S.E. & R.G. ROBINSON. 1997. Mechanism of disinhibition after brain lesions. J. Nerv. Ment. Dis. **185:** 108–114.
9. TRANEL, D., A. BECHARA & N.L. DENBURG. 2002. Asymmetric functional roles of right and left ventromedial prefrontal cortices in social conduct, decision-making, and emotional processing. Cortex **38:** 589–612.
10. CLARK, L. et al. 2003. The contributions of lesion laterality and lesion volume to decision-making impairment following frontal lobe damage. Neuropsychologia **41:** 1474–1483.
11. RORDEN, C. & H.O. KARNATH. 2004. Using human brain lesions to infer function: a relic from a past era in the fMRI age? Nat. Rev. Neurosci. **5:** 813–819.
12. BEAUREGARD, M., J. LEVESQUE & P. BOURGOUIN. 2001. Neural correlates of conscious self-regulation of emotion. J. Neurosci. **21:** 1–6.
13. ERNST, M. et al. 2002. Decision-making in a risk-taking task: a PET study. Neuropsychopharmacology **26:** 682–691.
14. FISHBEIN, D.H. et al. 2005. Risky decision making and the anterior cingulate cortex in abstinent drug abusers and nonusers. Brain Res. Cogn. Brain Res. **23:** 119–136.
15. LEVESQUE, J. et al. 2003. Neural circuitry underlying voluntary suppression of sadness. Biol. Psychiatry **53:** 502–510.
16. KERNS, J.G. et al. 2004. Anterior cingulate conflict monitoring and adjustments in control. Science **303:** 1023–1026.
17. SACK, A.T. et al. 2007. Imaging the brain activity changes underlying impaired visuospatial judgments: simultaneous FMRI, TMS, and behavioral studies. Cerebral Cortex Advance published online March 3.
18. ROBERTSON, E.M., H. THEORET & A. PASCUAL-LEONE. 2003. Studies in cognition: the problems solved and created by transcranial magnetic stimulation. J. Cogn. Neurosci. **15:** 948–960.
19. CHAMBERS, R.A., J.R. TAYLOR & M.N. POTENZA. 2003. Developmental neurocircuitry of motivation in adolescence: a critical period of addiction vulnerability. Am. J. Psychiatry **160:** 1041–1052.
20. BECHARA, A. et al. 1996. Failure to respond autonomically to anticipated future outcomes following damage to prefrontal cortex. Cereb. Cortex **6:** 215–225.

21. RAHMAN, S. *et al.* 2001. Decision making and neuropsychiatry. Trends Cogn. Sci. **5:** 271–277.
22. KNOCH, D. *et al.* 2006. Disruption of right prefrontal cortex by low-frequency repetitive transcranial magnetic stimulation induces risk-taking behavior. J. Neurosci. **26:** 6469–6472.
23. ROGERS, R.D. *et al.* 1999. Choosing between small, likely rewards and large, unlikely rewards activates inferior and orbital prefrontal cortex. J. Neurosci. **19:** 9029–9038.
24. FECTEAU, S. *et al.* Diminishing risk-taking behavior by increasing activity in the right prefrontal cortex. A direct current stimulation study: submitted.
25. RABIN, M. 1993. Incorporating fairness into game theory and economics. Am. Econ. Rev. **83:** 1281–1302.
26. FEHR, E. & K.M. SCHMIDT. 1999. A theory of fairness, competition, and cooperation. Q. J. Econ. **114:** 817–868.
27. GÜTH, W., R. SCHMITTBERGER & B. SCHWARZE. 1982. An experimental analysis of ultimatum bargaining. J. Econ. Behav. Organ. **3:** 367–388.
28. HENRICH, J. *et al.* 2001. In search of homo economicus: behavioral experiments in 15 small-scale societies. Am. Econ. Rev. **91:** 73–78.
29. CAMERON, L.A. 1999. Raising the stakes in the ultimatum game: experimental evidence from Indonesia. Econ. Inq. **37:** 47–59.
30. CAMERER, C.F. 2003. Behavioral game theory—Experiments in strategic interaction. Princeton University Press. Princeton, NJ.
31. SANFEY, A.G. *et al.* 2003. The neural basis of economic decision-making in the ultimatum game. Science **300:** 1755–1758.
32. CALDER, A.J., A.D. LAWRENCE & A.W. YOUNG. 2001. Neuropsychology of fear and loathing. Nat. Rev. Neurosci. **2:** 352–363.
33. MILLER, E.K. & J.D. COHEN. 2001. An integrative theory of prefrontal cortex function. Ann. Rev. Neurosci. **24:** 167–202.
34. KNOCH, D. *et al.* 2006. Diminishing reciprocal fairness by disrupting the right prefrontal cortex. Science **314:** 829–832.
35. DAMASIO, A.R. 1995. Descartes' error: Emotion, reason and the human brain. Penguin Books. New York.
36. SHAMAY-TSOORY, S.G. *et al.* 2003. Characterization of empathy deficits following prefrontal brain damage: the role of the right ventromedial prefrontal cortex. J. Cogn. Neurosci. **15:** 324–337.
37. STUSS, D.T., G.G. GALLUP JR. & M.P. ALEXANDER. 2001. The frontal lobes are necessary for 'theory of mind.' Brain **124:** 279–286.
38. DECETY, J. & P.L. JACKSON. 2004. The functional architecture of human empathy. Behav. Cogn. Neurosci. Rev. **3:** 71–100.
39. MYCHACK, P. *et al.* 2001. The influence of right frontotemporal dysfunction on social behavior in frontotemporal dementia. Neurology **56:** 11–15.
40. ADOLPHS, R. 2001. The neurobiology of social cognition. Curr. Opin. Neurobiol. **11:** 231–239.
41. EDWARDS-LEE, T.A. & R.E. SAUL. 1999. Neuropsychiatry of the right frontal lobe. *In* The Human Prefrontal Lobes: Functions and Disorders. B.L. Miller & J.L. Cummings, Eds.: 304–320. Guilford Press. New York.
42. ALEXANDER, M.P., D.T. BENSON & D.F. STUSS. 1989. Frontal lobe and language. Brain Lang. **37:** 656–691.
43. SHAMMI, P. & D.T. STUSS. 1999. Humour appreciation: a role of the right frontal lobe. Brain **122:** 657–666.

Adding Prediction Risk to the Theory of Reward Learning

KERSTIN PREUSCHOFF AND PETER BOSSAERTS

Computation and Neural Systems, California Institute of Technology, Pasadena, California, USA

ABSTRACT: This article analyzes the simple Rescorla–Wagner learning rule from the vantage point of least squares learning theory. In particular, it suggests how measures of risk, such as prediction risk, can be used to adjust the learning constant in reinforcement learning. It argues that prediction risk is most effectively incorporated by scaling the prediction errors. This way, the learning rate needs adjusting only when the covariance between optimal predictions and past (scaled) prediction errors changes. Evidence is discussed that suggests that the dopaminergic system in the (human and nonhuman) primate brain encodes prediction risk, and that prediction errors are indeed scaled with prediction risk (adaptive encoding).

KEYWORDS: reinforcement learning; learning rate; least squares learning; dopaminergic system; reward anticipation; prediction risk; uncertainty; adaptive encoding

INTRODUCTION

Major progress has been made in understanding the way the primate brain learns to anticipate uncertain rewards and about the crucial role of the dopaminergic system in such learning. Much of this work has been driven by reinforcement learning (RL) whereby prediction errors in trial t, e_t, lead to updates of the prediction x_t of the reward (payoff) p_t using the Rescorla–Wagner (RW) learning rule:

$$x_{t+1} = x_t + \theta + \kappa e_t$$

where κ is the learning constant ($e_t = p_t - x_t$; for generality, we added a constant θ to the usual formulation). The activation patterns of dopaminergic neurons in the nonhuman primate brain[1] and of subcortical dopaminoceptive areas of the human brain[2,3] have recently been formalized in terms of such RL models.

Address for correspondence: Peter Bossaerts, m/c 228-77 California Institute of Technology, Pasadena, CA 91125, USA. Voice: +1-626-395-4028; fax: +1-626-405-9841.
pbs@rioja.caltech.edu

Ann. N.Y. Acad. Sci. 1104: 135–146 (2007). © 2007 New York Academy of Sciences.
doi: 10.1196/annals.1390.005

In this article, we are interested in the learning constant of the RW rule. To keep things simple, we shall focus on prediction of a single random reward from a single prediction in pure Pavlovian conditioning. However, the approach is not limited to such a simple situation. The RW rule has been successfully applied to learning value functions. Value functions are discounted expected values of a sequence of potential future rewards that apply to complex multidimensional stimulus–action–reward situations. They can be obtained with or without potential instrumental interference, and with or without potential variable time delays. In the context of value functions, RL is referred to as *TD Learning*.[4] The dopamine system in the primate brain is thought to adapt the RW rule to handle TD learning.[5]

In standard expositions of the RW learning rule little attention is paid to the learning constant κ that is kept constant and often found empirically. But what determines κ? We are looking for general answers, which do not require much knowledge about the specific stochastic structure in each possible application, and therefore, answers that feature some amount of robustness.

Least squares learning theory[6] provides one possible answer. It suggests that κ is the projection coefficient of predictions onto past prediction errors. Least squares learning can be defended against more powerful learning rules, such as Bayes' law because it is agnostic about the model that generates the stimuli (on which predictions are based) and the rewards. In other words, it is model-free, and, as such, generates robustness.

As we shall see, the projection coefficient of predictions onto past prediction errors depends on two quantities: (i) the covariance between predictions and past prediction errors, (ii) the variance of the prediction errors. The variance of the prediction error is the expected size of this error. We shall refer to it as the prediction risk.

First, consider the covariance between predictions and past prediction errors. The following example demonstrates how least squares learning correctly determines κ based on this covariance:

> A fair coin is tossed repeatedly for a $1 win or loss. Imagine that our player predicts the reward on trial t to be zero, $x_t = 0$. If the player wins on the next coin toss ($p_t = 1$), then according to RL, which usually assumes a strictly positive learning rate κ (and $\theta = 0$), the prediction for the subsequent trial, x_{t+1}, should be larger than the previous prediction, $x_{t+1} > 0$. This does not make sense: since coin tosses are independent across trials, the best prediction continues to be 0. Only if the player uses κ = 0, does the RW rule set the prediction for the subsequent trial equal to zero: $x_{t+1} = x_t + 0\, e_t = x_t$.

> However, if our player studies the history of past trials, she will realize that the best prediction she could have used is to always predict that the expected reward is zero. This best prediction is not correlated with prior prediction errors. The zero covariance between predictions and prior prediction

errors implies a zero projection coefficient, and hence, $\kappa = 0$. Therefore, least squares learning theory correctly assigns a zero value to κ.

This example shows that retrospective analysis of the correlation between the best possible prediction and prior prediction errors, together with the least squares formula, allows one to determine the right value for the learning rate in the RL.

Second, consider the dependence of κ on the prediction risk. Compare the following two situations:

> *In the first situation, our player is in a stock market game and she is asked to predict the price change of a stock, say, IBM, over the next day. That is, her reward p_t depends on the IBM price change. Her past experience is that even with the best possible prediction, the size of the prediction error is large, and hence, the prediction risk is large, usually about \$3. On day t, she predicts the price change to be \$1 ($x_t = 1$). The actual price change on day t happens to be: $-\$2$ ($p_t = -2$). Should she revise her forecast for the next day much compared to the previous forecast? The answer is no, because the prediction error ($e_t = -2\text{-}1 = -3$) is within the range she expected. If she adjusts her forecast, it should be minimal. That is, the learning rate in the RW rule should be small.*

> *In the second situation, our player is predicting the sales of a store in town. Usually, sales are pretty steady, but they are subject to occasional large shifts, often indicating arrival or departure of a competitor. As such, the prediction risk is small, but forecasts need to be sensitive to outliers—changes in sales that are much larger than expected. Such sensitivity is implemented by making the learning rate in the RW rule larger.*

This comparison shows that the learning rate should change with the prediction risk. The least squares learning formula shows how the learning rate should be adjusted when there are changes in the prediction risk. In the two situations above, least squares learning theory would provide values that agree with the intuitive prescriptions given.

Consequently, least squares learning theory suggests that effective implementation of RL involves three tasks:

1. To keep track of prediction errors;
2. To encode correlation between the best possible predictions and past prediction errors;
3. To track prediction risk.

Evidence of involvement of the dopamine system in RL of rewards has been limited to the first of these tasks. The primary purpose of this article is to discuss recent evidence that the dopamine system is engaged in the third task as well. A secondary purpose of the article is to point out that tracking of prediction risk may simultaneously achieve optimal evaluation of options with random outcomes.

While no less interesting, we leave discussion of the second task, namely, encoding of the covariance between predictions and prior prediction errors, for a future occasion.

But before we discuss evidence of tracking of prediction risk in the dopamine system, let us first present the mathematics of least squares learning theory. It will suggest that the most effective way to accommodate prediction risk in RL is not by adjusting the learning rate, but by scaling the prediction errors. This ensures that the learning rate depends only on the covariance (between predictions and prior prediction errors), and hence, the learning rate needs to be adjusted only when the covariance changes.

TO SIMPLIFY DETERMINATION OF THE LEARNING CONSTANT, PREDICTION RISK SHOULD SCALE PREDICTION ERRORS

According to least squares learning theory, the learning constant κ should be the coefficient in a projection of predictions on past prediction errors. The projection coefficient is defined to be the ratio of a covariance and a variance. In our setting,

$$\kappa = \frac{\text{cov}(x_{t+1}, e_t)}{\text{var}(e_t)}.$$

The numerator is the covariance between predictions and past prediction errors. The denominator is the variance of prediction errors. Its square root is the standard deviation of prediction errors, and this quantity is what we refer to as *prediction risk*. We will use the symbol υ to represent it as

$$\upsilon = \sqrt{\text{var}(e_t)}.$$

As such, the RW learning rule becomes

$$x_{t+1} = x_t + \theta + \frac{\text{cov}(x_{t+1}, e_t)}{\upsilon^2} e_t,$$

which can be rewritten as

$$x_{t+1} = x_t + \theta + \text{cov}\left(x_{t+1}, \frac{e_t}{\upsilon}\right) \frac{e_t}{\upsilon}.$$

As such the learning rate no longer depends on prediction risk, but only on a covariance. We can now *define the new learning rate* $\overline{\kappa}$, which depends only on the covariance between the predictions and past *scaled* prediction errors as

$$\overline{\kappa} = \text{cov}\left(x_{t+1}, \frac{e_t}{\upsilon}\right).$$

In addition, these same-scaled prediction errors can be used to update the predictions—there is no need anymore to keep track of the raw prediction errors

$$x_{t+1} = x_t + \theta + \overline{\kappa e_t},$$

where

$$\overline{e_t} = \frac{e_t}{v}.$$

This adaptive encoding simplifies the learning rule, as the learning rate depends only on covariance (i.e., association), and no longer on prediction risk. In other words, the learning rate changes whenever the covariance changes but no longer changes when the prediction risk changes.

PREDICTION RISK AND THE DOPAMINERGIC SYSTEM

The above analysis suggests that any system that is engaged in effective RL should not only encode prediction errors, but also prediction risk, and that prediction risk should be used to scale prediction errors. Evidence has converged over the last 10 years that activation of the dopaminergic system in the primate brain reflects prediction errors. Is there any evidence that this system also encodes prediction risk and that prediction errors are scaled?

These issues have not been addressed directly in the empirical neuroscience literature, but there is indirect evidence. FIGURE 1, for instance, reproduces a well-known result.[7] It shows that dopamine neurons in the ventral tegmental area of the nonhuman primate brain increase firing in the anticipation period between cue presentation and outcome revelation (reward/no reward), and that the rate of increase correlates with prediction risk. This effect is referred to as "ramping." In the experiment, prediction risk (when measured as reward variance) is maximal when the reward probability equals 0.5; it is minimal for probabilities equal to zero or one. Correspondingly, ramping increases in probability for probabilities up to 0.5, and decreases for higher probabilities.

A delayed prediction–risk-related signal in subcortical dopaminoceptive areas was recently uncovered in the human brain as well.[8] In the experiments, subjects were playing a simple card gambling game that allowed the experimenter to vary predicted rewards and prediction risk independently and over a broad range. FIGURE 2 displays some of the results. It shows that the fMRI blood oxygen level dependent (BOLD) signal in bilateral ventral striatum and other areas increases quadratically in the probability of winning for probabilities less than 0.5, and decreases quadratically for probabilities above 0.5. This is exactly what one would expect if the signal encodes prediction risk. The evidence matches nicely the extent of risk-related "ramping" in firing of dopamine neurons of the nonhuman primate brain alluded to before. Closer inspection reveals that this signal is delayed. The response onset to risk is seen only 3–4 sec after cue onset. The delay is consistent with the "ramping" of dopaminergic neurons in the nonhuman primate brain: because firing of dopaminergic neurons increases

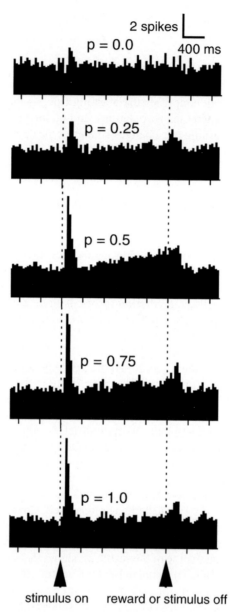

FIGURE 1. Relationship between firing of dopamine neurons in the ventral tegmental area and probability of reward. Firing increases gradually ("ramping") in the period of anticipation of reward; the increase is more pronounced the higher prediction risk (measured as reward variance) is. Prediction risk is highest for reward probability (P) equal to 0.5, and lowest for $P = 0.1$. (Reprinted with permission.[7])

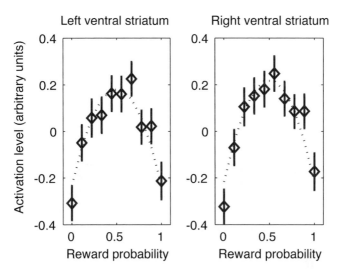

FIGURE 2. Modulation of BOLD signal in left and right ventral striatum of the human brain as a function of reward probability. The relationship is quadratic, implying that ventral striatum encodes prediction risk (measured in terms of reward variance). Relationship is modeled from one second after stimulus onset to reward revelation. (Reprinted with permission.[8])

only gradually, BOLD-related activation in dopaminoceptive regions will only become significant after a certain time.

Controversy exists about the origin of the correlation between prediction risk and "ramping" in dopaminergic neurons or delayed activation in dopaminoceptive areas. Such correlation could spuriously emerge as a result of the averaging of activation across trials on which FIGURES 1 AND 2 are based.[9] The spurious correlation emerges even in a standard TD learning model, with a constant learning rate, and hence, no account for prediction risk. Still, this explanation relies on specific aspects of the TD learning model (backpropagation of prediction errors in the anticipation period) for which there exists little physiological support. In addition, "ramping" is observed in single trials for single neurons as well.[10] But another way to settle the issues is to look at independent evidence for prediction risk encoding in the dopaminergic system.

Independent evidence comes from encoding of prediction errors at reward delivery. Unlike neuronal firing that reflects reward anticipation at stimulus onset, firing that correlates with prediction errors at reward delivery appears to scale with prediction risk.[11] If a stimulus unambiguously signals the size of an upcoming reward, firing rates at the time of reward do not correlate with the size of reward. However, firing rates at the time of the predictive stimulus do. FIGURE 3 displays the evidence. Three stimuli signal three different reward sizes, all occurring with the same probability. Firing at stimulus presentation

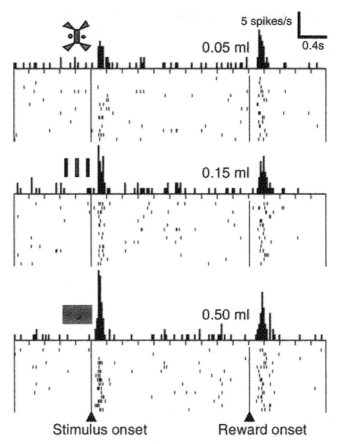

FIGURE 3. Firing of dopamine neurons in ventral tegmental area of nonhuman primate brain, as a function of the size of the prediction risk: low (top), medium (middle), and high (bottom). Upon reward delivery, the prediction error encoded in neuronal firing is scaled so that it is independent of prediction risk size. (Reprinted with permission.[11])

correlates with reward size. Upon reward delivery, however, the firing rate does not depend on reward size. The firing rate only reflects the sign of the prediction error, as when the prediction error is normalized with the prediction risk. The firing rate effectively reflects the scaled prediction error

$$\overline{e}_t = \frac{e_t}{\upsilon}.$$

Recent analysis in Wolfram Schultz's laboratory of imaging data of ventral striatum confirms this finding for the human brain (unpublished data).

Such adaptive encoding, as we discussed before, is an effective way to ensure that the learning rate in the RW model depends only on the covariance (between predictions and past scaled prediction errors), and not prediction risk.

As such, there is evidence of encoding of prediction risk in the dopamine system of the primate brain during the anticipation period, and of scaling of prediction errors with prediction risk at reward delivery. It is not known, however, to what extent the two phenomena are related, and what the origin is of the risk prediction signal. Future research should provide answers.

TRACKING PREDICTION RISK: A DUAL GOAL

There is ample behavioral evidence that primates, in particular humans, are sensitive to risk. Primates show a varying degree of aversion to risk[12] or, in some context, even a tendency to seek risk.[13] Since behavior shows sensitivity to risk, the primate brain must somehow be engaged not only in forecasting future rewards (forming expectations of rewards) but also in evaluating risk.

But any system that keeps track of prediction risk to facilitate learning simultaneously generates a signal with which risk itself can be evaluated. As such, a single prediction risk signal can serve two purposes: learning and choice. To date, it is not known to what extent the dopaminergic system is involved in determining choice.

When analyzing brain activation in the context of risky choice, neuroscientists have focused on insula. Activation of insula has been shown to be increasing in the amount of uncertainty. It is not known to what extent this activation is related to uncertainty-induced activation in the dopaminergic system. Physiologically, this is possible: anterior insula projects to midbrain dopaminergic nuclei, although projections to the insula are less conclusive.[14] Comparison of the time course of risk-related activation in insula[15] and in subcortical dopaminoceptive structures, such as ventral striatum[8] suggest that there is a link. More research is needed to determine the origin of the risk prediction signal in the dopaminergic system and its relationship with insula activity.

While the association between the risk signal in insula and reward learning has not been explored yet, this risk signal does appear to have behavioral implications. The risk signal can be used to predict to what extent subjects choose risky options over safe alternatives.[16] As such, to the extent that the insula signal is also used in learning (as mentioned before, this has not yet been determined), it serves two purposes: decision making as well as learning.

In contrast, the reward anticipation signal in dopaminoceptive structures is known to play a dual role. Its importance for learning is by now well established. Recent evidence indicates that it can also be used to predict to what extent subjects choose more rewarding options over poorer alternatives,[16] and as such, the anticipation signal does serve a decision-making as well as a learning purpose.

So far, we have taken the position that (prediction) risk is to be measured in terms of variance. From a learning point of view, this accords well with least squares learning. From a choice point of view, the position is consistent with standard financial decision theory.[17] Still, humans are known to be sensitive to additional features of risk as well, such as skewness and kurtosis.[18] It is not known to what extent brain activation in response to learning and choice under uncertainty reflects these additional features.

CONCLUSION

Major progress has been made in understanding the way the primate brain learns to anticipate uncertain reward. The crucial role of the dopaminergic system in such learning as well as the nature of the involved learning algorithm (RL) has become appreciated. There is some crucial element missing in the analyses, however. Foremost, least squares learning principles prescribe that prediction risk needs to be accounted for, so that forecasts react correctly to prediction errors. When prediction risk is sizeable, large prediction errors are to be expected, and hence, updates of forecasts should not be too sensitive to outcomes. Conversely, when prediction risk is small, forecasts need to be updated significantly whenever prediction errors turn out to be large. This is done most effectively, not by adjusting the learning rate, but by scaling prediction errors by prediction risk.

The evidence supports such adaptive encoding. Activation in midbrain dopaminergic neurons in the nonhuman primate brain and of dopaminoceptive structures in the human brain suggests that the dopamine system not only registers prediction risk during the reward anticipation period, but also that it scales the prediction errors at reward delivery. As such, the dopamine system parameterizes RL using principles of least squares learning.

Encoding that adapts to prediction risk, however, presupposes that an estimate of the prediction risk is available, which raises the issue of prediction risk learning. While beyond the scope of this article, suffice it to mention that online reinforcement learning algorithms can be used to effectively track prediction risk.[19] The brain mechanism engaged in tracking of prediction risk remains, however, unknown. Still, anterior insula is a prime candidate region for prediction risk learning, because of its known involvement in tracking of uncertainty.

We argued that encoding of prediction risk serves a dual role: we emphasized its benefits for optimal learning, but prediction risk encoding also serves to improve choices for risk-sensitive agents. The evidence we cited here converges on a separate representation (spatially, and even temporally) of expected reward and risk. In order to come to an informed evaluation of the trade-off between risk and reward, the separate representations have somehow to be integrated. To date, it is not known how this is accomplished in the brain.

Similarly, little is known about how the brain determines the strength of association (covariance) between predictions and (past, scaled) prediction errors. In least squares learning theory, this is a crucial parameter that determines the magnitude of the learning constant in RL. One can conjecture that it is obtained through RL, in the same way that RL can lead to an assessment of reward expectation and of reward prediction risk. Evidence is emerging[20] that dorsal anterior cingulate cortex is engaged in monitoring the salience of stimuli, and hence, the extent to which predictions should be adjusted as a function of prediction errors.

Our account is related to recent analysis of *expected uncertainty*,[21] but instead of exploring possible sources for a prediction risk signal, we studied here its impact on RL in the dopamine system. We have not addressed the issue of *unexpected uncertainty*,[21] which can be interpreted, in our context, as changes in the association (covariance) between predictors and (past, scaled) prediction errors. Unexpected uncertainty refers to situations where previously used predictors become suboptimal; predictors need to be updated differently as a function of past (scaled) prediction errors. We interpret this to mean that the covariance between future predictions and (scaled) prediction errors has changed, and hence, that the learning rate $\bar{\kappa}$ needs to be adjusted.

ACKNOWLEDGMENT

We thank John O'Doherty for detailed comments on an earlier draft, and Tim Behrens, Mark Walton, and Matthew Rushworth for further discussions on the link between uncertainty and the learning rate. Peter Bossaerts thanks the Swiss Finance Institute for financial support during his stay at the Université de Lausanne, where this article was written.

REFERENCES

1. SCHULTZ, W. 2004. Neural coding of basic reward terms of animal learning theory, game theory, microeconomics and behavioural ecology. Curr. Opin. Neurobiol. **14:** 139–147.
2. MCCLURE, S.M., G.S. BERNS & P.R. MONTAGUE. 2003. Temporal prediction errors in a passive learning task activate human striatum. Neuron **38:** 339–346.
3. O'DOHERTY, J.P., P. DAYAN, K. FRISTON, *et al*. 2003. Temporal difference models and reward-related learning in the human brain. Neuron **38:** 329.
4. MONTAGUE, P.R., P. DAYAN & T.J. SEJNOWSKI. 1996. A framework for mesencephalic dopamine systems based on predictive Hebbian learning. J. Neurosci. **16:** 1936–1947.
5. MONTAGUE, P.R., S.E. HYMAN & J.D. COHEN. 2004. Computational roles for dopamine in behavioural control. Nature **431:** 760.
6. MASANAO, A. 1986. State Space Modeling of Time Series. Springer-Verlag. New York.
7. FIORILLO, C.D., P.N. TOBLER & W. SCHULTZ. 2003. Discrete coding of reward probability and uncertainty by dopamine neurons. Science **299:** 1898–1902.

8. PREUSCHOFF, K. 2006. Neural differentiation of expected reward and risk in human subcortical structures. Neuron **51:** 381–390.
9. NIV, Y., M.O. DUFF & P. DAYAN. 2005. Dopamine, uncertainty and TD learning. Behav. Brain Funct. **1:** 6.
10. FIORILLO, C.D., P.N. TOBLER & W. SCHULTZ. 2005. Evidence that the delay-period activity of dopamine neurons corresponds to reward uncertainty rather than back-propagating TD errors. Behav. Brain Funct. **1:** 7.
11. TOBLER, P.N., C.D. FIORILLO & W. SCHULTZ. 2005. Adaptive coding of reward value by dopamine neurons. Science **307:** 1642–1645.
12. HOLT, C.A. & S.K. LAURY. 2002. Risk aversion and incentive effects. Am. Eco. Rev. **92:** 1644–1655.
13. MCCOY, A.N. & M.L. PLATT. 2005. Risk-sensitive neurons in macaque posterior cingulate cortex. Nat. Neurosci. **8:** 1220–1227.
14. FLYNN, F.G. 1999. Anatomy of the insula functional and clinical correlates. Aphasiology **13:** 55.
15. HUETTEL, S., A. SONG & G. MCCARTHY. 2005. Decisions under uncertainty: probabilistic context influences activation of prefrontal and parietal cortices. J. Neurosci. **25:** 3304–3311.
16. KUHNEN, C.M. & B. KNUTSON. 2005. The neural basis of financial risk taking. Neuron **47:** 763–770.
17. MARKOWITZ, H. 1991. Foundations of portfolio theory. J. Finance **46:** 469–477.
18. KROLL, Y., H. LEVY & H.M. MARKOWITZ. 1984. Mean-variance versus direct utility maximization. J. Finance **39:** 47–61.
19. BOSSAERTS, P. *In* New Encyclopedia of Neuroscience. Eds.: L. Squire, T. Albright, F. Bloom, F. Gage & N. Spitzer Elsevier. In press.
20. BEHRENS, T., M. WOOLRICH, M. WALTON & M. RUSHWORTH. 2006. Learning the value of information in an uncertain world. Oxford University working paper.
21. YU, A.J. & P. DAYAN. 2003. Expected and unexpected uncertainty: ACh and NE in the neocortex. *In* Advances in Neural Information Processing Systems 15. MIT Press. Cambridge, MA.

Still at the Choice-Point

Action Selection and Initiation in Instrumental Conditioning

BERNARD W. BALLEINE AND SEAN B. OSTLUND

Department of Psychology and the Brain Research Institute, University of California, Los Angeles, California, USA

ABSTRACT: Contrary to classic stimulus–response (S-R) theory, recent evidence suggests that, in instrumental conditioning, rats encode the relationship between their actions and the specific consequences that these actions produce. It has remained unclear, however, how encoding this relationship acts to control instrumental performance. Although S-R theories were able to give a clear account of how learning translates into performance, the argument that instrumental learning constitutes the acquisition of information of the form "response R leads to outcome O" does not directly imply a particular performance rule or policy; this information can be used both to perform R and to avoid performing R. Recognition of this problem has forced the development of accounts that allow the O and stimuli that predict the O (i.e., S-O) to play a role in the initiation of specific Rs. In recent experiments, we have used a variety of behavioral procedures in an attempt to isolate the processes that contribute to instrumental performance, including outcome devaluation, reinstatement, and Pavlovian–instrumental transfer. Our results, particularly from experiments assessing outcome–selective reinstatement, suggest that both "feed-forward" (O-R) and "feed-back" (R-O) associations are critical and that although the former appear to be important to response selection, the latter—together with processes that determine outcome value—mediate response initiation. We discuss a conceptual model that integrates these processes and its neural implementation.

KEYWORDS: goal-directed action; reward; habit; reinforcement; associative learning

INTRODUCTION

Whether framed in computational or, more explicitly, in psychological terms, contemporary theories of decision making commonly rely on the cognitive

Address for correspondence: Bernard W. Balleine, Department of Psychology, UCLA, Box 951563, Los Angeles, CA. Voice: 310-825-7560 (office), 310-825-2998 (lab); fax: 310-206-5895.
balleine@psych.ucla.edu

Ann. N.Y. Acad. Sci. 1104: 147–171 (2007). © 2007 New York Academy of Sciences.
doi: 10.1196/annals.1390.006

or executive capacities of the agent to explain choice.[1–6] As a consequence, these positions can slip into indeterminacy and it is easy to see why. Any information that takes the form "action A leads to outcome O" can be used both to perform A and to avoid performing A. It is simply not possible purely on the basis of the information presented to an agent, such as "A leads to O" and "B leads to P," to predict whether they will choose A or B because, whether derived from perceptual, cognitive, social, or fictive sources, information alone is not sufficient to determine a course of action. This criticism of cognitive, information-based theories of action was recognized early in analyses of animal behavior. The classic critique in that context was Guthrie's[7] jibe at the cognitive behaviorism of Edwin Tolman,[8] particularly the latter's contention that the performance of a rat learning to traverse a maze to find food was a matter of acquiring a belief about "what (action) leads to what (outcome)." As Guthrie[7] put it, on this view "the rat is left buried in thought" (p. 172). Merely believing that, say, "turning left at the choice point is necessary to get food" does not *entail* turning left.

Guthrie himself favored an account of animal action based on the formation of sensorimotor, so-called stimulus–response (S-R), associations—a theory that rendered actions homologous to reflexes and that explained performance by confounding action selection and initiation within the function of the stimulus. The most influential version of this account was later developed by Hull,[9] who proposed that S-R associations are strengthened by reinforcement; that is, an association between the situational stimuli (S) and a response (R) is strengthened when R is followed by a reinforcing event (such as food) thereby accounting for the observation that R becomes more probable in S. However, as has been well documented in the past,[10,11] this account means to claim that animals do not encode the consequences associated with their actions; that is, the reinforcer or outcome contingent on the performance of R does not itself form a part of the associative structure controlling the performance of that R. Indeed, it was recognition of this fact that produced some of the critical experimental tests of S-R theory, notably the *outcome devaluation test*.[11]

The outcome devaluation test, conducted after training and so after the formation of any S-R connection has been made, involves changing the value of the instrumental outcome using any of a variety of motivational manipulations, such as taste aversion learning[10,12] or specific satiety,[13,14] after which the tendency of the rat to press the lever is assessed in extinction, that is, in a test in which no outcomes are delivered. Performance of the devalued action is compared either against that of a nondevalued control or, in a choice situation, against the performance of another nondevalued action. S-R theories of instrumental performance predict that, because of the S-R association established during training, the presence of the training S guarantees that R will be performed on test irrespective of the change in value of the training outcome. In direct conflict with this prediction, however, numerous experiments have found that choice between actions respects the current value of the specific outcome associated with an action and not the presence or absence of some eliciting stimulus or other.[12,15,16]

These findings suggest that choice is at least partly determined by integrating action–outcome associations with outcome values. Nevertheless, although this kind of account can provide a good basis for action selection, it is not immediately clear how it determines performance; that is, how does believing that "A leads to O" and "O is valuable" induce the agent to perform A? One can imagine that knowing A→O and B→P and that O has greater value than P could lead one generally to prefer A over B. But does that necessarily result in greater performance of A? Is it sufficient that Jack prefers coffee to tea and knows their different means of production for Jack to make coffee? Although, on this account, something further than selection is required for action initiation, the relationship between selection and initiation is unexplored and remains implicit in the description of performance. Is there any alternative?

TWO-PROCESS THEORY

Various alternatives to the strict S-R account described earlier have been developed that, given certain assumptions, can predict the effects of outcome devaluation on performance without resorting to action–outcome learning. The most influential of these has been two-process theory. There have been two variants of this theory, a motivational version[17] and an expectancy version.[18] Both versions propose that instrumental learning is fundamentally S-R and that performance depends on the influence of predictions of the outcome (O) based on certain state or situational cues (i.e., S-O), but differ in the details of the influence of these predictions: on the motivational account, S-O associations modulate performance elicited by the S-R association, whereas on the expectancy account, the expectancy of O forms a component of the S with which the R becomes associated, that is, S-(O)-R.

The standard actor–critic formulation of reinforcement learning models has most in common with the first of these alternatives.[19,20] In that model, learning, that is, the acquisition of policies, is essentially a matter of forming (internal and/or external) state–response associations strengthened by the (reinforcing) feedback derived from the value of the state to which the response allows a transition. States have values based on the outcomes that they predict. Hence, together, states both motivate actions and reinforce them. Unfortunately, both the motivational version of two-process theory and its actor–critic implementation have difficulties explaining choice performance in an outcome devaluation test. Simply put, unless differential predictors can be established, a state, S_a, that is trained with two actions, S_a–R1; S_a–R2, for distinct but equally valued outcomes cannot explain the well-documented changes in choice performance that ensue when one or other outcome is subsequently devalued; the state predicts both a valued and a devalued outcome and cannot proscribe which of the two actions associated with the state should be performed. In one example,[21] hungry rats were trained to push a pole in one direction to get grain pellets and in the opposite direction to get sucrose. In this situation, both policies should be equally preferred; both actions are equally associated with all potential state cues, and the value of the state transitions, or outcomes of each action, are equal. Nevertheless, after outcome devaluation, animals immediately, that is, without further training, reduce their tendency to push in the direction that previously resulted in the now devalued outcome.

The alternative, expectancy formulation of two-process theory has had somewhat greater success because it allows the assumption that policies are determined by state–state transitions. For example, this account would propose that the policy R1 under S_a that results in a transition to S_i is controlled by $S_a S_i$ (i.e., $S_a S_i$–R1), and of R2 resulting in a transition to S_j is controlled by $S_a S_j$ (i.e., $S_a S_j$–R2). If changing the value of S_i relative to S_j is assumed to immediately alter the ability of S_a to retrieve S_i relative to S_j, then a basis might be established for predicting the relative change in performance induced by devaluation in the choice test in the study of Dickinson et al. among others.[12,21] In associative terms, this is the equivalent of the expectancy of an outcome O that is associated with a stimulus, that is, S-(O), controlling the production of the response, that is, S-(O)-R, an association that is strengthened by the reinforcement of R in the presence of S-(O) in a manner that accords with S-R theory. Psychologically, this account can explain the relative reduction in performance of an action associated with a devalued outcome either by supposing that the expectancy of an outcome is suppressed (or relatively so) when that outcome is (relatively) devalued, or that by retrieving O, S also retrieves an aversive emotional state reducing the performance of the response accordingly.

There is evidence that accords with this version of two-process theory; when the expected outcome and the delivered outcome are the same, animals learn more rapidly than when they are different. For example, consider the experimental design described in Design 1 taken from an experiment conducted by Trapold and Overmier.[18] In this study, rats were trained first to predict two distinct outcomes (O1 and O2) on the basis of two stimuli (S1 and S2) after which the stimuli were used to discriminate which of the two responses (R1 and R2) would be reinforced. Two groups were used in this discrimination phase: one in which the predicted and earned outcomes were congruent and one in which they were incongruent (see Design 1).

Congruent: [S1-O1; S2-O2] then [(S1: R1-O1; R2-); (S2: R1-; R2-O2)]

Incongruent: [S1-O1; S2-O2] then [(S1: R1-O2; R2-); (S2: R1-; R2-O1]

$$(1)$$

Trapold and Overmier[18] found that, when the outcome expected on the basis of S1 was also earned by R1, rats learned the discrimination more rapidly than when it differed from the outcome earned by R1; that is, in accord with predictions from the S-(O)-R account, Group Congruent learned faster than Group Incongruent. Similarly, it was subsequently found, as described in Design 2, that when rats were trained to perform two actions for distinct outcomes, presentation of a stimulus associated with one or the other outcome elevated the performance of the action trained with that outcome more than performance of the other action, an instance of an outcome selective form of a phenomenon referred to as Pavlovian–instrumental transfer.[22]

$$\text{If } [S1\text{-}O1, S2\text{-}O2] \text{ and } [R1\text{-}O1, R2\text{-}O2] \tag{2}$$
$$\text{then } [(S1: R1 > R2) \text{ and } (S2: R1 < R2)]$$

Thus, again in line with the S-(O)-R account, increasing the expectancy of O1 by presenting S1 was found to increase the performance of the R trained with O1, in this case R1 relative to R2; indeed performance of R2 did not differ from a baseline (no stimulus) period.

HIERARCHICAL THEORY

Despite this clear evidence in favor of the S-(O)-R account, there is even more compelling evidence against it as a general explanation of instrumental performance. First and foremost, studies assessing the neural bases of instrumental conditioning have found evidence that distinct structures mediate outcome devaluation and the influence of Pavlovian S-O associations on instrumental performance. For example, Corbit et al.[23] compared the effects of lesions of distinct nuclei in the ventral striatum, the core and shell subregions of the nucleus accumbens, on outcome devaluation effects induced by specific satiety and on the Pavlovian–instrumental transfer effect reported by Colwill and Rescorla.[22] Sham lesioned control subjects showed both devaluation and selective transfer effects. In contrast, and against predictions from the S-(O)-R account, lesions of the accumbens shell were found to abolish the excitatory effect of reward-related cues on instrumental performance without affecting outcome devaluation. Furthermore, although lesions of the accumbens core were found to abolish outcome devaluation, they had no effect on Pavlovian–instrumental transfer. This evidence that devaluation and transfer effects can be doubly dissociated at the level of the nucleus accumbens makes the argument that the explanation of one effect should be made in terms of the other particularly difficult to maintain (see also Ref. 24).

Furthermore, as described earlier, devaluation effects using a bidirectional manipulandum can only be explained by two-process theory if the expectancy based on state–state transitions argued to control performance is suppressed by outcome devaluation. Several studies have found that this is not so. For example, using a design similar to that described earlier by Trapold and Overmier, Rescorla[25] first trained explicit S1-O1, S2-O2 expectancies and then trained the rats to perform two actions, R1 and R2, one in the presence of each S, such that the reward delivered was incongruent with that predicted by S, that is, S1:R1-O2; S2:R2-O1. During this training, S1-O1 and S2-O2 continued to be presented intermixed with the instrumental contingencies. He then devalued O1, using a taste aversion procedure, and assessed the choice between R1 and R2 in the presence of both S1 and S2. On the S-(O)-R account, devaluation should have suppressed the control of S1 over R1 resulting in R1 < R2 on test. In direct contradiction of this prediction, Rescorla found that, in the presence of S1S2, R2 was performed less than R1; that is, the performance of the rats in this experiment respected the expectancy of the outcome based on their actions rather than that based on S1 and S2. Similarly, outcome devaluation after training on a biconditional discrimination, should not, on the two-process account, be predicted to generate differential performance. Rescorla[26] reports the results of a study along these lines using a design similar to that described in Design 3, in which two responses were trained for different outcomes in each of two stimuli after which O1 was devalued (i.e., O1-):

$$
\begin{array}{l|l}
\text{S1: R1-O1, R2-O2} & \text{S1: R1} < \text{R2} \\
\text{S2: R1-O2, R2-O1} & \text{S2: R1} > \text{R2}
\end{array}
\quad \text{O1-} \qquad (3)
$$

Although, on the S(O)-R account both S1 and S2 should be equally impaired in their control of R1 and R2 after the devaluation of O1, a choice test between R1 and R2 found that, under S1, R1 was reduced relative to R2 whereas, under S2, R2 was reduced relative to R1.

Together these kinds of findings have encouraged the development of a hierarchical theory of instrumental conditioning in which discriminative stimuli, such as S1 and S2, are proposed to control the association between response and outcome by reducing the threshold required for activation of the outcome representation by the action[12, 26] (see FIG. 1A). On this view, instrumental performance is not a simple matter of binary S-O, S-R, or R-O associations but reflects the hierarchical control of S on the R-O association; whenever the controlling S is present and the response is available, the S-(R-O) structure ensures that the instrumental response will be performed.

Although the hierarchical account is consistent with much of the data, recent tests of predictions derived from this position have raised doubts about its veracity. One particularly difficult set of observations for this view has come from what may be called "component discriminations," which can be used to model the discriminative properties of free operant schedules (see Design 4). In free operant conditioning, training is typically conducted in sessions in which only one response manipulandum is available at any one time. This arrangement ensures that the outcome earned by performing that response maintains a consistent relationship both as a consequence of the response but also as an antecedent or discriminative stimulus for the next response— refer to Design 4(a). It is, however, possible to arrange a situation in which the outcome that serves as the antecedent differs from that earned as a consequence

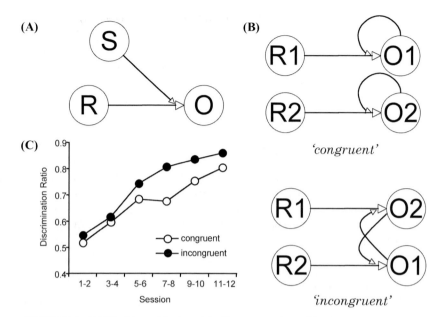

FIGURE 1. (**A**) The hierarchical model of instrumental conditioning according to which discriminative stimuli lower the threshold on the outcome representation increasing its activation by the associated response to generate performance. (**B**) Predictions from the hierarchical theory for the congruent and incongruent component discriminations described in the text. When the outcome used as the discriminative stimulus is congruent with that earned by the response, the discrimination is readily solved because O1 lowers the threshold of O1 resulting in R1 > R2, and O2 lowers the threshold of O2 resulting in R2 > R1. When the outcome used as the discriminative stimulus differs from that earned by the response, however, O1 will lower the threshold on O2, activation of which will lower the threshold on O1 making both R1 and R2 equally likely and the discrimination relatively more difficult. (**C**) Data collected at UCLA by Sanne de Wit and Sean Ostlund comparing the congruent and incongruent discriminations that replicate the finding that, in contrast to the predictions from hierarchical theory, the incongruent discrimination is acquired more rapidly than the congruent (the asterisk signals that $P < 0.05$ for the groups' comparison).

of responding, for example, in a situation in which two responses (R1 and R2) that earn different outcomes (O1 and O2) are performed in strict alternation—refer to Design 4(b).

$$\text{(a)} \quad \text{R1-O1-R1-O1-R1} \dots \text{etc.}$$
$$\text{R2-O2-R2-O2-R2} \dots \text{etc.} \qquad (4)$$
$$\text{(b)} \quad \text{R1-O1-R2-O2-R1} \dots \text{etc.}$$

As illustrated in Design 5, component discriminations establish the components of the two kinds of free-operant schedule described in Design 4 as discrete problems; that is, in the congruent case, the delivery of O1 signals that R1, not R2, will deliver O1 and O2 signals that R2, not R1, will deliver O2 whereas, in the incongruent case, O1 signals that R1, not R2, will deliver O2 and O2 signals that R2, not R1, will deliver O1.

$$\text{Congruent: } [O1: R1-O1, R2-] \text{ and } [O2: R2-O2, R1-]$$
$$\text{Incongruent: } [O1: R1-O2, R2-] \text{ and } [O2: R2-O1, R1-]$$

(5)

Although both of these discriminations model components of free-operant schedules, hierarchical theory predicts that the rate of acquisition of the congruent and incongruent discriminations will differ considerably (see FIG. 1B). Hierarchical theory has no difficulty predicting successful discrimination performance in the case of the congruent discrimination; delivery of O1 reduces the threshold on the R1-O1 association resulting in the performance of R1 whereas delivery of O2 reduces the threshold of the R2-O2 association resulting in the performance of R2. Hence, delivery of O1 will result in R1 > R2, whereas O2 will result in R2 > R1. In contrast, hierarchical theory predicts that the incongruent discrimination should be much harder to solve. Consider what should happen following the delivery of O1. Because O1 signals that R1 will be reinforced by O2, the delivery of O1 should allow R1 to more readily activate O2. On this account, however, activation of O2 should in turn lower the threshold for R2 to activate O1. As a consequence, hierarchical theory makes the prediction that the delivery of O1 (or O2) should result in an increased tendency to perform both R1 and R2, thereby making this discrimination much more difficult and perhaps unsolvable.

Experiments assessing the rate of discrimination learning in this situation have not supported hierarchical theory. For example, Dickinson and de Wit[27] trained rats on two different component-discrimination problems, one in which the delivery of a drop of sucrose signaled that a left lever press response would deliver a drop of sucrose and a right lever press nothing and another on which a food pellet signaled that a right lever press response would deliver a pellet and a left lever press nothing (the responses were counterbalanced across subjects). In this case, the discriminative (antecedent) cue was congruent with the outcome (or consequence) of responding. A second group of rats was also trained on two component discriminations, but in this case the antecedents and consequences were incongruent; that is, the sucrose signaled that a left lever press would deliver a pellet and a right lever press nothing, and a pellet signaled that a right lever press would deliver the sucrose and a left lever press nothing (again, counterbalanced). It is this latter pair of discriminations that hierarchical theory predicts should be relatively difficult to solve because the delivery of either outcome should provoke the tendency to press both the left and right levers, thus leading to *response conflict.*

In fact, despite this clear prediction of hierarchical theory, it was not upheld in the results of this experiment. Indeed, Dickinson and de Wit[27] reported precisely the opposite result: they found that the incongruent discrimination was acquired faster than the congruent one. This surprising finding has been replicated in our lab. During a visit to UCLA by Sanne de Wit, we collaborated on an experiment in the course of which we replicated the procedure described by Dickinson and de Wit and found, again, that the incongruent discrimination was acquired faster than the congruent one (see FIG. 1C). There are at least two reasons why this might be so in this case. First, it is possible that the sucrose (or pellet) delivery produced a very short-term satiety or negative priming effect reducing the processing of the subsequent sucrose (or pellet) delivery, thereby reducing the rate of acquisition of the lever press sucrose (or pellet) association. Alternatively, although the presentation of the outcomes that served as the discriminative stimuli were not explicitly paired with any stimulus, they were, of course, presented in a specific set of contextual cues, the same context in which the lever press response was performed and paired with the outcome. This sets up a kind of relative-validity problem, wherein the context–outcome (C-O) association competes with the response–outcome (R-O) association by virtue of being present when the response is performed (C-O, C+R-O). However, this competition between context and response must be at last partially localized to individual trials; the context is an equally good predictor of the outcomes in the incongruent discrimination. This second account reduces, therefore, to a variant of the negative priming account in which the C-D association interferes with processing the R-O association.

Finally, it is worth pointing out that the ability of the rats to solve the incongruent discrimination suggests that they have mechanisms able to resolve the response conflict inherent in this discrimination problem. Studies assessing the neural basis of response conflict in humans using tasks such as the Stroop, Simon, or flanker tasks suggest that a region of the dorsal medial prefrontal cortex plays a critical role in this capacity.[28] In a recent study, we have found evidence that the same is true of rats when solving the incongruent discrimination.[29] Indeed, temporary inactivation of the dorsomedial prefrontal cortex (using the GABA-A receptor agonist muscimol) not only made this discrimination much more difficult, it rendered performance inferior to the control discrimination, not by reducing performance of the correct response in the discrimination but by increasing performance of the incorrect response. It is interesting to note, therefore, that the result predicted by hierarchical theory for normal rats does in fact emerge but only in rats with a deficit in medial prefrontal cortical function.

Generally, however, it should be clear that hierarchical theory, as currently formulated, cannot account for these data. Unless a principled argument can be developed that allows the sensory and expectancy properties of the outcome to be represented differently, this theory is forced to predict that the ability of rats to solve the incongruent discrimination should be reduced relatively to the incongruent discrimination.

AN ALTERNATIVE TWO-PROCESS ACCOUNT

Although two-process theory provides a relatively simple explanation for certain features of instrumental performance, the prior analysis suggests that variants of this theory, whether framed in S-(O)-R or S-(R-O) terms, fail to describe accurately the nature of the discriminative properties of the outcome in response selection and, as a consequence, do not generate an adequate account of instrumental performance. This should not be taken to imply that all forms of two-process theory are untenable. Indeed, recent evidence from our lab investigating selective reinstatement effects has provided evidence for just such an alternative two-process account.[32]

It has been well documented, using both Pavlovian[30] and instrumental conditioning procedures,[31] that, after a period of extinction, brief reexposure to the outcome will reinstate performance of the conditioned response or instrumental action, respectively. We have recently been investigating the outcome specificity of instrumental reinstatement.[32] In these experiments, rats were given free-operant training on two levers for distinct outcomes (R1-O1, R2-O2) after which both actions were subjected to a period of extinction (R1-, R2-). During a final test phase, the rats were given a choice test on the two levers, again in extinction, immediately after the delivery of one or other instrumental outcome, for example, O1: R1 versus R2. In several studies, we have found that outcome delivery almost completely reinstated responding on the action with which it was associated during training.

The similarity of this effect to the component discriminations above raises the possibility that the selective reinstatement effect is mediated by the animals' learning, during free-operant training, that the outcome serves not only as a consequence but as an antecedent of the response with which it was paired, that is, O1:R1-O1. This is the same as the congruent discrimination arranged by Dickinson and de Wit.[27] Of course, the finding that reinstatement is selective not only accords with this account based on the antecedent O-R association, it also accords with the account that the free outcome restores performance though retrieval of the R-O association, perhaps through a backward R-O association. To examine these distinct explanations for selective reinstatement, we assessed the response specificity of reinstatement in a group given congruent training, in which the outcome signaling a response was always the same as the one earned by that response (i.e., O1:R1-O1 & O2:R2-O2), and a group given incongruent training, in which the outcome signaling a response differed from the outcome earned by that response (i.e., O1:R1-O2 & R2:R2-O1; Ostlund and Balleine[32]). The O-R and R-O accounts of reinstatement make distinct predictions with regard to the responses that will be reinstated on these two schedules. If the R-O association mediates reinstatement, then the delivery of a specific outcome (e.g., O1) should reinstate the same response (i.e., R1) in both groups. If, however, it is the antecedent O-R association that mediates the reinstatement effect then these two training procedures should generate opposing effects; the outcome should reinstate the response it signaled during training regardless of which response it followed. The results of this experiment, presented in FIGURE 2A, demonstrate that selective reinstatement effects are a product of the antecedent outcome–response association established during training; whereas O1 reinstated R1 in the congruent training group, it reinstated R2 after incongruent training. Of course, it is possible that animals in the incongruent condition failed to learn the specific R-O associations presented during training. In a subsequent outcome devaluation test, we found, against this claim, that both groups selectively suppressed performance of the action that had earned the now devalued outcome, indicating that both congruent and incongruent training had generated substantial R-O learning.

This experiment suggests that the antecedent O-R association is responsible for guiding the reinstatement of instrumental performance. Consistent with this claim, we have also found evidence that O1 will reinstate both actions equally when rats are trained with stable R-O, but unstable O-R, relationships, that is, one in which both R1 and R2 were preceded equally often by both O1 and O2 (see FIG. 2A). In this study, the stable R-O associations were maintained but were presented such that the opportunity to perform R1 to gain access to O1 and to perform R2 to gain O2 occurred equally often after both O1 and O2. Hence, although the consequent R-O relationships were consistent, the antecedent O-R associations were mixed such that O1 and O2 were equally often associated with both R1 and R2. If the O-R association mediates reinstatement, then the delivery of, say, O1 should have provoked reinstatement of both R1 and R2 and, indeed, this is what we found (see FIG. 2A).

These data make clear the potential for an alternative two-process account, not based on S-O and R-O processes, but based on the formation of O-R and R-O associations in the course of conditioning. It remains to be considered how these processes are formally related to each other, and, in what follows, we present both a model of this relationship and evidence supporting predictions derived from that model.

S-R versus O-R LEARNING

The suggestion that outcome–selective reinstatement is induced through the O-R association predicts, by analogy to other forms of S-R process, that reinstatement will not be affected if the reinstating outcome is devalued. We have assessed this prediction using a motivational shift from hunger to thirst[32] and, more recently, using devaluation by specific satiety—see Design 6. In both cases, we found evidence suggesting that although the influence of the

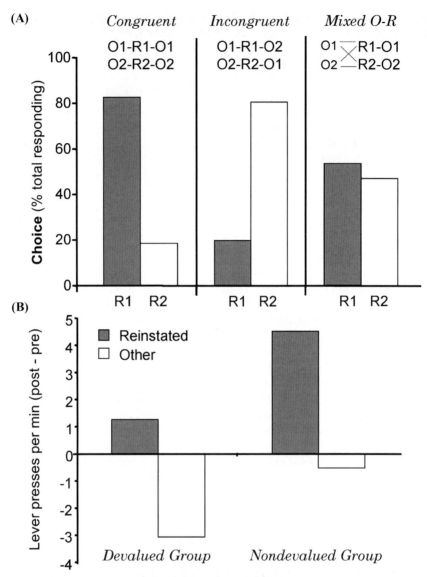

FIGURE 2. (A) Choice performance during a reinstatement test conducted after various types of training and the subsequent extinction of R1 and R2. (B) Choice between two levers during a reinstatement test after training on R1-pellet and R2-sucrose. Prior to extinction and test, one of the two outcomes was devalued for one group of rats (Devalued Group) but not the other (Nondevalued Group). Note that although the overall rate of performance during the test was strongly affected by devaluation, the proportional increase in responding generated by the devalued outcome was very similar to that generated by the nondevalued outcome.

reinstating outcome on response selection does not depend on outcome value, the overall degree of reinstatement is diminished by these treatments.

$$
\text{R1-O1, R2-O2} \quad \bigg| \quad \text{O1-} \quad \bigg| \quad \text{R1-, R2} \quad \bigg| \quad \frac{\text{(Dev) O1: R1 vs. R2}}{\text{(Non) O2: R1 vs. R2}} \tag{6}
$$

In the latter study, hungry rats were trained to press two levers, one for pellets and the other for a sucrose solution. Next, rats were allowed to consume freely one of the two outcomes for 1 h immediately before undergoing a reinstatement test. It is important to note that, after responding was extinguished on both levers, the Devalued Group was given noncontingent exposure to the outcome on which they were sated, whereas the Nondevalued Group received the other (valued) outcome. As predicted, whether devalued or nondevalued, outcome delivery resulted in similar shifts in subjects' choice of the action that had earned the reinstating outcome relative to their baseline choice (see FIG. 2B). Nevertheless, as is clear from this figure, and as we have found generally, the pattern of reinstatement that emerged also depended heavily on outcome value; whereas the Nondevalued Group performed substantially more responses for the reinstating outcome than at the extinction period, this elevation was strongly attenuated in the devalued group. Indeed, the reinstatement effect that emerges after devaluation is derived from the relative difference produced by a mild increase on the reinstated action and a strong decrease in responding on the other action (see FIG. 2B).

This pattern of results is in fact more generally consistent with the view that both O-R and R-O associations contribute to the overall pattern of responding during reinstatement. Although the O-R association is necessary to select the action associated with the outcome during training, the actual degree to which that association results in the initiation of the response depends on the R-O association and, hence, on the value of the associated outcome. This pattern of results, therefore, suggests two things. First, it suggests that the effect of the O-R association is, generally speaking, on response selection; it can induce a bias into choice performance that is multiplied by the R-O association to determine the overall rate of performance. Second, it suggests that if the O-R association is sufficiently strong, it can induce a degree of performance in and of itself and independently of the R-O association.

In fact, this latter suggestion is not new and has been advanced previously to explain the residual instrumental performance that is often observed in tests of outcome devaluation, although in that context it is usually thought to reflect the effect of more general S-R associations on performance rather than the specific O-R association proposed here.[33] Nevertheless, there is no doubting that the O-R association, as proposed, has much in common with standard S-R theoretical accounts and, indeed, can be seen as identifying one among a large number of other potential stimuli that could exert control over instrumental performance,

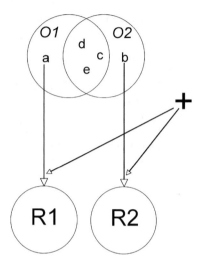

FIGURE 3. An elemental view of the formation of selective O-R associations during the training of two instrumental actions for distinct outcomes allowing for the selection of one response over another. Elements specific to R1 and R2 and that allow the selection of one response over another are shown as "a" and "b," whereas the other elements are common to both responses. Note that the reinforcement signal (+) functions to strengthen the connection between the specific stimulus element and the response.

particularly in the case of a single action. Of course, in the choice situation, the sources of differential response selection are more limited—the context or contextual elements being equally associated with both actions. Even so, in the choice situation, any stimulus that retrieves a specific outcome should, through the O-R association, be expected to select a specific action and, if owing to prior reinforcement (+) that association is sufficiently strong, to result in the initiation of that action irrespective of the value of the outcome associated with that action. This kind of structure is illustrated in FIGURE 3.

There is in fact evidence from a range of sources that supports this analysis. First, it is clear from a number of studies that the strength of S-R associations embedded in the instrumental training situation increases progressively with extended training.[33,34] Early in training, outcome devaluation effects are much larger than after continued training and, indeed, as Adams[34] has shown, with sufficient training the effects of devaluation can be abolished altogether. There is, therefore, good reason to believe that the amount of performance controlled by the O-R association increases with training as this association strengthens with each reinforced action and certainly sufficient evidence to propose that even when devaluation effects are quite strong, the O-R association may well be strong enough to induce the amount of reinstatement performance observed after devaluation, as illustrated in FIGURE 2B.

Furthermore, several studies have found that outcome devaluation does not necessarily affect the excitatory influence of stimuli associated with the instrumental outcome on instrumental performance in demonstrations of selective Pavlovian–instrumental transfer.[35–37] Just as we have found in our assessment of selective reinstatement, devaluing the outcome predicted by a stimulus can leave the excitatory influence of that stimulus on the performance of actions associated with that outcome relatively unaffected. Interestingly, this effect appears to be influenced by the degree of training. Holland[37] has reported that actions that have been given relatively little training are not much influenced by Pavlovian stimuli; much as one should propose for reinstatement, if the O-R association is relatively weak, both reinstatement and transfer should be correspondingly weak. As training progresses, however, the influence of Pavlovian cues and, potentially, the outcome itself on performance increases, presumably because the strength of the O-R association activated by the stimuli (or the outcome) increases with this training.

Generally, therefore, these results and, indeed, this overall analysis suggest that O-R associations have much in common with S-R learning and can be regarded as being acquired and strengthened in much the same way, that is, by contiguity.[38,39]

INTEGRATING O-R AND R-O ASSOCIATIONS: THE ASSOCIATIVE–CYBERNETIC MODEL

Some time ago, Dickinson and Balleine[40] advanced a general model of instrumental performance based on the interaction of S-R and R-O processes in what was referred to at the time as an associative–cybernetic model. Although this model was proposed in that paper as a vehicle for explaining differences between Pavlovian-conditioned responses and instrumental actions, particularly with respect to their sensitivity to degradation of the R-O contingency, it was also found that this model was able to offer a general account for the effects of overtraining on instrumental outcome devaluation and, more recently, has also been argued to have a degree of neural plausibility.[41,42] We contend that this model also provides a good architecture for understanding the interaction of O-R and R-O processes in the selection and initiation of instrumental actions, both in single-action training and in a choice situation. A revised version of this model that captures the various aspects of O-R and R-O associative processes described earlier and their integration is presented in FIGURE 4A.

In this model, action selection is largely controlled by outcome–response learning and hence by outcome–retrieval. Action selection initiates a process of action evaluation through the R-O association; that is, the value of the action is estimated on the basis of the predicted reward value of the outcome that is contingent on that action. Finally, the action selection and

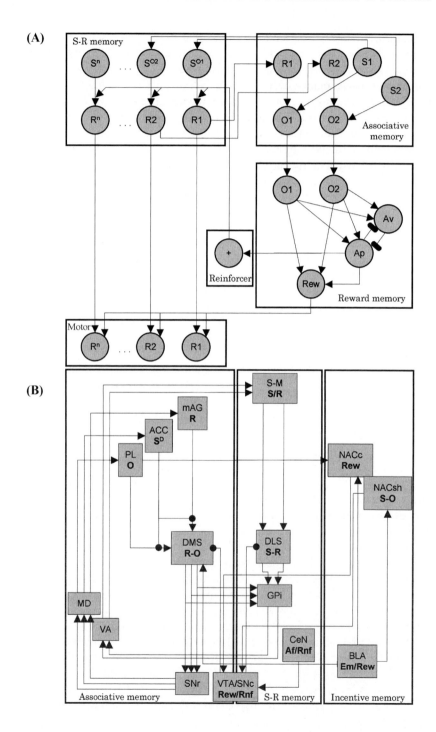

evaluation processes combine to initiate an action. The key to performance of any action is that response production will occur when the response representation in the motor system is activated above a low fluctuating threshold. Hence, performance of an action could occur without training when the threshold of this unit is randomly low or when activation of the representation is driven supra-threshold through activation by the response selection and evaluation processes. These sources of performance will, individually, result in a relatively low probability of a response and so highly variable rates of performance, particularly early in training. The probability of responding and the rate of performance generally is increased when both the feed-forward response selection process and the feedback evaluative process activate motor output together, guaranteeing supra-threshold levels of activation. On this view, although noise can result in a low fluctuating probability that a response will occur, consistent execution requires the integration of O-R and R-O processes at motor output. The occurrence of the response can result in the occurrence of its contingent outcome (O), and, when the latter occurs, this activates the outcome representation in the S-R memory, the associative memory, and the reward memory. The latter two update the strength of the R-O association and allow estimation of outcome value. At the same time, outcome delivery also gives rise to a reinforcement signal that strengthens contiguously active stimulus and response representations in the S-R memory, including the specific outcome–response association, as long as the outcome is appetitive. An aversive outcome or an outcome that has been devalued will inhibit the production

---

FIGURE 4. (A) A version of the associative cybernetic model. This version includes three memory modules: an S-R memory, composed of various S-R associations including the O-R association; an associative memory, composed of the S-O and R-O associations underlying Pavlovian and instrumental conditioning; and a reward memory that encodes the affective (appetitive [AP] or aversive [AV]) and emotional value of the outcome and separating its rewarding (Rew) and reinforcing (+) functions. As with earlier versions, performance is a product of the dual influence of S-R and associative memory modules on the motor system. Here, the latter influence is gated by the incentive memory. (B) Neural implementation of the associative–cybernetic model. Associative memory for instrumental conditioning is shown linking associative regions of prefrontal cortex—prelimbic area (PL), anterior cingulated (ACC), medial agranular area (mAG), and posterior dorsomedial striatum (DMS) with substantia nigra pars reticulata (SNr) output and feedback via mediaodorsal thalamus (MD) modulated by dopaminergic input to the DMS from the ventral tegmental area (VTA). The S-R memory is composed of sensorimotor cortex (S-M), dorsolateral striatum (DLS) pallidam (GPi), and thalamic feedback (via ventral anterior [VA] and the ventrolateral nuclei) again modulated by dopaminergic input from the substantia nigra pars compacta (SNc). The reward component of the incentive memory, barely sketched here, comprises the basolateral amygdala (BLA) and accumbens core (NACc). Reinforcement is mediated by affective modulation of SNc by the amygdala central nucleus (CeN). Further details can be found in Refs. 15, 40, 41, and 43.

of the reinforcement signal and, having a low value, will also ensure that action evaluation exerts very little if any excitatory impact on motor output.

Only the very general details of the reward system are provided here. For a more detailed discussion of the reward memory in this model, see Balleine.[15,41,43] One final point: it should be clear that outcome-specific transfer effects could be through the S-O connections in associative memory and activation of incentive and reward processes to ensure an increased activation of the motor response. Because these effects are resistant to outcome devaluation, however, it is likely that these effects are more usually produced by stimulus-induced retrieval of the O-R association in S-R memory, resulting in an increased probability of selection and hence performance of R.

THE ROLE OF CORTICOBASAL GANGLIA NETWORKS IN INSTRUMENTAL PERFORMANCE

When integrated with the existing literature, considerable recent research in our lab has allowed us to provide a reasonable working model of the neural network within which the associative–cybernetic model presented in FIGURE 4A could be implemented in the brain. This is presented in summary form in FIGURE 4B. Although we cannot describe this implementation in any detail here, its salient features and the evidence on which it is based can be quickly sketched; the interested reader is referred to the primary sources referenced later.

First a caveat: as should be clear from the above discussion, it is only possible to evaluate the effects of neural manipulations on instrumental performance when those manipulations are combined with tests that directly establish that the effects of the manipulation are specifically on instrumental conditioning and not due to changes in other forms of learning, notably Pavlovian conditioning, or due solely to changes in motor performance. Although a variety of theories have been advanced regarding the neural bases of the goal-directed aspects of instrumental conditioning, many of these have been derived using tasks that are only nominally instrumental and for which no direct evidence was provided to indicate that instrumental learning, rather than some other form of learning, was controlling performance (e.g., Refs. 44–47). Furthermore, many investigators have relied on only indirect measures of learning, for example, changes in performance or in the correlation between apparently goal-directed behavior and markers of neural activity, rather than on tests that provide direct assessments of the influence of a neural manipulation on learning.[20,48–50] For these reasons, much of the evidence behind these current theories is of little value in evaluating the putative mechanisms of instrumental performance.

The following summary of our working hypothesis regarding the neural bases of instrumental performance has been derived, however, from the results of experiments designed to assess the effects of neural manipulations on specific markers of instrumental learning, that is, on the instrumental outcome devaluation and contingency degradation tests. Using these tests, we have found strong evidence suggesting that the prelimbic region of the prefrontal cortex is involved in encoding the R-O contingency.[13,31,51] Cell body lesions of this structure made prior to training did not prevent the acquisition of an instrumental action but did abolish the ability of animals to choose a course of action after the outcome had been devalued or when a specific action–outcome contingency had been degraded. It is clear, however, that this region does not store the action–outcome association; after a period of training, lesions of this area no longer induced a deficit and performance appeared to be normal.[31] Furthermore, in line with the division of R-O and O-R associations, neither outcome-specific Pavlovian instrumental transfer[51] nor the specificity of instrumental reinstatement was affected by prelimbic lesions, although, similar to the influence of outcome devaluation, the latter was generally attenuated by this treatment.[31]

The time-limited influence of the prelimbic area on the acquisition of instrumental actions, coupled with its connectivity—particularly its amygdala, thalamic, and midbrain dopaminergic afferents—suggested that this region acts in a coordinated fashion with other prefrontal areas, notably anterior cingulate and medial agranular areas, that send efferents to common regions of thalamus and striatum to encode and modulate the encoding of the R-O association. In line with this suggestion, we found that a similar pattern of deficits was induced by lesions of the mediodorsal thalamus, a structure that maintains reciprocal connections with the prelimbic area, but not by lesions of the anterior thalamic nuclei.[52] The prelimbic prefrontal cortex also sends well-documented projections to the striatum, notably the dorsomedial striatum and the core of the nucleus accumbens. Considerable evidence suggests that the latter region is not involved in instrumental learning[23,53,54] although, as noted earlier, it does appear to play a role in establishing action values, likely through its connections with the basolateral amygdala, an area that we have found to be critical for establishing the incentive value of the instrumental outcome.[55,56]

In contrast, we have found evidence that lesions of the posterior dorsomedial striatum, whether made prior to or after training, abolished sensitivity to both outcome devaluation and contingency degradation.[57] Furthermore, temporary inactivation using muscimol or pretreatment with the N-Methyl-D-Aspartate (NMDA) antagonist 2-amino-5-phosphonopentanoic acid (APV) prior to learning were also found to abolish sensitivity to outcome devaluation but not infusions into adjacent dorsolateral striatum.[57,58] Indeed, in contrast to the previously dominant view that the dorsal striatum mediates the acquisition of skills or habits,[59,60] this evidence suggests that it is heterogeneous in function subserving both goal-directed and habitual actions. We have investigated this heterogeneity in several ways. First, on the basis of the observation that most studies assessing the role of the dorsal striatum in sensorimotor functions confined their manipulations of dorsolateral striatum,[61] we examined the effects of lesions of this area on overtraining-induced insensitivity to outcome devaluation. In this study we found that although the instrumental performance of sham lesioned rats was insensitive to outcome devaluation, lesions of this structure rendered performance once again goal directed; that is, lesioned rats showed greater control by the R-O association than sham rats.[62] In contrast, lesions of the dorsomedial striatum were without effect. Furthermore, in a second series of experiments we found evidence that, although overtrained rats were relatively insensitive to changes in the R-O contingency, muscimol-induced inactivation of the dorsolateral striatum increased the rats' sensitivity to this manipulation.[63]

Finally, it is important to note that the distinct reward and reinforcement functions of the outcome representation in the associative and S-R memory systems in the associative–cybernetic model overlay the distinct functions in associative and motor striatum. As mentioned earlier, the basolateral amygdala appears to play a direct role in establishing the reward value of the instrumental outcome by attaching emotional significance to its distinct sensory features. From the model, the reward memory establishes the reward value of the outcome and acts to gate the output of the associative memory on the motor system. This could be achieved either directly, through the direct connections that the basolateral area maintains with the dorsomedial striatum,[64] the prelimbic area and mediodorsal thalamus,[65] or indirectly through basolateral control of accumbens output onto the ventral tegmental area,[66] the source of the dopaminergic input to dorsomedial striatum.[67] Likewise, connections between the amygdala and accumbens shell exert control over Pavlovian–instrumental transfer (Corbit *et al.*[23]) likely through the shell's control of midbrain dopaminergic efferents on the dorsal striatum. In contrast, the role of the instrumental outcome as a reinforcer in the S-R memory is likely mediated by amygdala central nucleus efferents on the substantia nigra pars compacta, the source of the dopaminergic input to dorsolateral striatum.[68] Current evidence suggests that this input is modulated by affective processes in the central and extended amygdala,[69] and, as such, the amygdala as a whole appears in a position to parse the distinct reward and reinforcement signals produced by outcome delivery and to separate their impact onto the R-O and S-R associations in the parallel circuits that course through the dorsal striatum.[70]

Generally, therefore, although the dorsomedial region of the striatum mediates the encoding of R-O associations, the dorsolateral region appears to be critical for S-R learning. In terms of the associative–cybernetic model, therefore, current evidence suggests that the critical associative and S-R memory systems that contribute to instrumental performance course through corticostriatal circuits localized to adjacent regions of the dorsal striatum. How these two pathways interact, however, is currently a matter of debate. The generally accepted architecture of the basal ganglia emphasizes the operation of functionally distinct, closed parallel loops connecting prefrontal cortex, dorsal striatum, pallidum/substantial nigra, thalamus, and feeding back onto the originating area of prefrontal cortex.[71-73] There is, on this view, vertical integration within loops but not lateral integration across loops and, as a consequence, various theories have had to be developed to account for lateral integration, for example, the split loop[68] or spiraling midbrain–striatal integration[45,74] models. These models have not yet found wide acceptance. In contrast, older theories of striato-pallido-nigral integration proposed that, rather than being discrete, corticostriatal connections converge onto common target regions particularly in the globus pallidus and substantia nigra, a view that allows naturally for integration between various corticostriatal circuits.[75-77] Although anatomical studies challenge this view, recent evidence has emerged supporting a hybrid version: that, in addition to the segregated loops, there may also be integration through collateral projections from caudate (or dorsomedial striatum) converging with projections from the putamen (or dorsolateral striatum) onto common regions in both the internal and external globus pallidus.[78] Whether these converging projections underlie the integration of the O-R and R-O associations that the associative–cybernetic model identifies as critical for the initiation of instrumental performance remains an open question.

ACKNOWLEDGMENTS

The research reported here and the preparation of this manuscript were supported by a grant (no. MH56446) from the National Institute of Mental Health. The authors thank Sanne de Wit for permission to present the data in FIGURE 1C and Andrew Delamater for his comments on an earlier version of this manuscript.

REFERENCES

1. BECHARA, A. & M. VAN DER LINDEN. 2005. Decision-making and impulse control after frontal lobe injuries. Curr. Opin. Neurol. **18:** 734–739.
2. DALLEY, J.W., R.N. CARDINAL & T.W. ROBBINS. 2004. Prefrontal executive and cognitive functions in rodents: neural and neurochemical substrates. Neurosci. Biobehav. Rev. **28:** 771–784.
3. DAW, N.D., Y. NIV & P. DAYAN. 2005. Uncertainty-based competition between prefrontal and dorsolateral striatal systems for behavioral control. Nat. Neurosci. **8:** 1704–1711.
4. GLIMCHER, P.W. 2005. Indeterminacy in brain and behavior. Annu. Rev. Psychol. **56:** 25–56.
5. MA, W.J., J.M. BECK, P.E. LATHAM & A. POUGET. 2006. Bayesian inference with probabilistic population codes. Nat. Neurosci. **9:** 1432–1438.
6. MONTAGUE, P.R., B. KING-CASAS & J.D. COHEN. 2006. Imaging valuation models in human choice. Annu. Rev. Neurosci. **29:** 417–448.
7. GUTHRIE, E.R. 1935. The Psychology of Learning. Harpers. New York.

8. TOLMAN, E.C. 1932. Purposive Behavior in Animals. Century Books. New York.
9. HULL, C.L. 1943. Principles of Behavior. Appleton. New York.
10. ADAMS, C.D. & A. DICKINSON. 1981. Instrumental responding following reinforcer devaluation. Q. J. Exp. Psychol. 33B: 109–121.
11. HOLMAN, E.W. 1975. Some conditions for the dissociation of consummatory and instrumental behavior in rats. Learn. Motiv. 6: 358–366.
12. COLWILL, R.M. & R.A. RESCORLA. 1986. Associative structures in instrumental learning. In: The Psychology of Learning and Motivation. Vol. 20. G.H. Bower, Ed.: 55–104. Academic Press. Orlando, FL.
13. BALLEINE, B.W. & A. DICKINSON. 1998. Goal-directed instrumental action: contingency and incentive learning and their cortical substrates. Neuropharmacology 37: 407–419.
14. CORBIT, L.H. & B.W. BALLEINE. 2000. The role of the hippocampus in instrumental conditioning. J. Neurosci. 20: 4233–4239.
15. BALLEINE, B.W. 2001. Incentive processes in instrumental conditioning. In: Handbook of Contemporary Learning Theories. R.M.S. Klein, Ed.: 307–366. LEA. Hillsdale, NJ.
16. DICKINSON, A. & B.W. BALLEINE. 1994. Motivational control of goal-directed action. Anim. Learn. Behav. 22: 1–18.
17. RESCORLA, R.A. & R.L. SOLOMON. 1967. Two-process learning theory: relationships between Pavlovian conditioning and instrumental learning. Psychol. Rev. 74: 151–182.
18. TRAPOLD, M.A. & J.B. OVERMIER. 1972. The second learning process in instrumental conditioning. In: Classical Conditioning: II. Current Research and Theory. A.A. Black & W.F. PROKASY, Eds.: 427–452. Appleton-Century-Crofts. New York.
19. SUTTON, R.S. & A.G. BARTO. 1998. Reinforcement Learning. MIT Press. Cambridge, MA.
20. O'DOHERTY, J., et al. 2004. Dissociable roles of ventral and dorsal striatum in instrumental conditioning. Science 304: 452–454.
21. DICKINSON, A., J. CAMPOS, Z.I. VARGA & B. BALLEINE. 1996. Bidirectional instrumental conditioning. Q. J. Exp. Psychol. B. 49: 289–306.
22. COLWILL, R.M. & R.A. RESCORLA. 1988. Associations between the discriminative stimulus and the reinforcer in instrumental learning. J. Exp. Psychol. Anim. Behav. Process. 14: 155–164.
23. CORBIT, L.H., J.L. MUIR & B.W. BALLEINE. 2001. The role of the nucleus accumbens in instrumental conditioning: evidence of a functional dissociation between accumbens core and shell. J. Neurosci. 21: 3251–3260.
24. CORBIT, L.H. & B.W. BALLEINE. 2003. Instrumental and Pavlovian incentive processes have dissociable effects on components of a heterogeneous instrumental chain. J. Exp. Psychol. Anim. Behav. Process. 29: 99–106.
25. RESCORLA, R.A. 1992. Response-outcome versus outcome-response associations in instrumental learning. Anim. Learn. Behav. 20: 223–232.
26. RESCORLA, R.A. 1991. Associative relations in instrumental learning: the Eighteenth Bartlett Memorial Lecture. Q. J. Exp. Psychol. 43: 1–23.
27. DICKINSON, A. & S. DE WIT. 2003. The interaction between discriminative stimuli and outcomes during instrumental learning. Q. J. Exp. Psychol. B. 56: 127–139.
28. BOTVINICK, M.M., J.D. COHEN & C.S. CARTER. 2004. Conflict monitoring and anterior cingulate cortex: an update. Trends Cogn. Sci. 8: 539–546.

29. DE WIT, S., Y. KOSAKI, B.W. BALLEINE & A. DICKINSON. 2006. Dorsomedial pre-frontal cortex resolves response conflict in rats. J. Neurosci. **26:** 5224–5229.

30. DONEGAN, N.H., J.W. WHITLOW JR., & A.R. WAGNER. 1977. Posttrial reinstatement of the CS in Pavlovian conditioning: facilitation or impairment of acquisition as a function of individual differences in responsiveness to the CS. J. Exp. Psychol. Anim. Behav. Process. **3:** 357–376.

31. OSTLUND, S.B. & B.W. BALLEINE. 2005. Lesions of medial prefrontal cortex disrupt the acquisition but not the expression of goal-directed learning. J. Neurosci. **25:** 7763–7770.

32. OSTLUND, S.B. & B.W. BALLEINE. 2007. Instrumental reinstatement depends on sensory- and motivationally-specific features of the instrumental outcome. Learn. Behav. 35(1).

33. DICKINSON, A., B.W. BALLEINE, A. WATT, F. GONZALES & R.A. BOAKES. 1995. Overtraining and the motivational control of instrumental action. Anim. Learn. Behav. **22:** 197–206.

34. ADAMS, C.D. 1981. Variations in the sensitivity of instrumental responding to reinforcer devaluation. Q. J. Exp. Psychol. **34B:** 77–98.

35. COLWILL, R.M. & R.A. RESCORLA. 1990. Effect of reinforcer devaluation on dis-criminative control of instrumental behavior. J. Exp. Psychol. Anim. Behav. Pro-cess. **16:** 40–47.

36. RESCORLA, R.A. 1994. Transfer of instrumental control mediated by a devalued outcome. Anim. Learn. Behav. **22:** 27–33.

37. HOLLAND, P.C. 2004. Relations between Pavlovian-instrumental transfer and rein-forcer devaluation. J. Exp. Psychol. Anim. Behav. Process. **30:** 104–117.

38. DICKINSON, A. 1985. Actions and habits: the development of behavioural auton-omy. Philos. Trans. R. Soc. Lond. B. **308:** 67–78.

39. DICKINSON, A. 1994. Instrumental conditioning. *In*: Animal Cognition and Learn-ing. N.J. Mackintosh, Ed.: 4–79. Academic Press. London.

40. DICKINSON, A. & B.W. BALLEINE. 1993. Actions and responses: the dual psychology of behaviour. *In*: Spatial Representation. N. Eilan, R. McCarthy & M.W. Brewer, Eds.: 277–293. Basil Blackwell Ltd. Oxford.

41. BALLEINE, B.W. 2005. Neural bases of food seeking: affect, arousal and reward in corticostriatolimbic circuits. Physiol. Behav. **86:** 717–730.

42. DAYAN, P. & B.W. BALLEINE. 2002. Reward, motivation, and reinforcement learn-ing. Neuron. **36:** 285–298.

43. BALLEINE, B.W. 2004. Incentive Behavior. *In*: The Behavior of the Laboratory Rat: A Handbook with Tests. I.Q. Whishaw & B. Kolb, Eds.: 436–446. Oxford University Press. Oxford.

44. KELLEY, A.E. 2004. Ventral striatal control of appetitive motivation: role in inges-tive behavior and reward-related learning. Neurosci. Biobehav. Rev. **27:** 765–776.

45. HARUNO, M. & M. KAWATO. 2006. Heterarchical reinforcement-learning model for integration of multiple cortico-striatal loops: fMRI examination in stimulus-action-reward association learning. Neural. Netw. **19:** 1242–1254.

46. GOTO, Y. & A.A. GRACE. 2005. Dopaminergic modulation of limbic and cortical drive of nucleus accumbens in goal-directed behavior. Nat. Neurosci. **8:** 805–812.

47. ATALLAH, H.E., D. LOPEZ-PANIAGUA, J.W. RUDY & R. O'REILLY. 2007. C. Separate neural substrates for skill learning and performance in the ventral and dorsal striatum. Nat. Neurosci. **10:** 126–131.

48. KOECHLIN, E., A. DANEK, Y. BURNOD & J. GRAFMAN. 2002. Medial prefrontal and subcortical mechanisms underlying the acquisition of motor and cognitive action sequences in humans. Neuron **35:** 371–381.
49. HAMILTON, A.F. & S.T. GRAFTON. 2006. Goal representation in human anterior intraparietal sulcus. J. Neurosci. **26:** 1133–1137.
50. BUCCINO, G., F. BINKOFSKI & L. RIGGIO. 2004. The mirror neuron system and action recognition. Brain Lang. **89:** 370–376.
51. CORBIT, L.H. & B.W. BALLEINE. 2003. The role of prelimbic cortex in instrumental conditioning. Behav. Brain Res. **146:** 145–157.
52. CORBIT, L.H., J.L. MUIR & B.W. BALLEINE. 2003. Lesions of mediodorsal thalamus and anterior thalamic nuclei produce dissociable effects on instrumental conditioning in rats. Eur. J. Neurosci. **18:** 1286–1294.
53. DE BORCHGRAVE, R., J.N. RAWLINS, A. DICKINSON & B.W. BALLEINE. 2002. Effects of cytotoxic nucleus accumbens lesions on instrumental conditioning in rats. Exp. Brain. Res. **144:** 50–68.
54. BALLEINE, B. & S. KILLCROSS. 1994. Effects of ibotenic acid lesions of the nucleus accumbens on instrumental action. Behav. Brain Res. **65:** 181–193.
55. BALLEINE, B.W., A.S. KILLCROSS & A. DICKINSON. 2003. The effect of lesions of the basolateral amygdala on instrumental conditioning. J. Neurosci. **23:** 666–675.
56. WANG, S.H., S.B. OSTLUND, K. NADER & B.W. BALLEINE. 2005. Consolidation and reconsolidation of incentive learning in the amygdala. J. Neurosci. **25:** 830–835.
57. YIN, H.H., S.B. OSTLUND, B.J. KNOWLTON & B.W. BALLEINE. 2005. The role of the dorsomedial striatum in instrumental conditioning. Eur. J. Neurosci. **22:** 513–523.
58. YIN, H.H., B.J. KNOWLTON & B.W. BALLEINE. 2005. Blockade of NMDA receptors in the dorsomedial striatum prevents action-outcome learning in instrumental conditioning. Eur. J. Neurosci. **22:** 505–512.
59. POLDRACK, R.A. & M.G. PACKARD. 2003. Competition among multiple memory systems: converging evidence from animal and human brain studies. Neuropsychologia **41:** 245–251.
60. GRAYBIEL, A.M. 1995. Building action repertoires: memory and learning functions of the basal ganglia. Curr. Opin. Neurobiol. **5:** 733–741.
61. MCDONALD, R.J. & N.M. WHITE. 1993. A triple dissociation of memory systems: hippocampus, amygdala, and dorsal striatum. Behav. Neurosci. **107:** 3–22.
62. YIN, H.H., B.J. KNOWLTON & B.W. BALLEINE. 2004. Lesions of dorsolateral striatum preserve outcome expectancy but disrupt habit formation in instrumental learning. Eur. J. Neurosci. **19:** 181–189.
63. YIN, H.H., B.J. KNOWLTON & B.W. BALLEINE. 2006. Inactivation of dorsolateral striatum enhances sensitivity to changes in the action-outcome contingency in instrumental conditioning. Behav. Brain Res. **166:** 189–196.
64. KELLEY, A.E., V.B. DOMESICK & W.J. NAUTA. 1982. The amygdalostriatal projection in the rat–an anatomical study by anterograde and retrograde tracing methods. Neuroscience **7:** 615–630.
65. VERTES, R.P. 2006. Interactions among the medial prefrontal cortex, hippocampus and midline thalamus in emotional and cognitive processing in the rat. Neuroscience **142:** 1–20.
66. ALHEID, G.F. 2003. Extended amygdala and basal forebrain. Ann. N. Y. Acad. Sci. **985:** 185–205.

67. Sesack, S.R., D.B. Carr, N. Omelchenko & A. Pinto. 2003. Anatomical substrates for glutamate-dopamine interactions: evidence for specificity of connections and extrasynaptic actions. Ann. N. Y. Acad. Sci. **1003:** 36–52.

68. Joel, D. & I. Weiner. 2000. The connections of the dopaminergic system with the striatum in rats and primates: an analysis with respect to the functional and compartmental organization of the striatum. Neuroscience **96:** 451–474.

69. Fudge, J.L. & A.B. Emiliano. 2003. The extended amygdala and the dopamine system: another piece of the dopamine puzzle. J. Neuropsychiatry Clin. Neurosci. **15:** 306–316.

70. Balleine, B.W. & S. Killcross. 2006. Parallel incentive processing: an integrated view of amygdala function. Trends Neurosci. **29:** 272–279.

71. Alexander, G.E., M.R. DeLong & P.L. Strick. 1986. Parallel organization of functionally segregated circuits linking basal ganglia and cortex. Annu. Rev. Neurosci. **9:** 357–381.

72. Alexander, G.E. & M.D. Crutcher. 1990. Trends Neurosci. **13:** 266–271.

73. Nakahara, H., K. Doya & O. Hikosaka. 2001. Parallel cortico-basal ganglia mechanisms for acquisition and execution of visuomotor sequences—a computational approach. J. Cogn. Neurosci. **13:** 626–647.

74. Haber, S.N. 2003. The primate basal ganglia: parallel and integrative networks. J. Chem. Neuroanat. **26:** 317–330.

75. Percheron, G. & M. Filion. 1991. Parallel processing in the basal ganglia: up to a point. Trends Neurosci. **14:** 55–59.

76. Bar-Gad, I., G. Morris & H. Bergman. 2003. Information processing, dimensionality reduction and reinforcement learning in the basal ganglia. Prog. Neurobiol. **71:** 439–473.

77. Yelnik, J. 2002. Functional anatomy of the basal ganglia. Mov. Disord. **17** (Suppl. 3): S15–S21.

78. Nadjar, A., *et al.* 2006. Phenotype of striatofugal medium spiny neurons in parkinsonian and dyskinetic nonhuman primates: a call for a reappraisal of the functional organization of the basal ganglia. J. Neurosci. **26:** 8653–8661.

Plastic Corticostriatal Circuits for Action Learning

What's Dopamine Got to Do with It?

RUI M. COSTA

Section on In Vivo *Neural Function, Laboratory for Integrative Neuroscience, NIAAA, NIH, Bethesda, Maryland, USA*

ABSTRACT: Reentrant corticobasal ganglia circuits are important for voluntary action and for action selection. *In vivo* and *ex vivo* studies show that these circuits can exhibit a plethora of short- and long-lasting plastic changes. Convergent evidence at the molecular, cellular, and circuit levels indicates that corticostriatal circuits are involved in the acquisition and automatization of novel actions. There is strong evidence that activity in corticostriatal circuits changes during the learning of novel actions, but the plastic changes observed during the early stages of learning a novel action are different than those observed after extensive training. A variety of studies indicate that the neural mechanisms and the corticostriatal subcircuits involved in the initial acquisition of actions and skills differ from those involved in their automatization or in the formation of habits. Dopamine, a critical modulator of short- and long-term plasticity in corticostriatal circuits, is differentially involved in early and late stages of action learning. Changes in dopaminergic transmission have several concomitant effects in corticostriatal function, which may be important for action selection and action learning. These diverse effects may subserve different roles for dopamine in reinforcement and action learning.

KEYWORDS: dopamine; striatum; synchrony; oscillatory activity; skill learning; reinforcement; action selection; goal-directed actions; habits

INTRODUCTION

We frequently monitor the environment around us, and depending upon our internal state, we decide which actions to generate to obtain suitable outcomes. The execution of those actions depends on our knowledge of the environment, in encoding action–outcome relations, and on the capacity to execute the actions efficiently. Corticostriatal circuits play an important role in selecting the

Address for correspondence: Rui M. Costa, Section on *In Vivo* Neural Function, Laboratory for Integrative Neuroscience, NIAAA, NIH, 5625 Fishers Lane, Room TS-20D, MSC 9411, Bethesda, MD 20852-9411. Voice: 301-443-1196; fax: 301-480-0466.
costarui@mail.nih.gov

Ann. N.Y. Acad. Sci. 1104: 172–191 (2007). © 2007 New York Academy of Sciences.
doi: 10.1196/annals.1390.015

appropriate action, in encoding the action–outcome relation, and in optimizing the action. In the process of optimizing the action or making it more efficient, there is an initial stage of rapid increment in performance, which is followed by slow increment as performance reaches asymptotic levels. This process of optimization of the action may change the nature of the action, and after extensive practice the action may become automatic or habitual, instead of goal directed. Therefore, it is plausible that the mechanisms mediating the initial performance of a skill or action are different than those mediating overtrained skills or habitual actions. Dopaminergic transmission modulates several corticostriatal processes, which could have different roles in action selection, reinforcement learning, and habit formation; and dopamine seems to be differentially involved in the different stages of action learning. We will discuss convergent evidence at the molecular, cellular, and circuit level indicating that plasticity in corticostriatal circuits is important for action learning, that different neural mechanisms support the different stages of action learning, and that dopamine may modulate more than one process important for reinforcement learning.

CORTICOSTRIATAL CIRCUITS AND ACTION LEARNING

Corticostriatal circuits are critical for the performance of actions. Lesions or disorders affecting different nodes of these reentrant circuits lead to a variety of deficits in movement initiation, movement accuracy, and goal-directed movements.[1-4] These same reentrant circuits display a plethora of short-term and long-term plastic changes, ranging from changes in synaptic strength and excitability to changes in microcircuit states, as measured *in vivo* and *ex vivo*.[5-22] Therefore, it is not surprising that these corticostriatal loops are not only necessary for the initiation and performance of actions, but also necessary for the learning of new skills and procedures and for the formation of habits (or stimulus–response relationships, S–R).[23-30] Accumulated evidence from previous studies has shown changes in activity and connectivity during motor learning in the motor cortices[17,31-44] and the basal ganglia.[16,17,42,45-50] Consistently, learning of new skills appears to be affected in disorders affecting corticostriatal circuits, like Parkinson's[27,51] and Huntington's diseases.[52] Rats with diminished nigrostriatal dopamine release are impaired in learning a new skill in a rotarod.[29] Similarly, habit formation seems to also be impaired in Parkinson's disease[25] and in rodents with nigrostriatal lesions.[30] Finally, animal models with striatum-specific manipulations that impair synaptic plasticity show impaired skill learning[53] and S–R learning.[54]

However, many actions we perform are not skilled or habitual, and require the appropriate selection of strategies during a choice situation that can lead to different outcomes or require different efforts to access the same outcome. There is growing evidence that corticostriatal circuits are required for the appropriate selection of actions in a choice situation,[55-63] and that the encoding of the action–outcome contingencies during the learning of goal-directed actions is also dependent upon these circuits.[57,59,64,65] Therefore, corticostriatal circuits seem to have a wider involvement in action learning than previously thought, from the appropriate action selection, to the learning of goal-directed actions, and to the automatization of those actions and formation of habits.

DIFFERENT CORTICOSTRIATAL MECHANISMS FOR THE ACQUISITION AND THE AUTOMATIZATION OF ACTIONS

When we learn a new motor skill we experience fast improvement in motor performance during the initial training period and continue to display slow improvement with further training. Similarly, when we learn to perform a particular action to obtain a particular outcome there is an initial stage of rapid increment in performance, which is followed by slow increment as performance reaches asymptotic levels. Corticostriatal circuits play an important role in the acquisition and subsequent automatization of actions. However, the initial performance of a skill or an action appears to be mediated by different corticostriatal subcircuits and to engage different mechanisms than the performance of automatized skills or habitual actions.

Both striatum and motor cortex are activated throughout the different phases of motor learning.[50,66] However, imaging studies have revealed that different corticostriatal nodes may be differentially activated during different phases of motor learning.[50] Simultaneous recordings from striatum and motor cortex revealed that plastic changes in corticostriatal circuits happen during the early stages of training but continue even after animals have reached asymptotic performance.[16,17] The proportion of corticostriatal neurons with activity correlated with task performance increases rapidly during the early stage of skill learning,[16,17,43] indicating a fast expansion of the task-related corticostriatal network. This expansion could represent a rapid way to improve motor performance by incrementing the computational space available to control the motor response. Importantly, cortical and striatal neuronal ensembles continue to change with further refinement and automatization of the movement after extended training.[16,17,41] These slowly developing changes occurring after the initial improvement in performance differ from those observed during the fast phase, and may not necessarily be the same in cortex and striatum. For example, during the performance of a motor skill task, the accelerating rotarod, about 70% of movement-related neurons in striatum increased firing rate during performance of this task while 30% decreased firing rate[17] (FIG. 1). This proportion remained relatively unchanged throughout training, which is consistent with the existence of functionally different neuronal subpopulations in striatum that could correspond to the direct and indirect pathway.[67] However, in motor cortex, this proportion changed throughout training. At the onset of the task about 50% of neurons increase firing and 50% decrease firing frequency in relation to task performance, but as training progressed close to 90% of the

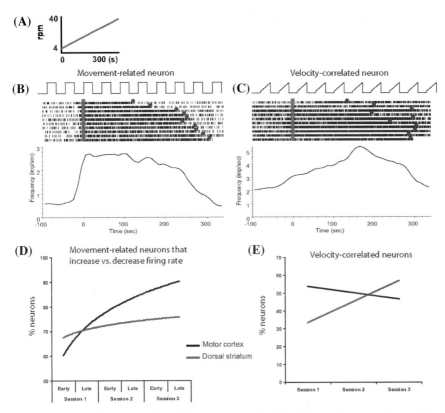

FIGURE 1. The neuronal activity in corticostriatal circuits continues to change after asymptotic performance during skill learning. During this late phase of skill learning motor cortex and dorsal striatum exhibit dissociable changes in activity. (**A**) Mice were trained to run in a rotarod accelerating from 4 to 40 rpm in 5 min, and given 10 trials per session with an intertrial interval of 5 min, across three sessions (one session per day). Animals reached asymptotic performance by the second session. (**B**) Example of a movement-related neuron that increased firing rate abruptly during the running of each trial. The perievent time histogram throughout one session is depicted at the bottom of the panel, and the raster plot of each trial during that session at the top; trials are presented top to bottom as they were presented during training. The beginning of each running period is aligned at time zero and marked by a square. The end of the running period is marked by a circle. (**C**) Example of a velocity-correlated neuron that gradually increases its firing rate during the running. (**D**) Percentage of movement-related neurons that increased versus decreased firing rate during movement across the different training sessions. In motor cortex, the proportion of neurons that increased firing rate increased gradually across sessions, while in dorsal striatum it remained unchanged. (**E**) Percentage of velocity-correlated neurons across the different sessions. In dorsal striatum the percentage of velocity-correlated neurons is low at the onset of training but increases slowly across sessions, while in motor cortex it remains unchanged. The figure appears in color online.

neurons in motor cortex increased firing rate during movement and only 10% decreased firing rate. These results are consistent with studies showing an increase in synaptic strength in motor cortex after skill learning.[68] On the other hand, if we consider the number of neurons that are correlated not just with movement, but with the speed of movement, many neurons in motor cortex (>50%) display activity that is related to the speed of movement, as one might expect. In striatum, fewer neurons (30%) display activity correlated with the speed of movement at the onset of training but slowly this number increases to be as large as in motor cortex, as if the cortex would be "training" the striatum.[17]

These data suggest different corticostriatal mechanisms mediating the initial acquisition of an action and the automatization of that same action. For example, although motor cortex is essential for the initial stages of motor skill learning,[39] synaptogenesis and reorganization of motor representations, as well as increases in signal-to-noise ratio of the neural activity in motor cortex, occur during the late but not the early phase of motor skill learning.[44,69] Similarly, other studies have shown that the activity in striatum during the initial encoding of a procedure or an action differs from the activity in striatum when that procedure becomes automatized or habitual.[11,70] These studies show that with training, task-related neural activity in the sensorimotor region of the striatum seems to decrease, reflecting perhaps a contraction of the corticostriatal network necessary to perform the task. Presumably, as S–R associations become stronger and the behavior less flexible, a suprathreshold sparse circuit could mediate the motor response, contrasting with the rapid expansion of corticostriatal task-related network observed during initial training.

Hence, the neural activity and the plastic changes in corticostriatal circuits differ between the initial acquisition of novel actions and during the automatization of those actions. But are the initial acquisition and the automatization encoded by the same corticostriatal loops? There is accumulated evidence that this is not the case. Studies in monkeys performing a sequential button press task show that during the early stage of acquisition the task-related activity is preferentially seen in the associative areas of the striatum (caudate and anterior putamen), while during the late stages of automatization the task-related activity is predominantly observed in the sensorimotor striatum (mainly posterior putamen).[71] In agreement, reversible inactivation of the associative striatum impaired the learning of new sequences but less so of well-learned sequences. Conversely, inactivation of the sensorimotor striatum affected the execution of well-learned sequences but not the acquisition of novel ones.[26] Similar results have been found in rodents. Rats trained on an instrumental task acquire goal-directed actions that depend on the action–outcome contingency relatively rapidly and only slowly acquire sensorimotor habits after a period of overtraining.[72–75] In rodents, lesions of the region corresponding to the associative striatum, that is, the medial portion of the dorsal striatum, render them unable to learn new goal-directed actions.[59] However, these animals can acquire habits. On the other hand, rodents with lesions in the lateral region of the dorsal striatum (sensorimotor striatum) fail to acquire habits but acquire goal-directed actions perfectly well.[28] These and other studies clearly suggest that different corticostriatal loops mediate the fast acquisition of goal-directed actions and the automatization of those actions.[30, 57,58, 64,76–82] It remains to be determined if such an anatomical dissociation in the function of corticostriatal circuits can be generalized to all actions and skills.

Do the molecular mechanisms necessary for the initial encoding and for the automatization of actions differ? Dopamine is a critical modulator of corticostriatal circuit function (see below), and hypodopaminergic function leads to difficulties in the initiation of voluntary actions. Not surprisingly, blockade of dopaminergic signaling during the early stages of training impairs the execution of the newly learned response.[83] However, with extended training these actions became less dopamine-dependent, in particular less dependent on dopamine receptor type 1 (D1) signaling.[83] Similar results were obtained in mice trained in a motor skill task. During the early stages of training dopamine receptor blockade severely impaired performance of the skill, but with extended training the performance of the skill became D1-independent, while D1 blockade still impaired voluntary movement (Clouse, White, and Costa, unpublished results). These studies suggest that dopaminergic signaling (at least D1 receptor-dependent signaling) is critical for performance during the early stages of action learning, but less critical after extended training. Accordingly, the activity of dopaminergic neurons induced by the presentation of reward-predicting stimuli decreases after overtraining.[84]

DOPAMINE, ACTIONS, AND REINFORCEMENT LEARNING

Dopamine is an important modulator of corticostriatal circuit function and has been implicated in reinforcement learning, motivation, and voluntary movement. Consequently, the progressive loss of the dopaminergic neurons from the substantia nigra pars compacta as observed in Parkinson's disease eventually leads to difficulties in the initiation of voluntary actions, motivation, and reinforcement learning[85–88] underscoring the importance of dopaminergic transmission for the multiple functions of corticostriatal circuits in behavior.[89] As we mentioned, loss of the nigrostriatal dopaminergic projection leads to impaired habit learning in humans[25] and animal models,[30,90] and downregulation of dopamine release from the substantia nigra impairs skill learning.[29] Conversely, amphetamine injections into the caudate can accelerate S–R learning,[91] and amphetamine sensitization leads to a faster transition from goal-directed to habitual responses.[92] Furthermore, electrical stimulation of the substantia nigra is able to reinforce lever pressing.[12] Therefore, dopamine seems to play an important role in reinforcement learning and in the acquisition of novel actions. But what is dopamine's role?

Dopaminergic neurons have been shown to change firing rate when presented with an unexpected or surprising outcome, and this change in phasic dopaminergic firing has been hypothesized to represent an error signaling during reinforcement learning.[93–96] Furthermore, it has been more recently proposed that phasic dopamine firing can code uncertainty and probability of reward,[97] and that dopamine release can code reward value.[98]

However, despite the correlation between phasic dopaminergic firing and reward value and probability, studies manipulating dopaminergic transmission suggest that dopamine may not be necessary to code for reward *per se*. For example, in a two-bottle preference test, dopamine-depleted mice still display a preference for hedonically positive sweet tastants over water.[99] Additionally, animals with slightly elevated levels of dopamine (70% of control) display similar liking reactions to sucrose,[100] and animals with elevated dopamine levels (500% of control) display normal preference for sweet and bitter stimuli.[101] Similarly, the role of phasic dopamine as an error prediction signal is not unambiguous. Studies showing that Parkinson's patients without medication are better at learning to avoid choices that lead to negative than positive outcomes, while patients with dopamine medication are more sensitive to positive than negative outcomes,[87,102] certainly point to a role of dopaminergic transmission in positive and negative reinforcement, but hypothesize a different effect of the error signal in direct and indirect pathways. Nonetheless, dopamine-depleted mice can still acquire morphine-induced conditioned place preference.[103] Also, animals with slightly (70%) elevated dopamine levels display normal Pavlovian and instrumental learning,[104,105] and normal facilitation of instrumental actions by discrete cues,[105] suggesting that dopaminergic neurons are part of a more distributed circuit that computes reward prediction error.[106–109]

In most of these studies manipulating dopaminergic transmission, both the phasic and tonic dopaminergic transmission are affected. Several studies have postulated that tonic dopaminergic transmission is important for modulating the goal-directed actions that lead to the reward, or in other words, how much animals "want" the reward.[100,104,110,111] Operationally, this means that dopamine can modulate the sensory-specific motivation to access that reward, or how much an animal is willing to work for that reward, which is different than how much an animal "likes" or prefers a reward. A recent study used an inducible approach in genetically modified mice to trigger changes in tonic dopaminergic transmission after learning, and showed that dopaminergic transmission indeed modulates how much animals are willing to work to get access to reinforcing stimuli.[112] These results are in agreement with the hypothesized role of tonic dopamine as a biological indicator of the opportunity cost of time.[113] In those conditions, tonic dopamine is a good predictor of average reward rate, and therefore of how much an animal should respond at a particular time.[113,114]

The sensory-specific motivation, or how much an animal is willing to work for a reward at a particular moment, is tightly coupled to the incentive value of the reward at that moment, and therefore to the mechanisms mediating the updating of reward value.[74] Hence, dopamine could modulate how much an animal "wants" to access a reward by virtue of modulating the updating of its incentive value. The role of dopamine in incentive learning has been less studied (see Ref. 115), but there are some indications that hyperdopaminergic animals (500%) update more positively the value of both sweet and bitter tastants.[101] Finally, it should be considered that dopamine could modulate changes in the value of reward-related actions,[63,116] and therefore influence the incentive salience of actions,[74,117] much as it has been proposed for reward-related stimuli.[111]

DOPAMINE MODULATES CORTICOSTRIATAL PLASTICITY AND CORTICOSTRIATAL CIRCUIT COORDINATION

As we have discussed, dopamine can act as a key regulator of corticostriatal circuit function, and has been implicated in reinforcement learning and voluntary movement. However, what is the impact of dopamine release on the target, that is, on the activity of corticostriatal circuits?

In dorsal striatum, concomitant dopaminergic release from the midbrain projection neurons and glutamatergic release from cortical terminals onto medium spiny neurons can result in changes in synaptic efficacy and synaptic strength.[118] Both long-term potentiation (LTP) and long-term depression (LTD) of corticostriatal synapses have been observed.[7–9,12,13] LTP at corticostriatal synapses has been shown to be mediated by postsynaptic changes and to be NMDA-dependent (N-methyl-D-aspartate glutamate receptor).[8,119] On the other hand, LTD is dependent on endocannabinoid-mediated retrograde signaling onto the cortical terminals and is expressed presynaptically.[14] Thus, coincident dopaminergic and glutamatergic input into the dorsal striatum may be important both to reinforce positively the appropriate actions and negatively irrelevant actions. But how does one know which ongoing actions to select and then reinforce? What happens to the activity of cortical and striatal neurons during the changes in dopaminergic transmission that occur during the online selection of voluntary actions?

Not surprisingly, studies recording the activity of single neurons in the dorsal striatum and primary motor cortex during rapid transitions between periods of hyperdopaminergia with hyperactivity and periods of dopamine depletion with hypoactivity show that alterations in dopaminergic transmission lead to changes in the firing rate of most striatal and cortical neurons.[20] However, what is more striking is that rapid changes in the levels of dopaminergic transmission result in profound changes in the coordinated activity of neurons in corticostriatal ensembles. It appears that during states of hyperdopaminergia less neurons display significant cross-correlated activity and fewer neurons fire entrained to the local field potential oscillation, indicating that the activity in corticostriatal circuits becomes less synchronous than in normal states (FIG. 2). Conversely, during hypodopaminergia, the synchronicity in these circuits

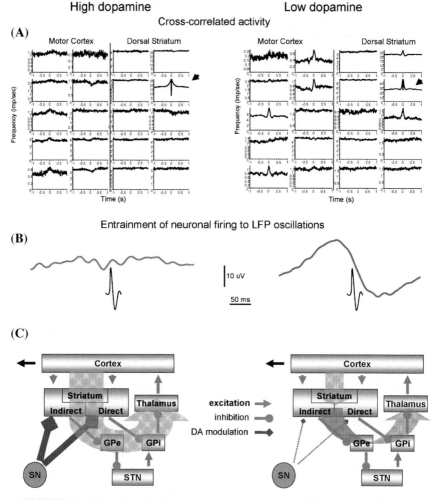

FIGURE 2. Alterations in dopaminergic transmission rapidly change corticostriatal synchrony. (**A**) Example of cross-correlations between pairs of cortical and striatal neurons during a state of hyperdopaminergia, and 20 min later during a state of hypodopaminergia. In each panel, the two left columns correspond to motor cortex neurons and the two right columns to dorsal striatum neurons. The activity of each neuron is correlated to the activity of the same striatal neuron (second from the top in the far right column indicated by the arrow and depicting the autocorrelation; time 0 is not depicted for the autocorrelation). During hyperdopaminergia corticostriatal circuits are largely asynchronous, and very few neurons display significant cross-correlated activity, while during hypodopaminergia synchronicity increases as indicated by an increase in cross-correlated firing. (**B**) Entrainment of a particular neuron to the local field potential oscillation during hyperdopaminergia and hypodopaminergia. The spike-triggered average of the local field potential oscillation, that is, the average oscillation in relation to the time of the spike (depicted as an action potential;

becomes high, with a high proportion of the neurons firing preferentially at a particular phase of the local field potential oscillation and showing cross-correlated firing.[20] These results are consistent with previous observations of increased synchronous neuronal activity in MPTP-treated primates.[120–122]

How can dopamine modulation of the entrainment of particular neurons to the local field potential be related to action selection in the striatum? Local field potential oscillations seem to be correlated with neuronal firing and with synchronous synaptic activity, even when this activity is subthreshold.[123,124] If striatal neurons are not entrained to the local field potential during states of hyperdopaminergia this means that when an input arrives to the striatum from the cortex, this input (synaptic input) results in an output (spike) irrespective of the phase of the local field potential oscillation. This means that the relationship between input and output is rather high, and that inputs received during high dopaminergic transmission are relayed forward (FIG. 2). On the other hand, during low dopaminergic transmission, neurons fire highly entrained to the local field potential that most inputs arriving from the cortex do not result in outputs, and therefore there is not a high correlation between input and output and most inputs are lost or "gated." This indicates that information in corticostriatal loops is relayed onward more readily during high dopaminergic transmission resulting in most actions being selected and in hyperactivity, while during low dopaminergic transmission inputs arriving out of phase with the local field potential oscillation would be "gated" resulting in poor input–output correspondence (FIG. 2).

How can dopamine modulate the entrainment of neurons to the local field potential oscillations in striatum? One possible mechanism is through dopamine's modulation of excitability in medium spiny neurons.[125–127] High dopaminergic transmission could lead to increased neuronal excitability, especially in D1 receptor-expressing neurons, thereby decreasing the firing threshold and allowing neurons to fire whenever they receive a synaptic input, that is, during

---

←

not to scale) is shown. During states of high dopaminergic transmission neurons do not fire entrained to the local field potential oscillation as shown by the relatively flat spike-triggered average function around the time of the spike. During low dopamine transmission neurons fire preferentially at a particular phase of the local field potential oscillation as shown by a significant fluctuation of the spike-triggered average function around the time of the spike. (C) During hyperdopaminergia neurons fire whenever the input arrives, irrespective of the phase of the local field potential oscillations. Therefore, the ratio between output and input is rather high, with most inputs being relayed forward and few selected. During low dopaminergic transmission neurons fire mostly during the negative slope of the local field potential (corresponding to intracellular depolarization, see above), and therefore most inputs arriving during other phases of the local field potential oscillation are not relayed forward. The ratio between output and input is very low, and the information relayed forward is highly synchronous.

any phase of the local field potential oscillation. Low dopaminergic trans-mission could reduce neuronal excitability, therefore elevating the threshold for medium spiny neurons to fire an action potential. In this case, neurons would fire only after enough depolarization; that is, during the negative slope or close to the trough of the extracellularly recorded oscillations (which would correspond to intracellular depolarization) (FIG. 2). Alternatively, increased excitability of D2 –receptor-expressing neurons[128] could lead to synchronous firing in striatum, potentially via their collateral connectivity, and drive directly the field potential oscillations. In addition to direct changes in the excitabil-ity of medium spiny neurons, dopamine could regulate the entrainment of these neurons to the local oscillations by modulating the function of choliner-gic interneurons,[129–131] or fast-spiking gabaergic interneurons.[132,133] Another possible mechanism would be that dopamine modulates the entrainment of neurons to the local field potential oscillation through rapid alterations in cor-ticostriatal synaptic efficacy.[12,13,19] Finally, these changes in synchrony and local field potential entrainment could emanate from changes in the relative frequency of medium spiny neuron upstates and downstates during hyper- and hypopaminergic periods.[13,134]

   Therefore, it appears that dopamine could have multiple roles during rein-forcement learning. For instance, it could rapidly modulate online selection of actions through a process of "gating" corticostriatal input–output (modulating the gain or salience), and reinforce the selected actions through the long-lasting modulation of synaptic strength. Furthermore, dopamine could embody more than the error signal in trial and error learning; it could also influence the diver-sity of actions tried. For instance, if during high tonic dopaminergic signaling corticostriatal circuits are more asynchronous and most inputs into striatum are relayed forward, this will increase the diversity of corticostriatal patterns generated and consequently the repertoire of actions to select from. Moreover, through the modulation of corticostriatal synchrony, dopamine could modu-late timing in corticostriatal circuits. However, these hypotheses are certainly simplistic because both the differential effects of dopamine in different cell populations in striatum (e.g., striatonigral and striatopallidal medium spiny neurons; cholinergic interneurons, fast-spiking neurons) and the contribution of thalamic inputs need to be considered.

DOPAMINE-DEPENDENT BRAIN STATES

   As we discussed, the phasic changes in the firing of dopaminergic neu-rons during the presentation of unexpected rewards or reinforcing stimuli happen in a background of tonic dopaminergic release. Therefore, the effect of the changes in phasic dopaminergic firing on corticostriatal function will depend on the particular state of the circuit during the encounter of unexpected reinforcing stimuli. The levels of dopamine released change throughout the circadian cycle and across different wake–sleep states.[135] Consequently, an increase in phasic firing (bursting) during a state of high tonic dopaminergic firing may have different impact than during a state of lower tonic dopaminergic transmission.[13,134] Considering that the levels of dopamine change the oscillatory behavior and synchrony in corticostriatal circuits, and therefore information processing and action selection in corticobasal ganglia circuits,[20] one could postulate the existence of dopamine-dependent brain states that set the relationship between the selection of voluntary actions and reinforcement learning, and other functions like attention and salience, sleep–wake transitions, and sensory perception.

It has been observed that during states of high dopaminergic transmission theta and gamma frequency oscillations are more prominent while during states of low dopaminergic transmission delta and beta frequencies are more apparent.[20,136–141] These changes are consistent with changes in the relative frequency of local field potential oscillations observed during movement initiation,[142–144] attention modulation,[145] sleep–wake states,[146,147] and vision,[148] suggesting that there could be a relation between different dopamine states and different information processing states in the brain.

CONCLUSION AND OUTSTANDING QUESTIONS

Plastic changes in corticostriatal circuits are important for different behavioral processes during action learning. To understand how these different behavioral processes are learned, we need to further understand the molecular and cellular processes mediating the acquisition of novel actions, and how those differ from habit formation. In doing so, we should consider not only the anatomy and connectivity of the different regions of these circuits, but also the role of different cell types and modulators. With the advent of more refined approaches to manipulate dopaminergic transmission and to measure or visualize plastic changes in corticostriatal circuits, some of the hypotheses and views reviewed here will become experimentally testable. These approaches can be applied to other neurotransmitter systems and to different nodes of the corticobasal ganglia circuitry.

ACKNOWLEDGMENTS

The writing of this article and the participation in the conference "Reward and Decision-Making in Cortico-basal Ganglia Networks" was supported by the NIAAA DICBR. I thank the organizers of the conference B. Balleine, K. Doya, J. O'Doherty, and M. Masamichi Sakagami. I also want to thank H. Yin and B. Balleine for valuable comments on the article, and my colleagues Miguel Nicolelis, Dana Cohen, and Shih-Chieh Lin.

REFERENCES

1. SAINT-CYR, J.A., A.E. TAYLOR & K. NICHOLSON. 1995. Behavior and the basal ganglia. Adv. Neurol. **65:** 1–28.
2. LALONDE, R. & T. BOTEZ-MARQUARD. 1997. The neurobiological basis of movement initiation. Rev. Neurosci. **8:** 35–54.
3. JELLINGER, K.A. 1998. Neuropathology of movement disorders. Neurosurg. Clin. N. Am. **9:** 237–262.
4. HERRERO, M.T., C. BARCIA & J.M. NAVARRO. 2002. Functional anatomy of thalamus and basal ganglia. Childs Nerv. Syst. **18:** 386–404.
5. LOVINGER, D.M. 1991. Trans-1-aminocyclopentane-1,3-dicarboxylic acid (t-ACPD) decreases synaptic excitation in rat striatal slices through a presynaptic action. Neurosci. Lett. **129:** 17–21.
6. KITAI, S.T. & D.J. SURMEIER. 1993. Cholinergic and dopaminergic modulation of potassium conductances in neostriatal neurons. Adv. Neurol. **60:** 40–52.
7. CALABRESI, P. *et al.* 1992. Long-term synaptic depression in the striatum: physiological and pharmacological characterization. J. Neurosci. **12:** 4224–4233.
8. CALABRESI, P. *et al.* 1992. Long-term potentiation in the striatum is unmasked by removing the voltage-dependent magnesium block of NMDA receptor channels. Eur. J. Neurosci. **4:** 929–935.
9. LOVINGER, D.M., E.C. TYLER & A. MERRITT. 1993. Short- and long-term synaptic depression in rat neostriatum. J. Neurophysiol. **70:** 1937–1949.
10. LOVINGER, D.M. & B.A. MCCOOL. 1995. Metabotropic glutamate receptor-mediated presynaptic depression at corticostriatal synapses involves mGLuR2 or 3. J. Neurophysiol. **73:** 1076–1083.
11. JOG, M.S. *et al.* 1999. Building neural representations of habits. Science **286:** 1745–1749.
12. REYNOLDS, J.N., B.I. HYLAND & J.R. WICKENS. 2001. A cellular mechanism of reward-related learning. Nature **413:** 67–70.
13. REYNOLDS, J.N. & J.R. WICKENS. 2000. Substantia nigra dopamine regulates synaptic plasticity and membrane potential fluctuations in the rat neostriatum, *in vivo*. Neuroscience **99:** 199–203.
14. GERDEMAN, G.L., J. RONESI & D.M. LOVINGER. 2002. Postsynaptic endocannabinoid release is critical to long-term depression in the striatum. Nat. Neurosci. **5:** 446–451.
15. NAKAHARA, H., S. AMARI SI & O. HIKOSAKA. 2002. Self-organization in the basal ganglia with modulation of reinforcement signals. Neural Comput. **14:** 819–844.
16. BRASTED, P.J. & S.P. WISE. 2004. Comparison of learning-related neuronal activity in the dorsal premotor cortex and striatum. Eur. J. Neurosci. **19:** 721–740.
17. COSTA, R.M., D. COHEN & M.A. NICOLELIS. 2004. Differential corticostriatal plasticity during fast and slow motor skill learning in mice. Curr. Biol. **14:** 1124–1134.
18. WANG, Z. *et al.* 2006. Dopaminergic control of corticostriatal long-term synaptic depression in medium spiny neurons is mediated by cholinergic interneurons. Neuron **50:** 443–452.
19. YIN, H.H. & D.M. LOVINGER. 2006. Frequency-specific and D2 receptor-mediated inhibition of glutamate release by retrograde endocannabinoid signaling. Proc. Natl. Acad. Sci. USA **103:** 8251–8256.

20. Costa, R.M. *et al.* 2006. Rapid alterations in corticostriatal ensemble coordination during acute dopamine-dependent motor dysfunction. Neuron **52:** 359–369.

21. Barnes, T.D. *et al.* 2005. Activity of striatal neurons reflects dynamic encoding and recoding of procedural memories. Nature **437:** 1158–1161.

22. Decoteau, W.E. *et al.* 2007. Oscillations of local field potentials in the rat dorsal striatum during spontaneous and instructed behaviors. J. Neurophysiol. Feb. 28 [Epub ahead of print]

23. Packard, M.G. & J.L. McGaugh. 1992. Double dissociation of fornix and caudate nucleus lesions on acquisition of two water maze tasks: further evidence for multiple memory systems. Behav. Neurosci. **106:** 439–446.

24. Graybiel, A.M. 1995. Building action repertoires: memory and learning functions of the basal ganglia. Curr. Opin. Neurobiol. **5:** 733–741.

25. Knowlton, B.J., J.A. Mangels & L.R. Squire. 1996. A neostriatal habit learning system in humans. Science **273:** 1399–1402.

26. Miyachi, S. *et al.* 1997. Differential roles of monkey striatum in learning of sequential hand movement. Exp. Brain Res. **115:** 1–5.

27. Doyon, J. *et al.* 1997. Role of the striatum, cerebellum, and frontal lobes in the learning of a visuomotor sequence. Brain Cogn. **34:** 218–245.

28. Yin, H.H., B.J. Knowlton & B.W. Balleine. 2004. Lesions of dorsolateral striatum preserve outcome expectancy but disrupt habit formation in instrumental learning. Eur. J. Neurosci. **19:** 181–189.

29. Akita, H. *et al.* 2006. Nigral injection of antisense oligonucleotides to synaptotagmin I using HVJ-liposome vectors causes disruption of dopamine release in the striatum and impaired skill learning. Brain Res. **1095:** 178–189.

30. Faure, A. *et al.* 2005. Lesion to the nigrostriatal dopamine system disrupts stimulus-response habit formation. J. Neurosci. **25:** 2771–2780.

31. Nudo, R.J. *et al.* 1996. Use-dependent alterations of movement representations in primary motor cortex of adult squirrel monkeys. J. Neurosci. **16:** 785–807.

32. Classen, J. *et al.* 1998. Rapid plasticity of human cortical movement representation induced by practice. J. Neurophysiol. **79:** 1117–1123.

33. Wise, S.P. *et al.* 1998. Changes in motor cortical activity during visuomotor adaptation. Exp. Brain Res. **121:** 285–299.

34. Kleim, J.A., S. Barbay & R.J. Nudo. 1998. Functional reorganization of the rat motor cortex following motor skill learning. J. Neurophysiol. **80:** 3321–3325.

35. Gandolfo, F. *et al.* 2000. Cortical correlates of learning in monkeys adapting to a new dynamical environment. Proc. Natl. Acad. Sci. USA **97:** 2259–2263.

36. Li, C.S., C. Padoa-Schioppa & E. Bizzi. 2001. Neuronal correlates of motor performance and motor learning in the primary motor cortex of monkeys adapting to an external force field. Neuron **30:** 593–607.

37. Karni, A. *et al.* 1995. Functional MRI evidence for adult motor cortex plasticity during motor skill learning. Nature **377:** 155–158.

38. Laubach, M., J. Wessberg & M.A. Nicolelis. 2000. Cortical ensemble activity increasingly predicts behaviour outcomes during learning of a motor task. Nature **405:** 567–571.

39. Muellbacher, W. *et al.* 2002. Early consolidation in human primary motor cortex. Nature **415:** 640–644.

40. Kleim, J.A. *et al.* 2002. Motor learning-dependent synaptogenesis is localized to functionally reorganized motor cortex. Neurobiol. Learn. Mem. **77:** 63–77.

41. KARGO, W.J. & D.A. NITZ. 2003. Early skill learning is expressed through selection and tuning of cortically represented muscle synergies. J. Neurosci. **23:** 11255–11269.
42. DEBAERE, F. *et al.* 2004. Changes in brain activation during the acquisition of a new bimanual coordination task. Neuropsychologia **42:** 855–867.
43. COHEN, D. & M.A. NICOLELIS. 2004. Reduction of single-neuron firing uncertainty by cortical ensembles during motor skill learning. J. Neurosci. **24:** 3574–3582.
44. KARGO, W.J. & D.A. NITZ. 2004. Improvements in the signal-to-noise ratio of motor cortex cells distinguish early versus late phases of motor skill learning. J. Neurosci. **24:** 5560–5569.
45. SEITZ, R.J. *et al.* 1990. Motor learning in man: a positron emission tomographic study. Neuroreport **1:** 57–60.
46. JENKINS, I.H. *et al.* 1994. Motor sequence learning: a study with positron emission tomography. J. Neurosci. **14:** 3775–3790.
47. GRAFTON, S.T., R.P. WOODS & T. MIKE. 1994. Functional imaging of procedural motor learning: relating cerebral blood flow with individual subject performance. Hum. Brain Map. **1:** 221–234.
48. DOYON, J. *et al.* 1996. Functional anatomy of visuomotor skill learning in human subjects examined with positron emission tomography. Eur. J. Neurosci. **8:** 637–648.
49. POLDRACK, R.A. *et al.* 1999. Striatal activation during acquisition of a cognitive skill. Neuropsychology **13:** 564–574.
50. UNGERLEIDER, L.G., J. DOYON & A. KARNI. 2002. Imaging brain plasticity during motor skill learning. Neurobiol. Learn Mem. **78:** 553–564.
51. HARRINGTON, D.L. *et al.* 1990. Procedural memory in Parkinson's disease: impaired motor but not visuoperceptual learning. J. Clin. Exp. Neuropsychol. **12:** 323–339.
52. HEINDEL, W.C. *et al.* 1989. Neuropsychological evidence for multiple implicit memory systems: a comparison of Alzheimer's, Huntington's, and Parkinson's disease patients. J. Neurosci. **9:** 582–587.
53. DANG, M.T. *et al.* 2006. Disrupted motor learning and long-term synaptic plasticity in mice lacking NMDAR1 in the striatum. Proc. Natl. Acad. Sci. USA **103:** 15254–15259.
54. PITTENGER, C. *et al.* 2006. Impaired bidirectional synaptic plasticity and procedural memory formation in striatum-specific cAMP response element-binding protein-deficient mice. J. Neurosci. **26:** 2808–2813.
55. HIKOSAKA, O. 1998. Neural systems for control of voluntary action—a hypothesis. Adv. Biophys. **35:** 81–102.
56. DOYA, K. 1999. What are the computations of the cerebellum, the basal ganglia and the cerebral cortex? Neural Netw. **12:** 961–974.
57. YIN, H.H. & B.J. KNOWLTON. 2004. Contributions of striatal subregions to place and response learning. Learn. Mem. **11:** 459–463.
58. YIN, H.H., B.J. KNOWLTON & B.W. BALLEINE. 2006. Inactivation of dorsolateral striatum enhances sensitivity to changes in the action-outcome contingency in instrumental conditioning. Behav. Brain Res. **166:** 189–196.
59. YIN, H.H. *et al.* 2005. The role of the dorsomedial striatum in instrumental conditioning. Eur. J. Neurosci. **22:** 513–523.
60. PRESCOTT, T.J. *et al.* 2006. A robot model of the basal ganglia: behavior and intrinsic processing. Neural Netw. **19:** 31–61.

61. ATALLAH, H.E. *et al.* 2007. Separate neural substrates for skill learning and performance in the ventral and dorsal striatum. Nat. Neurosci. **10:** 126–131.
62. BOGACZ, R. & K. GURNEY. 2007. The basal ganglia and cortex implement optimal decision making between alternative actions. Neural Comput. **19:** 442–477.
63. SAMEJIMA, K. & K. DOYA. 2007. Action value and action selection in the cortico-basal ganglia loop. Ann. N. Y. Acad. Sci.
64. YIN, H.H., B.J. KNOWLTON & B.W. BALLEINE. 2005. Blockade of NMDA receptors in the dorsomedial striatum prevents action-outcome learning in instrumental conditioning. Eur. J. Neurosci. **22:** 505–512.
65. BALLEINE, B.W. & S.B. OSTLUND. 2007. Still at the choice-point: action selection and initiation in instrumental conditioning. Ann. N. Y. Acad. Sci.
66. KARNI, A. *et al.* 1998. The acquisition of skilled motor performance: fast and slow experience-driven changes in primary motor cortex. Proc. Natl. Acad. Sci. USA **95:** 861–868.
67. GERFEN, C.R. 1992. The neostriatal mosaic: multiple levels of compartmental organization. J. Neural Transm. Suppl. **36:** 43–59.
68. RIOULT-PEDOTTI, M.S., D. FRIEDMAN & J.P. DONOGHUE. 2000. Learning-induced LTP in neocortex. Science **290:** 533–536.
69. KLEIM, J.A. *et al.* 2004. Cortical synaptogenesis and motor map reorganization occur during late, but not early, phase of motor skill learning. J. Neurosci. **24:** 628–633.
70. TANG, C. *et al.* 2007. Changes in activity of the striatum during formation of a motor habit. Eur. J. Neurosci. **25:** 1212–1227.
71. MIYACHI, S., O. HIKOSAKA & X. LU. 2002. Differential activation of monkey striatal neurons in the early and late stages of procedural learning. Exp. Brain Res. **146:** 122–126.
72. ADAMS, C. 1982. Variations in the sensitivity of instrumental responding to re-inforcer devaluation. Q. J. Exp. Psychol. Comp. Physiol. Psychol. **34B:** 77–98.
73. DICKINSON, A. 1985. Actions and habits: the development of behavioural auton-omy. Phil. Trans. Roy. Soc. Lond. B **308:** 67–78.
74. DICKINSON, A. & B. BALLEINE. 2002. The role of learning in the operation of motivational systems. In Steven's Handbook of Experimental Psychology. *In* Learning, Motivation, and Emotion, Vol. 3. H. Pashler & R. Gallistel, Eds.: 497–533. Wiley. New York.
75. DICKINSON, A. & B.W. BALLEINE. 1994. Motivational control of goal-directed action. Anim. Learn. Behav. **22:** 1–18.
76. COUTUREAU, E. & S. KILLCROSS. 2003. Inactivation of the infralimbic prefrontal cortex reinstates goal-directed responding in overtrained rats. Behav. Brain Res. **146:** 167–174.
77. YIN, H.H. & B.J. KNOWLTON. 2006. The role of the basal ganglia in habit forma-tion. Nat. Rev. Neurosci. **7:** 464–476.
78. CORBIT, L.H. & B.W. BALLEINE. 2003. The role of prelimbic cortex in instrumental conditioning. Behav. Brain Res. **146:** 145–157.
79. CORBIT, L.H., J.L. MUIR & B.W. BALLEINE. 2001. The role of the nucleus ac-cumbens in instrumental conditioning: evidence of a functional dissociation between accumbens core and shell. J. Neurosci. **21:** 3251–3260.
80. CORBIT, L.H., J.L. MUIR & B.W. BALLEINE. 2003. Lesions of mediodorsal thala-mus and anterior thalamic nuclei produce dissociable effects on instrumental conditioning in rats. Eur. J. Neurosci. **18:** 1286–1294.

81. CORBIT, L.H., S.B. OSTLUND & B.W. BALLEINE. 2002. Sensitivity to instrumental contingency degradation is mediated by the entorhinal cortex and its efferents via the dorsal hippocampus. J. Neurosci. **22:** 10976–10984.

82. OSTLUND, S.B. & B.W. BALLEINE. 2005. Lesions of medial prefrontal cortex disrupt the acquisition but not the expression of goal-directed learning. J. Neurosci. **25:** 7763–7770.

83. CHOI, W.Y., P.D. BALSAM & J.C. HORVITZ. 2005. Extended habit training reduces dopamine mediation of appetitive response expression. J. Neurosci. **25:** 6729–6733.

84. LJUNGBERG, T., P. APICELLA & W. SCHULTZ. 1992. Responses of monkey dopamine neurons during learning of behavioral reactions. J. Neurophysiol. **67:** 145–163.

85. CARLSSON, A. 1972. Biochemical and pharmacological aspects of Parkinsonism. Acta Neurol. Scand. Suppl. **51:** 11–42.

86. FREUND, H.J. 2002. Mechanisms of voluntary movements. Parkinsonism Relat. Disord. **9:** 55–59.

87. FRANK, M.J., L.C. SEEBERGER & C. O'REILLY R. 2004. By carrot or by stick: cognitive reinforcement learning in parkinsonism. Science **306:** 1940–1943.

88. AARSLAND, D., G. ALVES & J.P. LARSEN. 2005. Disorders of motivation, sexual conduct, and sleep in Parkinson's disease. Adv. Neurol. **96:** 56–64.

89. NIEOULLON, A. 2002. Dopamine and the regulation of cognition and attention. Prog. Neurobiol. **67:** 53–83.

90. ROBBINS, T.W. *et al.* 1990. Effects of dopamine depletion from the caudate-putamen and nucleus accumbens septi on the acquisition and performance of a conditional discrimination task. Behav. Brain Res. **38:** 243–261.

91. PACKARD, M.G. & N.M. WHITE. 1991. Dissociation of hippocampus and caudate nucleus memory systems by posttraining intracerebral injection of dopamine agonists. Behav. Neurosci. **105:** 295–306.

92. NELSON, A. & S. KILLCROSS. 2006. Amphetamine exposure enhances habit formation. J. Neurosci. **26:** 3805–3812.

93. SCHULTZ, W. 1998. Predictive reward signal of dopamine neurons. J. Neurophysiol. **80:** 1–27.

94. SCHULTZ, W., P. DAYAN & P.R. MONTAGUE. 1997. A neural substrate of prediction and reward. Science **275:** 1593–1599.

95. SURI, R.E. & W. SCHULTZ. 1998. Learning of sequential movements by neural network model with dopamine-like reinforcement signal. Exp. Brain Res. **121:** 350–354.

96. WAELTI, P., A. DICKINSON & W. SCHULTZ. 2001. Dopamine responses comply with basic assumptions of formal learning theory. Nature **412:** 43–48.

97. FIORILLO, C.D., P.N. TOBLER & W. SCHULTZ. 2003. Discrete coding of reward probability and uncertainty by dopamine neurons. Science **299:** 1898–1902.

98. TOBLER, P.N., C.D. FIORILLO & W. SCHULTZ. 2005. Adaptive coding of reward value by dopamine neurons. Science **307:** 1642–1645.

99. CANNON, C.M. & R.D. PALMITER. 2003. Reward without dopamine. J. Neurosci. **23:** 10827–10831.

100. PECINA, S. *et al.* 2003. Hyperdopaminergic mutant mice have higher "wanting" but not "liking" for sweet rewards. J. Neurosci. **23:** 9395–9402.

101. COSTA, R.M. *et al.* 2006. Dopamine levels modulate the updating of tastant values. Genes Brain Behav.

102. FRANK, M.J. 2005. Dynamic dopamine modulation in the basal ganglia: a neurocomputational account of cognitive deficits in medicated and nonmedicated Parkinsonism. J. Cogn. Neurosci. **17:** 51–72.

103. HNASKO, T.S., B.N. SOTAK & R.D. PALMITER. 2005. Morphine reward in dopamine-deficient mice. Nature **438:** 854–857.

104. CAGNIARD, B. *et al.* 2006. Mice with chronically elevated dopamine exhibit enhanced motivation, but not learning, for a food reward. Neuropsychopharmacology **31:** 1362–1370.

105. YIN, H.H., X. ZHUANG & B.W. BALLEINE. 2006. Instrumental learning in hyperdopaminergic mice. Neurobiol. Learn. Mem. **85:** 283–288.

106. KOBAYASHI, Y. & K. OKADA. 2007. Reward prediction error computation in the pedunculopontine tegmental nucleus neurons. Ann. N. Y. Acad. Sci.

107. HIKOSAKA, O. & M. MATSUMOTO. 2006. Role of the primate habenula in reward processing. II. Inhibitory effect on dopamine neurons. Soc. Neurosci. Abstracts.

108. MORRIS, G. *et al.* 2004. Coincident but distinct messages of midbrain dopamine and striatal tonically active neurons. Neuron **43:** 133–143.

109. MATSUMOTO, M. & O. HIKOSAKA. 2006. Role of the primate habenula in reward processing. I. Prediction of negative reward value. Soc. Neurosci. Abstracts.

110. ROBINSON, S. *et al.* 2005. Distinguishing whether dopamine regulates liking, wanting, and/or learning about rewards. Behav. Neurosci. **119:** 5–15.

111. BERRIDGE, K.C. 2007. The debate over dopamine's role in reward: the case for incentive salience. Psychopharmacology (Berl.) **191:** 391–431.

112. CAGNIARD, B. *et al.* 2006. Dopamine scales performance in the absence of new learning. Neuron **51:** 541–547.

113. NIV, Y. 2007. Cost, benefit, tonic, phasic: what do response rates tell us about dopamine and motivation? Ann. N. Y. Acad. Sci.

114. NIV, Y. *et al.* 2007. Tonic dopamine: opportunity costs and the control of response vigor. Psychopharmacology (Berl.) **191:** 507–520.

115. DICKINSON, A., J. SMITH & J. MIRENOWICZ. 2000. Dissociation of Pavlovian and instrumental incentive learning under dopamine antagonists. Behav. Neurosci. **114:** 468–483.

116. PESSIGLIONE, M. *et al.* 2006. Dopamine-dependent prediction errors underpin reward-seeking behaviour in humans. Nature **442:** 1042–1045.

117. MORRIS, G. *et al.* 2006. Midbrain dopamine neurons encode decisions for future action. Nat. Neurosci. **9:** 1057–1063.

118. ARBUTHNOTT, G.W. & J. WICKENS. 2007. Space, time and dopamine. Trends Neurosci. **30:** 62–69.

119. PARTRIDGE, J.G., K.C. TANG & D.M. LOVINGER. 2000. Regional and postnatal heterogeneity of activity-dependent long-term changes in synaptic efficacy in the dorsal striatum. J. Neurophysiol. **84:** 1422–1429.

120. GOLDBERG, J.A. *et al.* 2002. Enhanced synchrony among primary motor cortex neurons in the 1-methyl-4-phenyl-1,2,3,6-tetrahydropyridine primate model of Parkinson's disease. J. Neurosci. **22:** 4639–4653.

121. GOLDBERG, J.A. *et al.* 2004. Spike synchronization in the cortex/basal-ganglia networks of Parkinsonian primates reflects global dynamics of the local field potentials. J. Neurosci. **24:** 6003–6010.

122. RAZ, A. *et al.* 2001. Activity of pallidal and striatal tonically active neurons is correlated in MPTP-treated monkeys but not in normal monkeys. J. Neurosci. **21:** RC128.

123. KAUR, S., R. LAZAR & R. METHERATE. 2004. Intracortical pathways determine breadth of subthreshold frequency receptive fields in primary auditory cortex. J. Neurophysiol. **91:** 2551–2567.
124. LOPES DA SILVA, F. 1991. Neural mechanisms underlying brain waves: from neural membranes to networks. Electroencephalogr. Clin. Neurophysiol. **79:** 81–93.
125. NICOLA, S.M., J. SURMEIER & R.C. MALENKA. 2000. Dopaminergic modulation of neuronal excitability in the striatum and nucleus accumbens. Annu. Rev. Neurosci. **23:** 185–215.
126. HERNANDEZ-LOPEZ, S. *et al.* 1997. D1 receptor activation enhances evoked discharge in neostriatal medium spiny neurons by modulating an L-type Ca2+ conductance. J. Neurosci. **17:** 3334–3342.
127. HERNANDEZ-LOPEZ, S. *et al.* 2000. D2 dopamine receptors in striatal medium spiny neurons reduce L-type Ca2+ currents and excitability via a novel PLC[beta]1-IP3-calcineurin-signaling cascade. J. Neurosci. **20:** 8987–8995.
128. DAY, M. *et al.* 2006. Selective elimination of glutamatergic synapses on striatopallidal neurons in Parkinson disease models. Nat. Neurosci. **9:** 251–259.
129. RAZ, A. *et al.* 1996. Neuronal synchronization of tonically active neurons in the striatum of normal and parkinsonian primates. J. Neurophysiol. **76:** 2083–2088.
130. DING, J. *et al.* 2006. RGS4-dependent attenuation of M4 autoreceptor function in striatal cholinergic interneurons following dopamine depletion. Nat. Neurosci. **9:** 832–842.
131. MAURICE, N. *et al.* 2004. D2 dopamine receptor-mediated modulation of voltage-dependent Na+ channels reduces autonomous activity in striatal cholinergic interneurons. J. Neurosci. **24:** 10289–10301.
132. MALLET, N. *et al.* 2006. Cortical inputs and GABA interneurons imbalance projection neurons in the striatum of parkinsonian rats. J. Neurosci. **26:** 3875–3884.
133. MALLET, N. *et al.* 2005. Feedforward inhibition of projection neurons by fast-spiking GABA interneurons in the rat striatum *in vivo*. J. Neurosci. **25:** 3857–3869.
134. SURMEIER, D.J. & S.T. KITAI. 1997. State-dependent regulation of neuronal excitability by dopamine. Nihon Shinkei Seishin Yakurigaku Zasshi **17:** 105–110.
135. KHALDY, H. *et al.* 2002. Circadian rhythms of dopamine and dihydroxyphenyl acetic acid in the mouse striatum: effects of pinealectomy and of melatonin treatment. Neuroendocrinology **75:** 201–208.
136. SILBERSTEIN, P. *et al.* 2003. Patterning of globus pallidus local field potentials differs between Parkinson's disease and dystonia. Brain **126:** 2597–2608.
137. BROWN, P. *et al.* 2001. Dopamine dependency of oscillations between subthalamic nucleus and pallidum in Parkinson's disease. J. Neurosci. **21:** 1033–1038.
138. LEVY, R. *et al.* 2002. Dependence of subthalamic nucleus oscillations on movement and dopamine in Parkinson's disease. Brain **125:** 1196–1209.
139. WILLIAMS, D. *et al.* 2002. Dopamine-dependent changes in the functional connectivity between basal ganglia and cerebral cortex in humans. Brain **125:** 1558–1569.
140. CASSIDY, M. *et al.* 2002. Movement-related changes in synchronization in the human basal ganglia. Brain **125:** 1235–1246.
141. SILBERSTEIN, P. *et al.* 2005. Cortico-cortical coupling in Parkinson's disease and its modulation by therapy. Brain **128:** 1277–1291.
142. BAKER, S.N. *et al.* 1999. The role of synchrony and oscillations in the motor output. Exp. Brain Res. **128:** 109–117.

143. MASIMORE, B., J. KAKALIOS & A.D. REDISH. 2004. Measuring fundamental frequencies in local field potentials. J. Neurosci. Methods **138:** 97–105.
144. MASIMORE, B. *et al.* 2005. Transient striatal gamma local field potentials signal movement initiation in rats. Neuroreport **16:** 2021–2024.
145. FRIES, P. *et al.* 2001. Modulation of oscillatory neuronal synchronization by selective visual attention. Science **291:** 1560–1563.
146. GERVASONI, D. *et al.* 2004. Global forebrain dynamics predict rat behavioral states and their transitions. J. Neurosci. **24:** 11137–11147.
147. DZIRASA, K. *et al.* 2006. Dopaminergic control of sleep-wake states. J. Neurosci. **26:** 10577–10589.
148. RAGER, G. & W. SINGER. 1998. The response of cat visual cortex to flicker stimuli of variable frequency. Eur. J. Neurosci. **10:** 1856–1877.

Striatal Contributions to Reward and Decision Making

Making Sense of Regional Variations in a Reiterated Processing Matrix

JEFFERY R. WICKENS,[a,b] CHRISTOPHER S. BUDD,[b] BRIAN I. HYLAND,[c] AND GORDON W. ARBUTHNOTT[a,b]

[a]Neurobiology Research Unit, Okinawa Institute of Science and Technology, Okinawa, Japan

[b]Department of Anatomy and Structural Biology, University of Otago, Dunedin, New Zealand

[c]Department of Physiology, University of Otago, Dunedin, New Zealand

ABSTRACT: The striatum is the major input nucleus of the basal ganglia. It is thought to play a key role in learning on the basis of positive reinforcement and in action selection. One view of the striatum conceives it as comprising a reiterated matrix of processing units that perform common operations in different striatal regions, namely synaptic plasticity according to a three-factor rule, and lateral inhibition. These operations are required for reinforcement learning and selection of previously reinforced actions. Analysis of the behavioral effects of circumscribed lesions of the striatum, however, suggests regional specialization of learning and decision-making operations. We consider how a basic processing unit may be modified by regional variations in neurochemical parameters, for example, by the gradient in density of dopamine terminals from dorsal to ventral striatum. These variations suggest subtle differences between dorsolateral and ventromedial striatal regions in the temporal properties of dopamine signaling, which are superimposed on regional differences in connectivity. We propose that these variations make sense in relation to the temporal structure of activity in striatal inputs from different regions, and the requirements of different learning operations. Dorsolateral striatal (DLS) regions may be subject to brief, precisely timed pulses of dopamine, whereas ventromedial striatal regions integrate dopamine signals over a longer time course. These differences may be important for understanding regional variations in the contribution to reinforcement of habits, versus incentive processes that are sensitive to the value of expected rewards.

Address for correspondence: Jeffery R. Wickens, Ph.D., Principal Investigator, Neurobiology Research Unit, Okinawa Institute of Science and Technology, 12-22 Suzaki, Uruma City, Okinawa, Japan. Voice: +81-98-921-4097; fax: +81-98-921-4435.

wickens@oist.jp

Ann. N.Y. Acad. Sci. 1104: 192–212 (2007). © 2007 New York Academy of Sciences.
doi: 10.1196/annals.1390.016

KEYWORDS: striatum; reinforcement; dopamine; nucleus accumbens; learning

INTRODUCTION

The striatum is the major input nucleus of the basal ganglia, a site of convergence of glutamatergic inputs from almost every area of the cerebral cortex, and the major target of dopaminergic inputs from the midbrain. The striatum exerts a controlling influence over the cerebral cortex via a series of reentrant corticobasal ganglia circuits. A key issue for behavioral neurobiology is to elucidate the operations performed by the striatum on the inputs it receives from the cerebral cortex, and how these are modulated by the dopaminergic signals from the midbrain. We include in the striatum not only the dorsal region, which encompasses the caudate nucleus and putamen, but also the ventral region that includes the core and shell of the nucleus accumbens.[1]

Although segregation of the striatum into distinct dorsal and ventral zones has been proposed on functional grounds, there is no sharp line of demarcation between these regions anatomically. Once considered a motor center, the striatum is now recognized as playing a wider role in the brain's decision-making processes, extending to reward processing and motivation. This broadening of focus reflects several comparatively recent advances. These include recognition that the midbrain dopamine neurons are a crucial part of the brain's reward-signaling circuitry, that the neurons of the striatum are involved in associations of sensory stimuli with actions, and that circumscribed lesions of the striatum have highly specific effects on reinforcement and decision-making processes.

Two apparently contradictory principles of organization seem to operate in the striatum. Evidence of heterogeneity of striatal function seems to contradict striking uniformity in crucial aspects of cytology, local circuitry, and cellular physiology. For example, neurons modulated by reward and encoding action selections can be found throughout the striatum.[2–4] Similarly, at the local network level the striatum is characterized by a particular and repeating microcircuitry involving predominantly a single cell type with relatively constant physiological properties and interconnections.

Uniformity in cytology, local circuitry, and cellular physiology suggests circuit modules reiterated across the entire striatum, which perform a fundamental and common input–output operation.[5] Functional differentiation in such a system might be achieved by specific input and output connections of different modules. In addition, we propose that differences in dopamine terminal density and uptake capacity result in gradations in the time course of the dopamine signal, which is an additional important factor in determining how functional diversity across striatal regions arises from circuit uniformity.

THE BASIC INFORMATION PROCESSING OPERATIONS
OF THE STRIATUM

Based on neurobiological considerations, the basic canonical circuit of the striatum has been described by many authors, and has recently been reviewed.[6] The uniformity of this canonical circuit in different striatal regions is compatible with the idea of a reiterated processing unit, a phrase coined by Gerfen.[7] In brief, this unit is proposed to have two parts: a mechanism for reinforcement learning at the cellular level, and a mechanism for decision making at the network level.

We propose that the reinforcement learning mechanism is applied to the striatal inputs that project from almost every cortical region.[8] These cortical inputs provide the main source of excitation to the principal neurons of the striatum, the spiny projection neurons, and are necessary to make them fire.[9] Dopamine cells in the midbrain also send a major projection to the striatum, and the activity of these inputs is associated with reinforcing events that occur at the behavioral level during learning. Convergence of dopamine inputs from the midbrain with the glutamate inputs from the cortex occurs at the spines of the striatal output neurons.[10] At this site, dopamine appears to be involved in strengthening synaptic connections by a heterosynaptic effect on the corticostriatal synapses.[11,12] Together these properties provide a locus for a cellular mechanism for reinforcement according to a three-term contingency.[13]

The striatal substrate for decision making is assumed to be the network of spiny projection neurons. The spiny neurons are GABAergic projection neurons, which also make inhibitory synaptic contacts with other spiny neurons.[14,15] They may be considered to form a lateral inhibition type neural network. In such a network the spiny neurons that receive the most excitation suppress activity in less strongly excited spiny cells, and therefore control the outputs.[16] Although lateral inhibition in the striatum has been questioned,[17,18] many laboratories now report functional inhibitory interactions between spiny projection neurons.[19–22]

Combining the cellular mechanism for reinforcement with competitive interactions produces the proposed fundamental processing operation. In this mechanism, the strength of the excitatory synaptic inputs represents the history of dopamine-dependent plasticity, and thus the weighted sum of reinforcement associated with the activity of the particular synaptic connection. The activity of the striatal cells in a given context of cortical input thus represents the expected values of the alternative outputs. The competitive interactions are envisaged to select the outputs with the strongest inputs. This circuit may thus act to maximize expected rewards, in that the action chosen would be that represented by the spiny cell with the strongest excitatory inputs, reflecting the history of reinforcement of the cortical inputs to that cell.[12]

Remarkably, the important anatomical and biochemical elements of this mechanism are present throughout all regions of the striatum. The cytology of the striatum suggests similarity in the local circuits of the dorsal and ventral striatum, as well as medial and lateral parts. Projection neurons of the nucleus accumbens have a similar appearance to spiny projection neurons of the dorsal striatum.[23,24] Subtle differences in the degree of spinyness of projection neurons in different accumbal areas have been reported,[25,26] but the differences are quantitative rather than qualitative. At the cellular level there are few differences in receptor distribution between the accumbens and the dorsal striatum.[24] Similarly, very few differences in electrophysiological properties have been described. For example, the up/down states first described in dorsal striatum[9,27] have also been described in ventral striatum.[28]

The convergence of cortical inputs with dopamine inputs, which is crucial for the reinforcement learning mechanism, is also very similar in different striatal regions. Like in the dorsal striatum,[10] dopamine inputs to spiny neurons of the nucleus accumbens converge with hippocampal inputs[29] often converging on the same dendritic spine,[30] with similar patterns for prefrontal[31] and amygdalar[32] inputs to the nucleus accumbens. Also, at the local circuit level, lateral inhibitory interactions between the spiny projection neurons have been described in both dorsal[19,21,33] and ventral[20] striatal areas, where they have similar properties.

Since dopamine modulation of plasticity and lateral inhibition are regarded as the key properties underlying the fundamental information processing operations of the striatal units, their presence in all regions argues that in terms of these properties there is regional homogeneity. If as proposed this circuit does represent a central organizing principle of the striatal network, then we must acknowledge that it is repeated throughout the extent of the striatum. On the other hand, if this is what the striatum does, then it is necessary to explain the evidence for regional variation in the behavioral functions of the striatum, which we review below.

REGIONAL SPECIFICITY OF LEARNING IN THE STRIATUM

Current behavioral studies are leading to a new map of striatal function. With modern behavioral techniques distinct changes in behavioral capacities have been identified after circumscribed lesions. Earlier behavioral studies examining the effects of dorsal versus ventral lesions of the striatum led to the consensus that the ventral striatum subserves reward-related behavior and reinforcement processes,[34,35] with the dorsal striatum responsible for more habitual sensorimotor functions.[36] More recently, it has become evident that learning is affected in different ways by lesions of the dorsolateral versus dorsomedial striatum (DMS).

Since Konorski[37] and Divac *et al.*[38] first identified the importance of the striatum in instrumental learning, a number of behavioral studies in rodents have involved large lesions of the dorsal striatum. Potential medial to lateral gradients in anatomical and functional properties may be overlooked when there are large lesions. A lateral bias of damage in many of these studies has propagated the idea that the dorsal striatum is only involved with stimulus–response learning.[39,40] In contrast, because of its role in the reinforcing effects of psychostimulants and natural rewards,[34,35] the ventral striatum is often assumed to be involved in more complex learning processes.

The ventral striatum does appear to mediate the ability of rewards and reward-associated cues to affect appetitive behavior.[34,41] Recently, however, more localized dorsal lesions have shown that the DMS is responsible for associating specific actions with rewarding outcomes.[39,42,43]

Dorsolateral Striatum (DLS) in Habit Formation

As training progresses, interval schedules normally promote a rapid progression from goal-directed to habitual responding.[44,45] Goal-directed action is operationally defined by the sensitivity of responses to the outcome. Yin et al.[43] showed that the progression from goal-directed to habitual responding was disrupted by lesions of the DLS. The DLS is the rat equivalent of the putamen and is roughly equivalent to the area termed the sensorimotor striatum. The DLS-lesioned animals remained sensitive to devaluation of the outcome throughout training. In contrast, the experiment produced habitual responding in rats with lesions to the DMS and controls. This rendered them insensitive to the devaluation. The same results were obtained using a LiCl devaluation paradigm[43] or an omission procedure[46] to test for goal-directed responding.

Such evidence suggests that the DLS, but not the DMS, is responsible for the stimulus–response (S–R) associations, which are thought to underlie habitual responding. Further support that the DLS is critical in habitual responding comes from evidence that dopamine (DA) innervation of the sensorimotor striatum is necessary for habit formation in instrumental conditioning[47] and cue-controlled drug seeking.[48]

DMS in Goal-Directed Action

The rat equivalent of the caudate nucleus, the DMS, or associative striatum, on the other hand, appears crucial for the learning and expression of goal-directed actions.[42,46,49,50] Pre- and post-training lesions of the posterior DMS (pDMS), but not the anterior DMS, were equally effective in reducing the sensitivity of the animal to devaluation and contingency degradation treatments.[42] A related experiment showed that NMDA receptors are involved in encoding action–outcome associations in instrumental conditioning, an effect that is independent of performance.[50]

The findings above support the hypothesis that the DLS and the pDMS can be functionally dissociated. Early clues for such a dissociation came from the "place and response" literature.[39] In a situation in which food is available in a fixed location, place strategies endow an animal with a flexible representation of an environment. This allows them to find the food regardless of their start point. In contrast, response strategies rely on the recall of an inflexible sequence of turns from the start point. Response strategies are believed to be supported by S–R associations in the dorsal striatum.[40] Yin and Knowlton[49] found evidence that the pDMS plays a role in the spatially guided behavior of place strategies, leading to the possible interpretation that the pDMS may form a functional circuit with the hippocampus.[39] Supported by recent imaging studies of the caudate in humans[51] these findings all suggest that the DMS is involved in the selection of actions that lead to reward.

Nucleus Accumbens/Ventral Striatum

Lesions of the accumbens reduce instrumental performance,[52] as does selective blockade with glutamate antagonists[53] or dopamine antagonists.[54] However, the accumbens does not appear necessary for goal-directed action. Lesions of the whole accumbens do not affect sensitivity to the value of reward or to the action–outcome contingency.[41,52] Animals with lesions of either the nucleus accumbens shell (AcbS) or core (AcbC) subcomponents[55] maintain the ability to reduce performance when the reward is devalued. However, AcbC-lesioned animals reduce performance in a manner that is not selective for the particular reward that has been devalued. In an experiment by Corbit *et al.*[56] two separate rewards were associated with two different actions and then one reward was devalued. While AcbS-lesioned animals only reduced the execution of the devalued action, animals with AcbC lesions produced a decrement in the performance of both actions. This is consistent with the shell serving to potentiate behavior directed toward stimuli associated with reward (conditioned stimuli; CS). This potentiation appears to be directed to a particular stimulus by the action of the core.[57]

To understand the contribution of the ventral striatum to behavior, it is important to distinguish between two concepts. Incentive motivation is the process by which Pavlovian cues stimulate the vigor of ongoing behavior. It is also termed incentive salience, wanting,[58] or Pavlovian incentive learning (PIL).[59] Alternatively, instrumental incentive learning (IIL) is the process by which goals are assigned value, based on the prior experience of the reward in a relevant state of deprivation. PIL can be dissociated from IIL using DA antagonists, as only PIL is sensitive to dopamine manipulations.[59] Also, in a sequence of actions that lead to a reward (heterogeneous instrumental chain), PIL has a greater influence on actions more proximal to the delivery of reward.[60] Dopamine in the ventral striatum seems to mediate how stimuli that are associated with reward, and that are close to the delivery of reward, invigorate behavior. The ventral striatum and ventral striatal DA do not appear to be involved in attaching incentive value to the mental representation of a rewarding outcome.[41,61]

The AcbS and AcbC both contribute to the Pavlovian processes that influence instrumental behavior. The interaction between the two areas is complex and continues to be investigated. Stimuli that are paired with reward (CS) acquire the ability to activate a central motivational process[62,63] that can enhance approach behaviors and subsequent responses to the CS. Core lesions impair such conditioned approach (autoshaping),[64] but this behavior is unaffected by shell lesions.[57] Neither the core nor the shell appears to be essential for conditioned reinforcement, where a stimulus becomes reinforcing because of pairings with a reward.[57]

The presentation of a CS that has been paired with reward can enhance the ongoing performance of an independently acquired instrumental response. This process is termed Pavlovian to instrumental transfer (PIT).[65] If multiple reinforcers are presented contingently upon the execution of different actions, response-specific PIT can also be observed.[66] Response-specific PIT is a selective potentiation of an instrumental action, which results in the same reward as was paired with the CS. Interestingly, Holland[67] found the magnitude of PIT is unaffected by devaluation of the outcome associated with the CS. When a single reinforcer was used, extended training increased PIT, but reduced sensitivity to devaluation effects. In comparison, when multiple reinforcers were used, the amount of training had little effect on either PIT or devaluation.[67] It remains to be investigated whether such differential sensitivity of these two measures, observed with extended training, maps on to the neural dissociation between action–outcome (A–O) and (S–R).

PIT appears to engage the circuitry that is artificially activated by the selective action of psychostimulants on conditioned reinforcers.[34] Taylor and Robbins[68] found that dopaminergic lesions of the accumbens, but not the caudate nucleus, attenuated the enhanced responding to reward-related stimuli caused by intra-Acb amphetamine. Shell lesions stop this enhanced responding altogether, whereas core lesions stop any enhancement from being specific to the CS or conditioned reinforcer.[57] A disruption in conditioned reinforcement has been put forward as a possible explanation for why core lesions prevent the ability to choose a large delayed reinforcer, over a smaller immediate reinforcer.[69,70]

ANATOMICAL BASIS FOR STRIATAL SUBDIVISIONS AND HETEROGENEITY

The specific behavioral effects of anatomically circumscribed lesions of the striatum suggest regional specialization of function, but it is unclear whether differences in anatomy and physiology underlie the different operations involved in these behavioral functions. Anatomical considerations have led to many different ways to divide the striatum into different areas. These include: classical architectonics (Brockhaus[71] describes seven different areas in the caudate), neurochemical specificity,[7,72–75] cortical input,[76–78] and functional criteria.[39,79–82] However, none of these divisions imply a qualitative difference in the circuitry underlying the reiterated processing unit defined in the "Introduction" section, nor is it understood how any such differences could lead to a qualitatively different algorithm for reinforcement and decision-making operations. We argue instead that graded differences in parameters affecting dopamine signaling may play a key role in the functional specialization of different areas.

Heimer and Wilson[76] proposed a similar and parallel organization of dorsal and ventral striatal connectivity. This concept of a unified striatal complex with dorsal and ventral subdivisions has advanced understanding of basal ganglia function enormously. Anatomically a sharp distinction between the two areas is unfounded. Despite the greater complexity of the ventral striatum's structural and chemical architecture,[83–85] such measures fail to identify a definite point of transition to the dorsal region. Instead, a mediolateral zonation appears to determine the function of the largely equivalent striatal neuronal circuitry. This zonation is imposed on the structure by its excitatory cortical, thalamic, and amygdaloid inputs.[83] Under this scheme the shell-core division of the Acb can be seen as a continuation of a diagonal ventromedial–dorsolateral anatomofunctional gradient, imposed by the input connectivity of the striatum.

The major explosion in subdivision of the striatum was a continuation of the chemical neuroanatomy that had started the exercise. As antibody techniques developed the striatum could be seen to be subdivided again into compartments. In cat brain Malach and Graybiel[86] talks about striosomes while in the rat Gerfen *et al.*[73] and Herkenham and Pert[87] talk about "patches." These divisions are important in that they coincide with different inputs in some cases[88,89] and the "limbic" inputs to the striatum, although mainly directed to the accumbens, may expand into striosomes.[74,90] They also have a specific subset of nigral cells as their input,[91,92] although these are not within the ventral tegmental area (VTA), which is the main source of accumbens dopamine.

The innervation of the striatum by glutamatergic inputs from "limbic" cortical areas (hippocampus, amygdala, medial prefrontal cortex) coincides with inputs from more medial dopamine cell groups.[93] The exact boundaries depend a little upon the limits of the "limbic" cortex but the inputs from hippocampus, amygdala, lateral, and medial prefrontal areas are mainly restricted to the accumbens and the most medial edge of the striatum along the ventricle. In cat at least and probably monkey too the striosomes are also innervated from these "limbic" cortical areas although there has been some controversy in the rat where the patches seem to be innervated from different cortical layers rather than different cortical areas.[89] It is clear, however, that there is little if any innervation of striosomes from sensory cortical areas where there is a clear division between layer III and layer V.[94]

HOMOGENEITY OF DOPAMINE CELL ACTIVITY IN RELATION TO BEHAVIOR

One anatomical division that has been linked to functional differences between dorsal and ventral striatum has been a differential source of dopaminergic input to the two areas. Historically, the Swedish groups looking at the fluorescence of monoamines against a dark background saw the whole "nigrostriatal system" as one. Cell bodies formed a butterfly-like shape in the tegmentum with wings of cells lying along the "zona compacta" of the substantia nigra and the "body" over the midline in an area that they called the VTA of Tsai.[95] The midbrain DA cell fields defined by tyrosine hydroxylase immunohistochemistry form groups labeled A8, A9, and A10, corresponding roughly to the boundaries of the retrorubral field, substantia nigra (mainly pars compacta [SNc], but with some in reticulata), and VTA (including several subnuclei), respectively.[96] The terminal areas are continuous although they include several named nuclei. Despite the anatomical continuity in the cell fields, it has become standard to differentiate medial and lateral groups, on the assumption that these are functionally different. Thus the substantia nigra dopamine neurons are held to largely project to the DLS, and to be more involved in motor control, while VTA neurons project more ventromedially and to be more involved in cognitive or affective function.[97,98] To what extent do the physiological data on dopamine cell activity in relation to behavior provide support for this anatomically inspired division?

Over the past 20 years there has been a steady growth in the number of studies in which the behavioral characteristics of neurons in the SNc and VTA have been examined. Unfortunately, data relating to possible differences in the behavioral associations of neurons in the different areas are extremely sparse. In studies that have examined both SNc and VTA in the same paradigm, sample sizes for VTA are often small, making fair comparison difficult. This possibly relates to the difficulty accessing neuronal groups close to the midline, and the fact that while VTA contains many DA neurons, they form a smaller proportion of cells in the region.[99]

One implication of the proposed functional division is that DA cells of A9, in the SNc, might be more likely to exhibit "movement-related" activity. In one early study,[100] cells with activity considered "movement-related" were found in A8, A9, and A10. There was no significant difference in the proportion of DA cells showing this kind of activity between the regions. On the other hand, division of the recording area arbitrarily into three mediolateral zones did reveal a significant difference in proportions, with the lateral zone having the highest (68%) and the medial zone the lowest (30%) proportion of cells with such activity.[100] However, the movement-related modulations of DA cells has attracted little attention in subsequent studies due to their modest amplitude and rather tonic nature. The impressive phasic modulations in activity that are seen in other motor control regions are not seen in the DA neurons.

The converse approach to the question of possible functional differentiation between DA cells in VTA, and SNc, relates to the possibility that there may be differences in responsiveness to reward or reward-related cues. It might, for example, be hypothesized that VTA DA cells would be more influenced by such cues, than the presumed "motoric" SNc.

Evidence relating to this possibility comes from a series of papers from the Schultz group in which A10 cells were sampled (although the SNc was the main stereotaxic target in these experiments) during performance of various rewarded tasks. The same study noted above[100] found a converse relation for responses to CS, with the lateral zone having a significantly lower proportion of responding cells. Once again, however, there was no significant difference when analyzed by traditional nuclear boundaries.

A later study by this group[101] recorded larger numbers of cells (A8, 9, 10 $n = 38, 162, 39$, respectively) and did find a significant regional variation when analyzed by traditional boundaries. During the learning phases of three operant tasks, a significantly higher proportion of A10 (about 50%) than A8 (10%) or A9 (20%) DA cells responded to reward delivery. Furthermore, the average magnitude of the reward response measured from each neuron was greater in the A10 group. After learning of the task, proportions of (predicted) reward-responsive neurons decreased in all groups, but by more in A10, such that once the tasks were well learned, similar low proportions (around 9%) of neurons responded; and the amplitudes of responses were also no longer significantly different. Similarly, in these experiments, a significantly higher proportion of A10 cells ($\approx$90%) than A8 or A9 cells (both around 50%) responded to trigger or instruction stimuli that indicated the task requirements (which thus predicted reward if the monkey performed the task correctly). Perhaps surprisingly, the proportions of these conditioned responses fell after training, again such that differences in proportions were not significant (around 40–60% across the regions). For these stimuli there were never any differences in the amplitudes of the individual cell responses. These data show that all regions contain DA neurons that respond to reward and reward-predictive cues, but raise the possibility that VTA neurons may be more responsive to such stimuli early in learning, which could relate, for instance, to a greater sensitivity to novelty.

However, in contradistinction to this result, several other studies in which reasonable numbers of cells were recorded from each region found no significant differences in the distribution of cells showing responses to reward-related stimuli, in a range of contexts and task designs. Ljungberg *et al.*[102] reported that A8, A9, and A10 DA neurons respond to a trigger light and to reward in a delayed alternation task, but found no significant difference in proportions of responsive cells across the regions. Similarly, another study by the same group[103] found no significant difference in proportions of neurons responding to these stimuli in a simple reaction time task, whether analyzed according to nuclear boundaries or mediolateral zones.

Waelti et al.[104] recorded responses of 201 and 77 DA cells in A9 and A10, respectively and a smaller number of A8 cells, to compound stimuli-predicting reward in a study investigating whether DA cell activity conforms to classical theory with regard to prediction error signaling, using a "blocking" paradigm. Neurons in all regions responded to compound stimuli-predicting reward, as did those in another paper[105] showing that DA cell activity is consistent with the reward prediction error signal, although no formal statistical analysis was reported.

In another study very few DA cells were found to respond to a mild aversive stimulus, and the proportions doing so were similar in VTA and SNc.[106] On the other hand, there was some suggestion in this study that the A8 region may have an overrepresentation of such responses, relative to the other two. Several other studies from the Schultz group[107–110] and one from another laboratory,[111] have examined reward responsiveness of cells in VTA and SNC. In all these studies data from VTA and dSNc were pooled, implying that the responses were similar.

Unfortunately, studies from other species are able to contribute very little to this debate, as there have been few studies combining examination of responses to reward-related stimuli with a survey of different DA cell fields. Our own studies in the rat, for example, did not yield sufficient-sized groups of cells that could be clearly allocated to VTA and SNc.[112–114] Miller et al.[115] reported responses to reward-predicting stimuli from both nigral and tegmental dopamine cell populations. Tegmental cells seemed overall less responsive (in that most of the unaffected cells found were from this region), but the sample size was small and the analysis restricted to only three or four trials.

Thus overall the impression from the available literature is that while there may be subtle differences in proportion of presumed DA cells showing reward-related activity, this is insufficient to suggest that neuronal populations in these regions exhibit clearly demarcated functional relationships during behavior that would indicate significant differences in the functional associations of their afferent inputs. Therefore, it appears that the regions of striatum receiving inputs from different DA cell groups are in fact receiving a rather similar pattern of dopamine input, at least that component determined by action potential traffic in the DA cell axons. Functional differences in those target regions, insofar as they exist, may thus more relate to how the action potential firing of the DA cell axons is transformed into the neurochemical signal in the striatum, by release, reuptake, and diffusion, and by the postsynaptic neurons.

GRADIENTS IN DOPAMINE SIGNALING MAY INFLUENCE
TEMPORAL PROPERTIES

Although there are strong qualitative similarities in the behavioral associations of dopaminergic inputs to different striatal areas, heterogeneity in the density of the dopamine projection has been recognized for some time. These differences can be quantified in terms of the density of dopamine varicosities. In the DLS[116] the number of DA varicosities per mm^3 is 1.1×10^8 compared to the ventromedial striatum[116] where it is 0.6×10^8. These measurements apply to the matrix regions. The density of dopamine islands in DLS is 1.7 times higher. Thus there is a descending gradient in the density of dopamine innervation from higher dorsolateral densities to lower ventromedial densities.

Volume transmission is thought to play a key role in dopamine signaling in the striatum. The significance of innervation density for volume transmission is probably most closely related to the temporal precision of signaling. High innervation densities are necessary for rapid signaling, as such densities reduce the distance between release and receptor sites. However, release is not the only important parameter, and clearance also plays an important role in determining the characteristics of the dopamine signal. The plasmalemmal dopamine transporter (DAT) plays a critical role in regulating the clearance and hence time course and distance of volume transmission. Dopaminergic axons in the shell contain lower mean densities of DAT than those in the core.[117] The half-life of dopamine released by electrical stimulation is significantly shorter in the dorsal striatum than in the nucleus accumbens.[118–120] Thus, gradients in innervation densities are amplified by similar gradients in DAT densities, leading to a more precise temporal control over dopamine release in dorsolateral versus ventromedial regions, as depicted in FIGURE 1.

Differences in the temporal character of the dopamine signal would be likely to affect the requirements for dopamine-dependent synaptic plasticity, and the effects of dopamine on the excitability of postsynaptic neurons in different regions. Dopamine is implicated in different forms of activity-dependent synaptic plasticity in different regions.[12, 121, 122] Dopamine also modulates the function of ion channels in the dorsal and ventral striatum.[123] Dopamine-dependent plasticity has been described in the dorsal striatum using intracellular recording methods[11, 124] and in the ventral striatum using single unit activity measures.[125] Some studies have not shown dopamine-dependent plasticity in the ventral striatum.[126] However, very few studies have made regional comparisons in the requirements for synaptic plasticity using the same experimental paradigms in different areas.

The few studies to undertake regional comparisons of synaptic plasticity in slices from the striatum have produced some interesting results, especially when control slices are compared with dopamine-depleted slices. Using field potential recordings Partridge *et al.*[127] found that the DLS exhibited a tendency to display long-term depression (LTD) after high-frequency stimulation

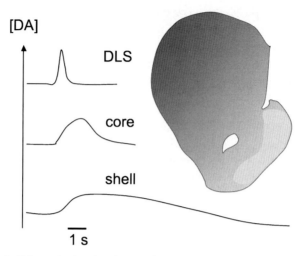

FIGURE 1. Schematic showing the postulated gradients in dopamine innervation density, and the effect of this on the time course of dopamine clearance. High density of release sites and DAT results in fast clearance in the dorsolateral striatum (DLS). Lower density of release sites and DAT in the core of the nucleus accumbens results in slower clearance. Clearance is slower again in the shell region of the accumbens. The fast time course may be more related to discrete events involving reinforcement; the slower time course may be related to action–outcome evaluation.

(HFS) in P17-34 animals; whereas the DMS had a tendency to display LTP. Smith *et al.*[128] also reported regional variation in activity-dependent plasticity. More examples of short-term potentiation (STP) and LTP were seen in DMS. Conversely, more examples of STD and LTD were seen in the lateral striatum. These differences were eliminated by 6-OHDA lesions,[128] suggesting they may be related to dopaminergic gradients. However, field potential studies in the striatum can be difficult to interpret, and these studies need to be repeated using intracellular recording techniques to establish which cells are expressing plasticity changes, and to separate changes in excitability from changes in synaptic efficacy. Also, there may be regional variations that affect experimentally induced plasticity, such as the physical arrangements of afferent fibers close to the stimulating electrode, which do not necessarily reflect different rules for synaptic plasticity at the cellular level.

CONCLUSIONS

The cytology and local circuitry of the striatum retains some key properties in all striatal regions, including the convergence of dopamine and glutamate inputs at the spiny projection neurons, and lateral inhibitory interactions between spiny projection neurons. In a simplistic way, these properties seem to underlie reinforcement and decision-making operations of the striatum. Heterogeneity in the morphology of spiny neurons, in immunohistochemical properties, and in the sources of dopamine input may be significant, but do not suggest qualitative differences in information processing operations. Recordings from presumed dopaminergic cells projecting to different striatal regions suggest that these regions are receiving a similar pattern of dopamine input.

We have suggested that one of the bases for regional specialization in behavioral functions may be differential dopamine innervation, and DAT expression. Dopamine systems that have high density of dopamine release sites and that express high levels of DAT would be expected to have high temporal resolution of signaling in the innervated regions. These regions may be more sensitive to phasic activity in dopamine cells. On the other hand, regions with less dense innervation and lower levels of DAT would not be so sensitive to phasic activity, and would tend to integrate the dopamine signal over a longer time period. The former type of signaling may be more closely related to dopamine-dependent plasticity and reinforcement processes, while the latter may be more closely related to incentives. Such differences in dopamine time course may parallel differences in behavioral function of different areas. The differences in dopamine signaling we discuss are, however, subtle. Far greater differences in dopamine signaling exist between the striatum and the prefrontal cortex, where the innervation density is on the order of a hundredfold less. Further work is needed to clarify whether the predicted differences in temporal properties of dopamine signaling exist in awake and behaving animals, and whether these are associated with different temporal requirements for postsynaptic effects of dopamine.

Our final conclusion is that regional anatomical and physiological differences do not suggest obvious regional differences in the algorithm implemented in the striatum. Perhaps the same algorithm applied to different sets of inputs and outputs might account for the functional differences described. Alternatively, differences in parameter values, such as those we have discussed in relation to the temporal properties of dopamine signaling, may play a role. An important caveat must be that few experiments have addressed regional differences in anatomy or physiology directly. Particular striatal regions, for example, the dorsal striatum or the nucleus accumbens, have attracted interest because of their associations with certain functions, and subsequent experiments have tended to be conducted in a particular striatal region. This has led to a regional specialization in experimental protocols, which may obscure a common, reiterated information processing operation.

REFERENCES

1. HEIMER, L. 2003. A new anatomical framework for neuropsychiatric disorders and drug abuse. Am. J. Psychiatry **160:** 1726–1739.

2. TAKIKAWA, Y., R. KAWAGOE & O. HIKOSAKA. 2002. Reward-dependent spatial selectivity of anticipatory activity in monkey caudate neurons. J. Neurophysiol. **87:** 508–515.
3. SCHULTZ, W., L. TREMBLAY & J.R. HOLLERMAN. 2003. Changes in behavior-related neuronal activity in the striatum during learning. Trends Neurosci. **26:** 321–328.
4. SAMEJIMA, K. *et al.* 2005. Representation of action-specific reward values in the striatum. Science **310:** 1337–1340.
5. WICKENS, J.R. 1990. Striatal dopamine in motor activation and reward-mediated learning. Steps towards a unifying model. J. Neural. Transm. **80:** 9–31.
6. BOLAM, J.P. *et al.* 2000. Synaptic organisation of the basal ganglia. J. Anat. **196:** 527–542.
7. GERFEN, C.R. 1985. The neostriatal mosaic: the reiterated processing unit. *In* Neurotransmitter Interactions in the Basal Ganglia. C.F.B. Scatton & M. Sandler, Eds. Raven Press. New York.
8. MCGEORGE, A.J. & R.L. FAULL. 1989. The organisation of the projections from the cerebral cortex to the striatum in the rat. Neuroscience **29:** 503–537.
9. WILSON, C.J. & Y. KAWAGUCHI. 1996. The origins of two-state spontaneous membrane potential fluctuations of neostriatal spiny neurons. J. Neurosci. **16:** 2397–2410.
10. SMITH, Y. *et al.* 1994. Synaptic relationships between dopaminergic afferents and cortical or thalamic input in the sensorimotor territory of the striatum in monkey. J. Comp. Neurol. **344:** 1–19.
11. REYNOLDS, J.N.J., B.I. HYLAND & J.R. WICKENS. 2001. A cellular mechanism of reward-related learning. Nature **413:** 67–70.
12. WICKENS, J.R., J.N. REYNOLDS & B.I. HYLAND. 2003. Neural mechanisms of reward-related motor learning. Curr. Opin. Neurobiol. **13:** 685–690.
13. WICKENS, J.R. & R. KOTTER. 1995. Cellular models of reinforcement. *In* Models of Information Processing in the Basal Ganglia. J.C. Houk, J.L. Davis & D.G. Beiser, Eds.: 187–214. MIT Press. Cambridge, MA.
14. SOMOGYI, J.P., J.P. BOLAM & A.D. SMITH. 1981. Monosynaptic cortical input and local axon collaterals of identified striatonigral neurons. A light and electron microscope study using the Golgi-peroxidase transport degeneration procedure. J. Comp. Neurol. **195:** 567–584.
15. YUNG, K.K.L. *et al.* 1996. Synaptic connections between spiny neurons of the direct and indirect pathways in the neostriatum of the rat: evidence from dopamine receptor and neuropeptide immunostaining. Eur. J. Neurosci. **8:** 861–869.
16. WICKENS, J.R., G.W. ARBUTHNOTT & T. SHINDOU. 2006. Simulation of GABA function in the basal ganglia: computational models of GABAergic mechanisms in basal ganglia function. Prog. Brain Res. In press.
17. JAEGER, D., H. KITA & C.J. WILSON. 1994. Surround inhibition among projection neurons is weak or nonexistent in the rat neostriatum. J. Neurophys. **72:** 2555–2558.
18. STERN, E.A., D. JAEGER & C.J. WILSON. 1998. Membrane potential synchrony of simultaneously recorded striatal spiny neurons *in vivo*. Nature **394:** 475–478.
19. TUNSTALL, M.J. *et al.* 2002. Inhibitory interactions between spiny projection neurons in the rat striatum. J. Neurophys. **88:** 1263–1269.
20. TAVERNA, S. *et al.* 2004. Direct physiological evidence for synaptic connectivity between medium-sized spiny neurons in rat nucleus accumbens *in situ*. J. Neurophysiol. **91:** 1111–1121.

21. Koos, T., J.M. Tepper & C.J. Wilson. 2004. Comparison of IPSCs evoked by spiny and fast-spiking neurons in the neostriatum. J. Neurosci. **24:** 7916–7922.

22. Czubayko, U. & D. Plenz. 2002. Fast synaptic transmission between striatal spiny projection neurons. Proc. Natl. Acad. Sci. USA **99:** 15764–15769.

23. Chang, H.T. & S.T. Kitai. 1985. Projection neurons of the nucleus accumbens: an intracellular labeling study. Brain Res. **347:** 112–116.

24. Meredith, G.E. 1999. The synaptic framework for chemical signaling in nucleus accumbens. Ann. N. Y. Acad. Sci. **877:** 140–156.

25. Meredith, G.E. *et al.* 1992. Morphological differences between projection neurons of the core and shell in the nucleus accumbens of the rat. Neuroscience **50:** 149–162.

26. O'Donnell, P. & A.A. Grace. 1993. Physiological and morphological properties of accumbens core and shell neurons recorded *in vitro*. Synapse **13:** 135–160.

27. Wilson, C.J. & P.M. Groves. 1981. Spontaneous firing patterns of identified spiny neurons in the rat neostriatum. Brain Res. **220:** 67–80.

28. Leung, L.S. & C.Y. Yim. 1993. Rhythmic delta-frequency activities in the nucleus accumbens of anesthetized and freely moving rats. Can. J. Physiol. Pharmacol. **71:** 311–320.

29. Totterdell, S. & A.D. Smith. 1989. Convergence of hippocampal and dopaminergic input onto identified neurons in the nucleus accumbens of the rat. J. Chem. Neuroanat. **2:** 285–298.

30. Sesack, S.R. & V.M. Pickel. 1990. In the rat medial nucleus accumbens, hippocampal and catecholaminergic terminals converge on spiny neurons and are in apposition to each other. Brain Res. **527:** 266–279.

31. Sesack, S.R. & V.M. Pickel. 1992. Prefrontal cortical efferents in the rat synapse on unlabeled neuronal targets of catecholamine terminals in the nucleus accumbens septi and on dopamine neurons in the ventral tegmental area. J. Comp. Neurol. **320:** 145–160.

32. Johnson, L.R. *et al.* 1994. Input from the amygdala to the rat nucleus accumbens: its relationship with tyrosine hydroxylase immunoreactivity and identified neurons. Neuroscience **61:** 851–865.

33. Venance, L., J. Glowinski & C. Giaume. 2004. Electrical and chemical transmission between striatal GABAergic output neurones in rat brain slices. J. Physiol. **559:** 215–230.

34. Cardinal, R.N. *et al.* 2002. Emotion and motivation: the role of the amygdala, ventral striatum, and prefrontal cortex. Neurosci. Biobehav. Rev. **26:** 321–352.

35. Kelley, A.E. 2004. Ventral striatal control of appetitive motivation: role in ingestive behavior and reward-related learning. Neurosci. Biobehav. Rev. **27:** 765–776.

36. Mogenson, G.J., D.L. Jones & C.Y. Yim. 1980. From motivation to action: functional interface between the limbic system and the motor system. Prog. Neurobiol. **14:** 69–97.

37. Konorski, J. 1967. Integrative Activity of the Brain: An Interdisciplinary Approach University of Chicago Press. Chicago.

38. Divac, I., H.E. Rosvold & M.K. Szwarcbart. 1967. Behavioral effects of selective ablation of the caudate nucleus. J. Comp. Physiol. Psychol. **63:** 184–190.

39. Yin, H.H. & B.J. Knowlton. 2006. The role of the basal ganglia in habit formation. Nat. Rev. Neurosci. **7:** 464–476.

40. PACKARD, M.G. & B.J. KNOWLTON. 2002. Learning and memory functions of the basal ganglia. Annu. Rev. Neurosci. **25**: 563–593.

41. DE BORCHGRAVE, R. *et al.* 2002. Effects of cytotoxic nucleus accumbens lesions on instrumental conditioning in rats. Exp. Brain Res. **144**: 50–68.

42. YIN, H.H. *et al.* 2005. The role of the dorsomedial striatum in instrumental conditioning. Eur. J. Neurosci. **22**: 513–523.

43. YIN, H.H., B.J. KNOWLTON & B.W. BALLEINE. 2004. Lesions of dorsolateral striatum preserve outcome expectancy but disrupt habit formation in instrumental learning. Eur. J. Neurosci. **19**: 181–189.

44. DICKINSON, A. *et al.* 1995. Motivation control after extended instrumental training. Anim. Learn. Behav. **23**: 197–206.

45. DICKINSON, A., D.J. NICHOLAS & C.D. ADAMS. 1983. The effect of instrumental training contingency on susceptibility to reinforcer devaluation. Quart. J. Exper. Psychol. Sec. B. Comp. Physio. Psych. **35**: 35–51.

46. YIN, H.H., B.J. KNOWLTON & B.W. BALLEINE. 2006. Inactivation of dorsolateral striatum enhances sensitivity to changes in the action-outcome contingency in instrumental conditioning. Behav. Brain Res. **166**: 189–196.

47. FAURE, A. *et al.* 2005. Lesion to the nigrostriatal dopamine system disrupts stimulus-response habit formation. J. Neurosci. **25**: 2771–2780.

48. VANDERSCHUREN, L.J., P. DI CIANO & B.J. EVERITT. 2005. Involvement of the dorsal striatum in cue-controlled cocaine seeking. J. Neurosci. **25**: 8665–8670.

49. YIN, H.H. & B.J. KNOWLTON. 2004. Contributions of striatal subregions to place and response learning. Learn. Mem. **11**: 459–463.

50. YIN, H.H., B.J. KNOWLTON & B.W. BALLEINE. 2005. Blockade of NMDA receptors in the dorsomedial striatum prevents action-outcome learning in instrumental conditioning. Eur. J. Neurosci. **22**: 505–512.

51. TRICOMI, E.M., M.R. DELGADO & J.A. FIEZ. 2004. Modulation of caudate activity by action contingency. Neuron **41**: 281–292.

52. BALLEINE, B. & S. KILLCROSS. 1994. Effects of ibotenic acid lesions of the nucleus accumbens on instrumental action. Behav. Brain Res. **65**: 181–193.

53. KELLEY, A.E., S.L. SMITH-ROE & M.R. HOLAHAN. 1997. Response-reinforcement learning is dependent on N-methyl-D-aspartate receptor activation in the nucleus accumbens core. Proc. Natl. Acad. Sci. USA **94**: 12174–12179.

54. SOKOLOWSKI, J.D. & J.D. SALAMONE. 1998. The role of accumbens dopamine in lever pressing and response allocation: effects of 6-OHDA injected into core and dorsomedial shell. Pharmacol. Biochem. Behav. **59**: 557–566.

55. BROG, J.S. *et al.* 1993. The patterns of afferent innervation of the core and shell in the "accumbens" part of the rat ventral striatum: immunohistochemical detection of retrogradely transported fluoro-gold. J. Comp. Neurol. **338**: 255–278.

56. CORBIT, L.H., J.L. MUIR & B.W. BALLEINE. 2001. The role of the nucleus accumbens in instrumental conditioning: evidence of a functional dissociation between accumbens core and shell. J. Neurosci. **21**: 3251–3260.

57. PARKINSON, J.A. *et al.* 1999. Dissociation in effects of lesions of the nucleus accumbens core and shell on appetitive Pavlovian approach behavior and the potentiation of conditioned reinforcement and locomotor activity by D-amphetamine. J. Neurosci. **19**: 2401–2411.

58. ROBINSON, T.E. & K.C. BERRIDGE. 1993. The neural basis of drug craving: an incentive-sensitization theory of addiction. Brain Res. Brain Res. Rev. **18**: 247–291.

59. DICKINSON, A., J. SMITH & J. MIRENOWICZ. 2000. Dissociation of Pavlovian and instrumental incentive learning under dopamine antagonists. Behav. Neurosci. **114:** 468–483.

60. CORBIT, L.H. & B.W. BALLEINE. 2003. Instrumental and Pavlovian incentive processes have dissociable effects on components of a heterogeneous instrumental chain. J. Exp. Psychol. Anim. Behav. Process. **29:** 99–106.

61. AHN, S. & A.G. PHILLIPS. 2006. Dopamine efflux in the nucleus accumbens during within-session extinction, outcome-dependent, and habit-based instrumental responding for food reward. Psychopharmacology (Berl.) **191:** 641–651.

62. BERRIDGE, K.C. & T.E. ROBINSON. 1998. What is the role of dopamine in reward: hedonic impact, reward learning, or incentive salience? Brain Res. Brain Res. Rev. **28:** 309–369.

63. BINDRA, D. 1974. A motivational view of learning, performance, and behavior modification. Psychol. Rev. **81:** 199–213.

64. PARKINSON, J.A. *et al.* 2000. Disconnection of the anterior cingulate cortex and nucleus accumbens core impairs Pavlovian approach behavior: further evidence for limbic cortical-ventral striatopallidal systems. Behav. Neurosci. **114:** 42–63.

65. LOVIBOND, P.F. 1983. Facilitation of instrumental behavior by a Pavlovian appetitive conditioned stimulus. J. Exp. Psychol. Anim. Behav. Process. **9:** 225–247.

66. COLWILL, R.M. & R.A. RESCORLA. 1988. Associations between the discriminative stimuli and the reinforcer in instrumental learning. J. Exp. Psychol. Anim. Behav. Process. **11:** 120–132.

67. HOLLAND, P.C. 2004. Relations between Pavlovian-instrumental transfer and reinforcer devaluation. J. Exp. Psychol. Anim. Behav. Process. **30:** 104–117.

68. TAYLOR, J.R. & T.W. ROBBINS. 1986. 6-Hydroxydopamine lesions of the nucleus accumbens, but not of the caudate nucleus, attenuate enhanced responding with reward-related stimuli produced by intra-accumbens d-amphetamine. Psychopharmacology (Berl.) **90:** 390–397.

69. CARDINAL, R.N. 2001. Neuropsychology of reinforcement processes in the rat, University of Cambridge. Unpublished Ph.D. thesis.

70. CARDINAL, R.N. *et al.* 2001. Impulsive choice induced in rats by lesions of the nucleus accumbens core. Science **292:** 2499–2501.

71. BROCKHAUS, H. 1942. Zur feineren Anatomie des Septum und des Striatum. J. Psychol. Neurol. (Leipzig) **51:** 1–56.

72. GERFEN, C.R. 1984. The neostriatal mosaic: compartmentalization of corticostriatal input and striatonigral output systems. Nature **311:** 461–464.

73. GERFEN, C.R., M. HERKENHAM & J. THIBAULT. 1987. The neostriatal mosaic: II Patch and matrix-directed mesostriatal dopaminergic and non-dopaminergic systems. J. Neurosci. **7:** 3915–3934.

74. GRAYBIEL, A.M. & C.W. RAGSDALE. 1978. Histochemically distinct compartments in the striatum of human, monkey and cat demonstrated by acetylcholinesterate staining. Proc. Natl. Acad. Sci. USA **75:** 5723–5726.

75. GRAYBIEL, A.M. 1984. Correspondence between the dopamine islands and striosomes of the mammalian striatum. Neuroscience **13:** 1157–1187.

76. HEIMER, L. & R.D. WILSON. 1975. The subcortical projections of the allocortex: similarities in the neuronal associations of the hippocampus, the pyriform cortex and the neocortex. *In* Golgi Centennial Symposium: Perspectives in Neurobiology. M. Santini, Ed.: 177–193. Raven Press. New York.

77. NAUTA, N.J.W., M.B. PRITZ & R.J. LASEK. 1974. Afferents to the rat striatum studied with horseradish peroxidase. An evaluation of a retrograde neuro-anatomical research method. Brain Res. **67:** 219–238.

78. RAGSDALE, C.W., Jr. & A.M. GRAYBIEL. 1990. A simple ordering of neocortical areas established by the compartmental organization of their striatal projections. Proc. Natl. Acad. Sci. USA **87:** 6196–6199.

79. ALEXANDER, G.E., M.R. DELONG & P.L. STRICK. 1986. Parallel organization of functionally segregated circuits linking basal ganglia and cortex. Ann. Rev. Neurosci. **9:** 357–381.

80. CANALES, J.J. & S.D. IVERSEN. 1998. Behavioural topography in the striatum: differential effects of quinpirole and D-amphetamine microinjections. Eur. J. Pharmacol. **362:** 111–119.

81. HABER, S.N., J.L. FUDGE & N.R. MCFARLAND. 2000. Striatonigrostriatal pathways in primates form an ascending spiral from the shell to the dorsolateral striatum. J. Neurosci. **20:** 2369–2382.

82. ZAHM, D.S. 1999. Functional-anatomical implications of the nucleus accumbens core and shell subterritories. Ann. N. Y. Acad. Sci. **877:** 113–128.

83. VOORN, P. *et al.* 2004. Putting a spin on the dorsal-ventral divide of the striatum. Trends Neurosci. **27:** 468–474.

84. HEIMER, L. 2000. Basal forebrain in the context of schizophrenia. Brain Res. Rev. **31:** 205–235.

85. PRENSA, L., S. RICHARD & A. PARENT. 2003. Chemical anatomy of the human ventral striatum and adjacent basal forebrain structures. J. Comp. Neurol. **460:** 345–367.

86. MALACH, R. & A.M. GRAYBIEL. 1988. Mosaic architecture of the somatic sensory-recipient sector of the cat's striatum. J. Neurosci. **6:** 3436–3458.

87. HERKENHAM, M. & C.B. PERT. 1981. Mosaic distribution of opiate receptors, parafascicular projections and acetylcholinesterase in rat striatum. Nature **291:** 415–418.

88. GERFEN, C.R. 1984. The neostriatal mosaic: compartmentalization of corticostriatal input and striatonigral output systems. Nature **311:** 461–464.

89. GERFEN, C.R. 1989. The neostriatal mosaic: striatal patch-matrix organization is related to cortical lamination. Science **246:** 385–388.

90. CHESSELET, M.F., C. GONZALES & L.P. 1991. Heterogeneous distribution of the limbic system-associated membrane protein in the caudate nucleus and substantia nigra of the cat. Neuroscience **40:** 725–733.

91. WRIGHT, A.K. & G.W. ARBUTHNOTT. 1981. The pattern of innervation of the corpus striatum by the substantia nigra. Neuroscience **6:** 2063–2067.

92. JIMENEZ-CASTELLANOS, J. & A.M. GRAYBIEL. 1987. Subdivisions of the dopamine-containing A8-A9-A10 complex identified by their differential mesostriatal innervation of striosomes and extrastriosomal matrix. Neuroscience **23:** 223–242.

93. UNGERSTEDT, U. 1971. Stereotaxic mapping of the monoamine pathways in the rat brain. Acta Physiol. Scand. Suppl. **367:** 1–48.

94. WRIGHT, A.K. *et al.* 1999. Double anterograde tracing of outputs from adjacent "barrel columns" of rat somatosensory cortex. Neostriatal projection patterns and terminal ultrastructure. Neuroscience **88:** 119–133.

95. DAHLSTROM, A. & K. FUXE. 1964. Evidence for the existence of monoamine containing neurons in the central nervous system. I. Demonstration of monoamines in the cell bodies of brainstem neurons. Acta Physiol. Scand. **62:** 1–55.

96. PHILLIPSON, O.T. 1979. The cytoarchitecture of the interfascicular nucleus and ventral tegmental area of TSAI in the rat. J. Comp. Neurol. **187:** 85–98.
97. LINDVALL, O. & A. BJORKLUND. 1974. The organisation of the ascending catecholamine neuron systems in the rat brain: as revealed by the glyoxylic acid flourescence method. Acta Physiol. Scand. Suppl. **412:** 1–48.
98. UNGERSTEDT, U. 1971. Stereotaxic mapping of the monoamine pathways in the rat brain. Acta Physiol. Scand. Suppl. **367:** 1–48.
99. MARGOLIS, E.B., *et al.* 2006. The ventral tegmental area revisited: is there an electrophysiological marker for dopaminergic neurons? J. Physiol. **577:** 907–924.
100. SCHULTZ, W. 1986. Responses of midbrain dopamine neurons to behavioral trigger stimuli in the monkey. J. Neurophysiol. **56:** 1439–1461.
101. SCHULTZ, W., P. APICELLA & T. LJUNGBERG. 1993. Responses of monkey dopamine neurons to reward and conditioned stimuli during successive steps of learning a delayed response task. J. Neurosci. **13:** 900–913.
102. LJUNGBERG, T., P. APICELLA & W. SCHULTZ. 1991. Responses of monkey midbrain dopamine neurons during delayed alternation performance. Brain Res. **567:** 337–341.
103. LJUNGBERG, T., P. APICELLA & W. SCHULTZ. 1992. Responses of monkey dopamine neurons during learning of behavioral reactions. J. Neurophys. **67:** 145–163.
104. WAELTI, P., A. DICKINSON & W. SCHULTZ. 2001. Dopamine responses comply with basic assumptions of formal learning theory. Nature **412:** 43–48.
105. HOLLERMAN, J.R., L. TREMBLAY & W. SCHULTZ. 1998. Influence of reward expectation on behavior-related neuronal activity in primate striatum. J. Neurophysiol. **80:** 947–963.
106. MIRENOWICZ, J. & W. SCHULTZ. 1996. Preferential activation of midbrain dopamine neurons by appetitive rather than aversive stimuli. Nature **379:** 449–451.
107. MIRENOWICZ, J. & W. SCHULTZ. 1994. Importance of unpredictability for reward responses in primate dopamine neurons. J. Neurophys. **72:** 1024–1027.
108. TOBLER, P.N., A. DICKINSON & W. SCHULTZ. 2003. Coding of predicted reward omission by dopamine neurons in a conditioned inhibition paradigm. J. Neurosci. **23:** 10402–10410.
109. FIORILLO, C.D., P.N. TOBLER & W. SCHULTZ. 2003. Discrete coding of reward probability and uncertainty by dopamine neurons. Science **299:** 1898–1902.
110. TOBLER, P.N., C.D. FIORILLO & W. SCHULTZ. 2005. Adaptive coding of reward value by dopamine neurons. Science **307:** 1642–1645.
111. SATOH, T. *et al.* 2003. Correlated coding of motivation and outcome of decision by dopamine neurons. J. Neurosci. **23:** 9913–9923.
112. HYLAND, B.I. *et al.* 2002. Firing modes of midbrain dopamine cells in the freely moving rat. Neuroscience **114:** 475–492.
113. PAN, W.X. *et al.* 2005. Dopamine cells respond to predicted events during classical conditioning: evidence for eligibility traces in the reward-learning network. J. Neurosci. **25:** 6235–6242.
114. PAN, W.X. & B.I. HYLAND. 2005. Pedunculopontine tegmental nucleus controls conditioned responses of midbrain dopamine neurons in behaving rats. J. Neurosci. **25:** 4725–4732.
115. MILLER, J.D. *et al.* 1983. Activity of mesencephalic dopamine and non-dopamine neurons across stages of sleep and walking in the rat. Brain Res. **273:** 133–141.

116. DOUCET, G., L. DESCARRIES & S. GARCIA. 1986. Quantification of the dopamine innervation in adult rat neostriatum. Neuroscience **19:** 427–445.
117. NIRENBERG, M.J. *et al.* 1997. The dopamine transporter: comparative ultrastructure of dopaminergic axons in limbic and motor compartments of the nucleus accumbens. J. Neurosci. **17:** 6899–6907.
118. SUAUD-CHAGNY, M.F. *et al.* 1995. Uptake of dopamine released by impulse flow in the rat mesolimbic and striatal systems *in vivo.* J. Neurochem. **65:** 2603–2611.
119. GARRIS, P.A. & R.M. WIGHTMAN. 1994. Different kinetics govern dopaminergic neurotransmission in the amygdala, prefrontal cortex, and striatum: an *in vivo* voltammetric study. J. Neurosci. **14:** 442–450.
120. GARRIS, P.A. & R.M. WIGHTMAN. 1995. Regional differences in dopamine release, uptake and diffusion measured by fast-scan cyclic voltammetry. *In* Neuromethods, Vol. 25. A. Boulton, G. Baker & R.N. Adams, Eds.: 179–220. Humana Press Inc. Totowa, NJ.
121. JAY, T.M. 2003. Dopamine: a potential substrate for synaptic plasticity and memory mechanisms. Prog. Neurobiol. **69:** 375–390.
122. REYNOLDS, J.N. & J.R. WICKENS. 2002. Dopamine-dependent plasticity of corticostriatal synapses. Neural Netw. **15:** 507–521.
123. NICOLA, S.M., D.J. SURMEIER & R.C. MALENKA. 2000. Dopaminergic modulation of neuronal excitability in the striatum and nucleus accumbens. Ann. Rev. Neuroscience. **23:** 185–215.
124. KERR, J.N.D. & J.R. WICKENS. 2001. Dopamine D-1/D-5 receptor activation is required for long-term potentiation in the rat neostriatum in vitro. J. Neurophys. **85:** 117–124.
125. FLORESCO, S.B. *et al.* 2001. Modulation of hippocampal and amygdalar-evoked activity of nucleus accumbens neurons by dopamine: cellular mechanisms of input selection. J. Neurosci. **21:** 2851–2860.
126. PENNARTZ, C.M.A. *et al.* 1993. Synaptic plasticity in an *in vitro* slice preparation of the rat nucleus accumbens. Eur. J. Neurosci. **5:** 107–117.
127. PARTRIDGE, J.G., K.C. TANG & D.M. LOVINGER. 2000. Regional and postnatal heterogeneity of activity-dependent long-term changes in synaptic efficacy in the dorsal striatum. J. Neurophysiol. **84:** 1422–1429.
128. SMITH, R. *et al.* 2001. Regional differences in the expression of corticostriatal synaptic plasticity. Neuroscience **106:** 95–101.

Multiple Representations of Belief States and Action Values in Corticobasal Ganglia Loops

KAZUYUKI SAMEJIMA[a] AND KENJI DOYA[b]

[a]Tamagawa University Brain Science Institute, Machida, Tokyo, Japan

[b]Okinawa Institute of Science and Technology, Uruma, Okinawa, Japan

ABSTRACT: Reward-related neural activities have been found in a variety of cortical and subcortical areas by neurophysiological and neuroimaging experiments. Here we present a unified view on how three subloops of the corticobasal ganglia network are involved in reward prediction and action selection using different types of information. The motor/premotor-posterior striatum loop is specialized for action-based value representation and movement selection. The orbitofrontal–ventral striatum loop is specialized for object-based value representation and target selection. The lateral prefrontal–anterior striatum loop is specialized for context-based value representation and context estimation. Furthermore, the medial prefrontal cortex (MPFC) coordinates these multiple value representations and actions at different levels of hierarchy by monitoring the error in predictions.

KEYWORDS: action value; reward; model-based reinforcement learning; decision making; dopamine; striatum; belief state

INTRODUCTION

Recent neurophysiological and neuroimaging experiments have revealed reward-related neural activity in a variety of cortical areas, such as orbitofrontal (OFC), prefrontal (PFC), and parietal cortices, as well as in subcortical areas, such as the amygdala, the striatum, and the thalamus. How do these areas differ in their functions? And how do they work together? Based on the theoretical framework of reinforcement learning (RL), we present a hypothesis on how reward prediction and action selection can be realized in the corticobasal ganglia loop, consisting of the cortex, the striatum, the pallidum, and the thalamus. We then review recent experimental literature and point out how different subloops

Address for correspondence: Kazuyuki Samejima, Tamagawa University Brain Science Institute, 6-1-1 Tamagawa-gakuen, Machida, Tokyo, Japan 195-8610. Voice: +81-42-739-8668; fax: +81-42-739-8663.

samejima@lab.tamagawa.ac.jp

Ann. N.Y. Acad. Sci. 1104: 213–228 (2007). © 2007 New York Academy of Sciences.
doi: 10.1196/annals.1390.024

of the corticobasal ganglia loop are specialized for particular types of reward prediction and action selection.

MODEL-FREE AND MODEL-BASED RL

RL[1] is a theoretical framework that describes the way an agent acquires behavior based on reward feedback, and has been successfully used for modeling animal and human behavior and its neural basis (see e.g., Daw & Doya[2]). There are two major classes of algorithms in RL: model-free learning, in which the agent simply associates the given state, its action, and the resulting outcome, and model-based learning, in which the agent uses the knowledge of the rule of the game, or the state-action-next state dynamics. Here we review the basic concepts in model-free and model-based RL.

Model-Free RL and Action Values

A popular model-free RL algorithm is called SARSA learning, in which the agent learns a so-called "action value function" from experiences of state-action-reward-state-action sequences. The action value function Q(state, action) estimates how much reward the agent is going to get by taking an action at a given state

$$Q(\text{state}(t), \text{action}(t)) = E[\text{reward}(t)$$
$$+ \gamma \text{ reward}(t+1) + \gamma^2 \text{ reward}(t+2) + \ldots],$$

where E[] means expectation and γ is the temporal discount factor that controls whether to focus on immediate reward ($\gamma = 0$) or take into account future rewards (γ close to 1). Once the action value function is learned, the optimal behavior, or policy, is to select an action that maximizes the action value Q(state, action) for the given state.

Learning of this action value function involves minimizing the so-called temporal difference (TD) error

$$\delta(t) = \text{reward}(t) + \gamma \ Q(\text{state}(t+1), \text{action}(t+1)) - Q(\text{state}(t), \text{action}(t)),$$

which measures the inconsistency between the reward expected, Q(state(t), action(t)), and the sum of the reward acquired, reward(t), and those further expected, Q(state(t+1), action(t+1)), discounted by γ.

Can these processes be realized in the brain? Accumulating pieces of evidence suggest that the network linking the cortex and the basal ganglia can implement such computation (see e.g., Daw & Doya[2]). First, the midbrain dopamine neurons, which project most heavily to the striatum, show a firing profile that resembles that of the TD signal.[3] Action value-like neural activity is also found in cortical and striatal neurons.[4,5] Furthermore, the plasticity of the corticostriatal synapses is regulated by the dopamine input.[6,7] These suggest

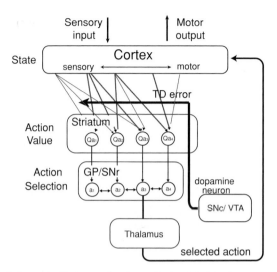

FIGURE 1. Schematic diagram of action selection model of corticobasal ganglia network in Doya's hypothesis. Cortex represents estimate state of environment. The corticobasal ganglia synaptic weights represent state action value for possible actions. The striatal projection neurons output action value (Qa_n) for action candidate for comparison between actions to select one in downstream of basal ganglia, including globus pallisdus (GP) or substantia nigra pars reticulate (SNr). Dopamine neuron in substantia nigra pars compacta (SNc) and ventral tegmental area (VTA) projected to thestriatum. The dopamine-dependent plasticity change the corticostriatal synaptic weights to adapt action values by temporal difference (TD) error signal.

that the corticobasal ganglia loop can realize basic model-free RL, with the striatum as the major candidate for action value learning[8] (FIG. 1). Specifically, based on the current environmental state represented in the cortex, subgroups of striatal neurons represent action values for different actions. Their outputs are compared in the structure downstream of the striatum including the globus pallidus and the subgroups of neurons coding the action with the highest action value will be put into execution. Any discrepancy between the predicted and actual rewards is signaled by dopamine neurons and used for learning of action value coding neurons in the striatum by dopamine-dependent plasticity.

Model-Based RL and Belief States

A major drawback of the model-free RL is its inflexibility. When the condition of the reward is changed, for example, the agent has to experience many failures to learn new action values for all the relevant states and actions. This is why model-based RL algorithms were developed, in which the predictive model of the state–action–state dynamics is learned and utilized.

$$\text{Prob}(\text{state}(t + 1)|\text{state}(t), \text{action}(t))$$

An important use of a state prediction model is to disambiguate noisy or missing sensory inputs, in a form called "belief states." A typical example is walking in a dark room; your vision does not tell you where you are, but by combining your previous position and the steps you took, you can estimate approximately where you are in the room and by combining it with any sensation, such as touching a wall, allows you to narrow down the estimate. A belief state can represent a detailed physical state, or a higher-level context of the task, which is critical for context-dependent behaviors and modular control.[9] The right way of combining noisy observation and dynamic prediction is by the framework of Bayesian inference.

Another important use of the predictive model is the so-called look-ahead planning, in which the predicted new state reached by taking a hypothetical action is used for action selection and learning. For example, if the agent notices that the rewarding state has changed, it can try and evaluate hypothetical actions and find the right sequence of actions without actually performing physical actions.[10]

There are accumulating pieces of evidence suggesting that the cortical network can implement Bayesian inference.[11] Activities of prefrontal and rostral premotor cortices have been reported in tasks that involve manipulation of belief states. Especially, the lateral prefrontal cortex (LPFC) is involved in sequential planning[12] and context-dependent switching behaviors.[13–15] Thus we postulate that the corticobasal ganglia network can realize model-based RL, with the cortex playing a major role in updating the belief state using dynamic probabilistic models. Furthermore, for deterministic motor control and planning, the "forward models" in the cerebellum may also contribute to the prediction through the corticocerebellar loop.[16]

NEURAL CORRELATES OF ACTION VALUES

Based on the theoretical framework presented above, let us now review recent experimental findings related to action values. Reward predictive activities in the striatum have often been reported. Discharge rate in caudate nucleus neurons changes adaptively during memory-guided saccade in the rewarded or nonreward direction.[17] Learning-related activities in stimulus–action association learning are also reported from different groups.[18,19] Different kinds,[20] amount,[21] and probability of reward[5,22] modulate projection neuronal activity in dorsal striatum when animals could predict reward value. Reward-related activities are also reported from cortical areas, including the OFC,[23–25] PFC,[26–30] and parietal cortices.[4,31,32]

Action Value Coding by Striatal Neurons

To test the hypothesis that striatal neurons encode action values, we trained two monkeys to perform a free-choice task with stochastic reward delivery

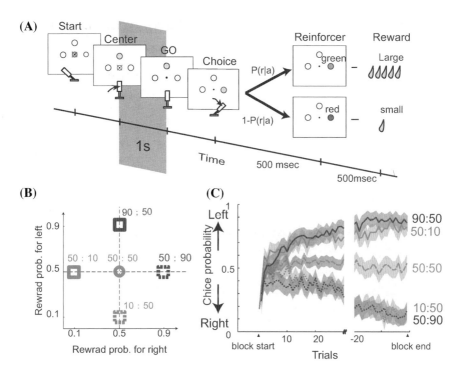

FIGURE 2. Reward-based free choice task and monkey performance. (**A**) Time chart of events during the task. (**B**) Diagram of large reward probabilities for left-, $P(r \mid a = \text{L})$, and right-handle turn, $P(r \mid a = \text{R})$, in five types of trial blocks. (**C**) Average curves of choice probability P_L (solid line) and its 95% confidence interval (shaded band) in five trial blocks in monkey RO.

(FIG. 2A), and recorded striatal neuronal activity.[5] After centering and holding the handle for 1 sec, monkeys turned the handle in one of two directions. An LED on the side chosen was illuminated stochastically in either green or red. A green LED predicted that a large reward (0.2 mL water) would follow, and a red LED predicted a small reward (0.07 mL). The probability of a large reward after left- and right-turns were fixed during a block of 30 to 150 trials, and varied between five types of trial blocks. In the "90-50" block, for example, the probability of a large reward was 90% for the left-turn and 50% for the right-turn. In this case, by taking the small reward as the baseline ($r = 0$) and the large reward as unity ($r = 1$), the action value for the left-turn Q_L was 0.9 and the action value for the right-turn Q_R was 0.5. We used four asymmetrically rewarded blocks, "90-50," "50-90," "50-10," and "10-50," and one symmetrically rewarded block, "50-50" (FIG. 2B).

To test whether the action value-based RL model could explain the monkey's adaptive choice behavior (FIG. 2C), we compare model predictions of choice behavior with actual monkey choice. The model has two free parameters, the update rate α and the inverse temperature β. These hidden parameters are estimated by a Bayesian estimate with observable data of reward and choice history.

FIGURE 3A shows an example of the time courses of the estimated action values for left-turn $Q_L(t)$ (lower panel, solid line) and for right-turn $Q_R(t)$ (dotted line), the predicted choice probability (upper panel solid line), and the actual monkey's choices (broken line). These results show that the RL model using an action value function well estimated actual, adaptive processes of choices of the monkey.

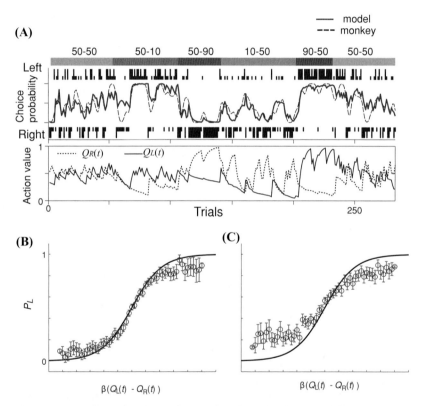

FIGURE 3. Estimated action values and action selection probability of monkey's choice. (**A**) Example sequence of action selection probability predicted by the RL model (solid line) and monkey's choice history smoothed by Gaussian filter ($\Sigma = 2.5$, broken line) are plotted in upper panel. Individual choice is indicated by vertical short bar (small rewarded) and long bar (large rewarded). The upper horizontal bar indicates blocks of large reward probability for left choice and right choices. The estimated action values $Q_L(t)$ (solid line) and $Q_R(t)$ (dotted line) are plotted in lower panel. The summary of choice probability predicted by the RL model (solid line) and monkey's choice ratio (circle and error bar) sorted by normalized value difference $\beta(Q_L(t)-Q_R(t))$ in (**B**) monkey RO and (**C**) monkey AR.

We plotted the probability of choice $p(a(t)|Q_a(t), \beta(t))$ predicted from the history of actions and reward up to trial $t-1, \{a(k), r(k), k = 1,\ldots, t-1\}$.

FIGURE 3B, C shows that the action choices are successfully predicted by the estimated action values in two monkeys. Logistic nonlinear relation between the estimated value difference and the action choice probability predicted by the model was observed. There were small deviations of the model prediction from monkey's choices at large action value differences, $\beta(Q_L(t) - Q_R(t))$. This was probably because, in the trials where choice was stabilized, monkeys, especially monkey AR, sometimes chose the actions opposite to those with higher reward probabilities in anticipation of change of trial block.

We recorded 504 striatal projection neurons in the right putamen and caudate nucleus of two monkeys. In this study, we focused on the 142 neurons that displayed increased discharges during at least one task event, and had discharge rates higher than 1 spike/sec during the delay period. We compared the average discharge rates during the delay period from two asymmetrically rewarded blocks. FIGURE 4A, B shows a representative neuron in which the delay period discharge rate was significantly higher in the 90-50 block (dark solid line) than in the 10-50 block (light dotted line) ($P = 0.003$, two tailed Mann–Whitney U test). On the other hand, delay period discharges were not significantly different ($P = 0.70$) in the 50-10 and 50-90 blocks (FIG. 4B), for which preferred actions were the opposite. Thus, the neuron encodes the action value for left turn, Q_L, but not the action itself. Another neuron showed a significantly higher discharge rate in the 50-10 block than in the 50-90 block ($P < 0.001$; FIG. 4C, D), but there was no significant difference between the 50-90 and 50-10 blocks ($P = 0.67$). This neuron may code right action value, Q_R. We also found neurons, as shown in FIG. 4E, F, which discharged more in the 90-50 block than in the 10-50 block ($P = 0.028$), but less in the 50-90 block than in the 50-10 block ($P = 0.003$). They may code the difference in action values, Q_L-Q_R, or left-turn action.

We found 24 left-turn value-type neurons (Q_L type, 17%) with discharge rates correlated to Q_L but not to Q_R, and 31 right-turn value-type neurons (Q_R type, 22%) correlated to Q_R but not to Q_L. There were 16 differential value-type (ΔQ type, 12%) neurons correlated with both Q_L and Q_R, but in opposite directions.

We next examined whether the neuronal activity encoding action values predict a monkey's action choices. The action values $Q_L(t)$ and $Q_R(t)$ at the t-th trial of a single block of trials were estimated based on a standard RL model and the past action $a(j)$ and reward $r(j)$ ($j = 1,\ldots, t-1$).[33] We performed regression analysis of neuronal activity with estimated action values fluctuated trial-by-trial. We also found 28 (20%) "Q_L-type" neurons that had a significant regression coefficient to $Q_L(t)$ (t-test, $P < 0.05$) but not to $Q_R(t)$ and 26 (19%) "Q_R-type" neurons that had a significant regression coefficient to $Q_R(t)$ but not to $Q_L(t)$. There were 9 (7%) "differential action value-type" (ΔQ) neurons correlated with the difference between $Q_L(t)$ and $Q_R(t)$. These results suggest

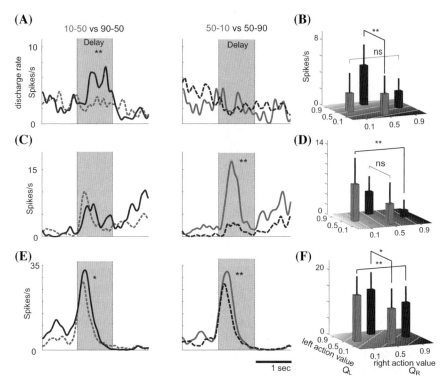

FIGURE 4. Three representative reward value-coding neurons in the striatum. (**A**) A left-action value (Q_L-type) neuron in the anterior striatum. Average discharge rates during 10-50 and 90-50 blocks (left panel) and during 50-10 and 50-90 blocks (right panel). (**B**) Three-dimensional bar graph of average magnitudes and standard deviation of activity during delay period (shaded period in A). Floor gradient shows the regression surface of neuronal activity by large reward probability after left and right turns. (**C, D**) A right-action value (Q_R-type) neuron in anterior putamen. (**E, F**) A differential action value (Q & m-type) neuron with correlation to action choice.

representation of action values in the striatum, which can guide action selection in the downstream of basal ganglia circuit.[8]

Action Value and Action Coding in the Globus Pallidus

Recently, Pasquereau *et al.*[22] studied premovement activities of the globus pallidus internal segment (GPi) and also found their correlation to action values. During movement, GPi neurons were tuned to the selected action rather than action value, whereas the striatal projection neurons still retained information of action value. This suggests that areas downstream of striatum, particularly in the globus pallidus, are more involved in action selection than in action value estimation. However, it is still an open question how an action is selected in the corticobasal ganglia loop using the action value information in the striatum.

Action Value Coding in Parietal Cortex

Action value-like neural activities were also found in lateral intraparietal region (LIP) during an oculomotor decision-making task,[4] in which the monkey Sugrue et al.[4] showed the local matching law, which is an online version of a descriptive model of animal choice behaviors (matching law[34]), could explain the monkey's choice behavior influenced by the history of actions and rewards. By using choice-triggered averaging method, which is the average reward histories triggered by particular choice to described quantitative influence weights for reward existence in the past particular trial to predict the value for the choice (probability of reward by making a saccade to the target color), they found the LIP neuronal activity were not only modulated by eye-movement direction but also by the value for the choice that was filted to the actual animal choice behavior related to reward amount or probability is specialized for the representation of space-based value representation, which could be used for spatial attention shifting or eye movement direction triggered by reward or the reward prediction error.

Object-Based Predictive Reward Coding in OFC

Different food rewards or visual stimuli associated with reward kind are represented in activity during food presentation and delay period activity[24] in OFC. Reward-predicting activity cued by visual stimuli and anticipatory build-up activity to reward delivery have also been reported in monkey OFC.[23,34] These reward representations for particular objects could be used for the selection of an object as a target.

Recently, Padoa-Schioppa and Assad[25] examined orbitofrontal neuron activities in decision making between targets with different amounts and kinds of rewards. They determined the subjective values of different rewards by finding a balance point, for example, choice between one drop of juice and three drops of water. Their results showed that orbitofrontal neuron firing encodes the subjective values for the target. Orbitofrontal reward prediction activity is also modulated by the delay time until reward delivery,[36] suggesting encoding of the delay-discounted value for each target.

While the size, kind, and delay of reward associated with a visual stimulus or an object are quantitatively represented in orbitofrontal neurons, their activity related to actions and movements is rarely reported. We hypothesize that the OFC–ventral striatum system learns to represent object-based values, which are useful for target selection, regardless of how to actually achieve the target.

Reward Prediction Error Coding by Dopamine

TD signal-like activities of dopamine neurons[3] have been further verified in experiments that varied the reward magnitude,[37] probability,[38] and contextual information, such as order of visual stimuli[39] and previous actions.[40]

In functional magnetic resonance imaging (fMRI) studies, TD error-like activity is often reported from the striatum,[41,44] which is consistent with the TD error coding in the dopamine neurons because the striatum is the major target of dopamine neurons and the fMRI signal is known to respond strongly to the presynaptic inputs. Pessiglione *et al.*[42] showed in an instrumental conditioning task that the activity of the striatum that correlated with reward prediction error was enhanced under medication of l-DOPA and reduced under haloperidol. This is convincing evidence linking reward-based action learning and dopamine-dependent striatal activity in a human fMRI study.

Recently, it was reported that dopamine D1 receptor blockade affects switching reward bias in a visually guided saccade task with an asymmetrical reward contingency.[45] Electrical microstimulation in the striatum during the outcome presentation period, which is supposed to facilitate dopamine release in the striatum, could enhance oculomotor saccadic eye movement in the stimulated direction,[46] and stimulus–action association learning.[19] These observations offer further support for the involvement of dopamine feedback to the striatum for action value learning.

NEURAL CORRELATES OF BELIEF STATES AND PLANNING

Belief State Coding in the LPFC

Several lines of evidence suggest that the LPFC is involved in the maintenance and manipulation of belief states. This idea extends the classical notion of working memory and executive functions of LPFC. In fMRI, Yoshida and Ishii[47] showed that the lateral prefrontal activity correlated with the uncertainty in estimating current position in a maze navigation task, where the subjects' view was limited to close proximity.

Monkey physiological experiments also show PFC activity related to estimates of a variety of hidden states in decision-making tasks. Postreward firing of LPFC neurons in memory-guided and visually guided oculomotor decision tasks provides information about reward existence, selected action, and also task context (memory-guided or visually guided).[29] When a monkey plays a matching pennies game against a computer opponent that takes an action in reference to actions and rewards in previous trials, the prefrontal activity also has the information about the action and the reward in previous trials.[28] In path planning task in a multiple-step maze, the PFC activity correlates with path information to reach a goal.[12]

Reward or punishment predictive activities are also reported in the LPFC[26,27] as is the correlation with action value computed by a RL model.[28] Stimulus–reward–action association learning in a reversal task revealed that the PFC activity changes during the acquisition of a new reward contingency but that learning is relatively slower than that of striatal neurons.[48]

From these lines of evidence, we speculate that LPFC neurons calculate belief state representation that is estimated by a Bayesian framework, in which the state prediction model and sensory observation are combined to estimate the hidden state of the environment. The same belief state mechanisms in LPFC should also be useful for inferring future states for hypothetical actions. This can be used for model-based RL, in which multistep forward prediction of states is used to evaluate states reached by a sequence of imaginary actions.

Medial Prefrontal Cortex (MPFC) and Hierarchy of Prediction and Decision

MPFC activity correlated with reward prediction error has been observed in a human fMRI task.[49–51] The cingulate motor area (CMA) is activated when a monkey shifts its preference due to the reduced amount of reward.[52] The muscimol injection into the CMA delays such a preference switch.[52] Lesions of the anterior cingulate cortex[53] alter the ability of the accumulated reward history to change the motor selection. Matsumoto et al.[54] recorded neurons from monkey anterior cingulate cortex and found phasic activation for particular stimulus–reward and action–reward associations.

One possible function of the MPFC is to detect environmental changes by updating higher-level belief states that encode different environmental modes.[9] Hampton and O'Doherty[55] introduced a higher-order structure in a decision task in an fMRI experiment. The activation of the ventromedial prefrontal area is more correlated with difference between prior and posterior probabilities of the higher hidden state inferred by Bayesian estimate, than the reward expectation error estimated by the action value-based RL model.

The most challenging inference task in our life would be the inference of other's mind, known as the theory-of-mind, which is essential for complex social interaction.[56] MPFC may have a hierarchical structure in which the anterior part of MPFC estimates hidden states in the higher-order structure of the environment, while the posterior part of MPFC estimates lower-order environmental state or predicts directly sensory or motor evidence.

HIERARCHICAL RL MODEL OF CORTICOBASAL GANGLIA LOOP

In summary, we propose a hypothesis that the multiple subloops of the corticobasal ganglia network are specialized for the selection of various levels of "actions," such as the selection of a strategy, a target object, a target in space, and a physical action, based on various levels of "states," such as the goal of the task, the hidden state of the environment, the history of past actions and outcomes, and sensory observations. This multiple-level hierarchy of states and actions might be embedded within the anterior–posterior axis of the frontal cortex (FIG. 5).

The hierarchical organization of lateral frontal cortices, including lateral pre-frontal, premotor, and motor cortex, generally provides a state representation and state prediction model for predicting sensory stimuli, abstract action, and concrete movement. These state codings can also be used for offline updating of state action values to predict the outcome of an action, and maintenance of the hidden state of the environment. The medial frontal cortex (MFC) also has a hierarchical organization, including CMA, supplemental motor area, presup-plemental motor area, and anterior cingulate, or the posterior part of medial MFC, and anterior part of MFC categorized in human brain. In the MFCs the hierarchical structure of hidden state representations could be updated by comparing predicted state with the actual observation.

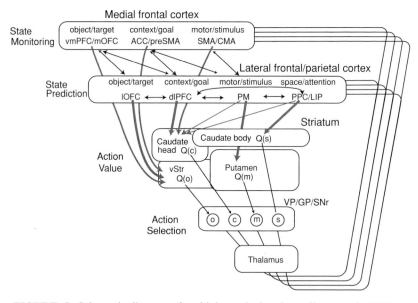

FIGURE 5. Schematic diagram of multiple corticobasal ganglia network. OFC = or-bitofrontal cortex; lOFC = lateral orbitofrontal cortex; mOFC = medial orbitofrontal cortex; ACC = anterior cingulate cortex; preSMA = presupplemental motor area; SMA = supple-mental motor area; CMA = cingulate motor area; dlPFC = dorsolateral prefrontal cortex; vmPFC = ventromedial prefrontal cortex; PM = premotor cortex; PPC = posterior parietal cortex; LIP = lateral intraparietal area; vStr = ventral striatum; Q() indicate "action" value function for object (o), context (c), space (s), and motor (m).

In this model, four different corticobasal ganglia subloops are specialized for different levels of action selection. The premotor/motor–putamen network is specialized for movement-based action or movement selection. The posterior parietal region–caudate body network is specialized for space-based value representations and spatial attention or eye/head direction selection. The orbitofrontal–ventral striatal loop is specialized for object-based value representations and target selection, regardless of how the goal is actually achieved. The lateral prefrontal–caudate head loop is specialized for context-based value representations and context selection. The context could be a representation of goal or subgoal representation or working memory manipulation dependent on task. The multiple and hierarchical structure of corticobasal ganglia loops could be organized by different levels of predictions and prediction errors calculated in multiple levels of lateral and MPFC.

Haber[57,58] proposed an interesting hypothesis regarding anatomical connection and function of corticobasal ganglia loops in which there are spiral structures in prefrontal–striatal connections and SNc–striatal connections, from the ventral striatum to the dorsal striatum. The anatomical description of these corticostriatal and SNc–striatum networks supports the suggested hierarchical organization and representation of model-based and model-free RL.

CONCLUSION

We reviewed reward-related activities in various regions of the cortex and the basal ganglia, especially in the striatum. These structures might implement different kinds of RL algorithm, such as model-free and model-based RL. The computational framework of multiple model-based RL[9] can be a theoretical guide in asking how such multiple representations of value learning and action selection can be coordinated by use of multiple prediction models.

REFERENCES

1. SUTTON, R.S. & A.G. BARTO. 1998. Reinforcement Learning. MIT Press. Cambridge, MA.
2. DAW, N.D. & K. DOYA. 2006. The computational neurobiology of learning and reward. Curr. Opin. Neurobiol. **16:** 199–204.
3. SCHULTZ, W., P. DAYAN & P.R. MONTAGUE. 1997. A neural substrate of prediction and reward. Science **275:** 1593–1599.
4. SUGRUE, L.P., G.S. CORRADO & W.T. NEWSOME. 2004. Matching behavior and the representation of value in the parietal cortex. Science **304:** 1782–1787.
5. SAMEJIMA, K. et al. 2005. Representation of action-specific reward values in the striatum. Science **310:** 1337–1340.
6. REYNOLDS, J.N., B.I. HYLAND & J.R. WICKENS. 2001. A cellular mechanism of reward-related learning. Nature **413:** 67–70.

7. REYNOLDS, J.N. & J.R. WICKENS. 2002. Dopamine-dependent plasticity of corticostriatal synapses. Neural Netw. **15:** 507–521.
8. DOYA, K. 2000. Complementary roles of basal ganglia and cerebellum in learning and motor control. Curr. Opin. Neurobiol. **10:** 732–739.
9. DOYA, K. *et al.* 2002. Multiple model-based reinforcement learning. Neural Comput. **14:** 1347–1369.
10. DAW, N.D., Y. NIV & P. DAYAN. 2005. Uncertainty-based competition between prefrontal and dorsolateral striatal systems for behavioral control. Nat. Neurosci. **8:** 1704–1711.
11. DOYA, K. *et al.* 2007. Bayesian Brain: Probablistic Approaches to Neural Coding (Computational Neuroscience). MIT Press. Cambridge, MA.
12. MUSHIAKE, H. *et al.* 2006. Activity in the lateral prefrontal cortex reflects multiple steps of future events in action plans. Neuron **50:** 631–641.
13. KONISHI, S. *et al.* 1998. Transient activation of inferior prefrontal cortex during cognitive set shifting. Nat. Neurosci. **1:** 80–84.
14. NAKAHARA, K. *et al.* 2002. Functional MRI of macaque monkeys performing a cognitive set-shifting task. Science **295:** 1532–1536.
15. AMEMORI, K. & T. SAWAGUCHI. 2006. Rule-dependent shifting of sensorimotor representation in the primate prefrontal cortex. Eur. J. Neurosci. **23:** 1895–1909.
16. DOYA, K. 1999. What are the computations of the cerebellum, the basal ganglia and the cerebral cortex? Neural Netw. **12:** 961–974.
17. KAWAGOE, R., Y. TAKIKAWA & O. HIKOSAKA. 1998. Expectation of reward modulates cognitive signals in the basal ganglia. Nat. Neurosci. **1:** 411–416.
18. TREMBLAY, L., J.R. HOLLERMAN & W. SCHULTZ. 1998. Modifications of reward expectation-related neuronal activity during learning in primate striatum. J. Neurophysiol. **80:** 964–977.
19. WILLIAMS, Z.M. & E.N. ESKANDAR. 2006. Selective enhancement of associative learning by microstimulation of the anterior caudate. Nat. Neurosci. **9:** 562–568.
20. HASSANI, O.K., H.C. CROMWELL & W. SCHULTZ. 2001. Influence of expectation of different rewards on behavior-related neuronal activity in the striatum. J. Neurophysiol. **85:** 2477–2489.
21. CROMWELL, H.C. & W. SCHULTZ. 2003. Effects of expectations for different reward magnitudes on neuronal activity in primate striatum. J. Neurophysiol. **89:** 2823–2838.
22. PASQUEREAU, B. *et al.* 2007. Shaping of motor responses by incentive values through the basal ganglia. J. Neurosci. **27:** 1176–1183.
23. TREMBLAY, L. & W. SCHULTZ. 1999. Relative reward preference in primate orbitofrontal cortex. Nature **398:** 704–708.
24. HIKOSAKA, K. & M. WATANABE. 2000. Delay activity of orbital and lateral prefrontal neurons of the monkey varying with different rewards. Cereb. Cortex **10:** 263–271.
25. PADOA-SCHIOPPA, C. & J.A. ASSAD. 2006. Neurons in the orbitofrontal cortex encode economic value. Nature **441:** 223–226.
26. WATANABE, M. *et al.* 2002. Coding and monitoring of motivational context in the primate prefrontal cortex. J. Neurosci. **22:** 2391–2400.
27. KOBAYASHI, S. *et al.* 2006. Influences of rewarding and aversive outcomes on activity in macaque lateral prefrontal cortex. Neuron **51:** 861–870.
28. BARRACLOUGH, D.J., M.L. CONROY & D. LEE. 2004. Prefrontal cortex and decision making in a mixed-strategy game. Nat. Neurosci. **7:** 404–410.

29. TSUJIMOTO, S. & T. SAWAGUCHI. 2005. Neuronal activity representing temporal prediction of reward in the primate prefrontal cortex. J. Neurophysiol. **93:** 3687–3692.
30. AMEMORI, K. & T. SAWAGUCHI. 2006. Contrasting effects of reward expectation on sensory and motor memories in primate prefrontal neurons. Cereb. Cortex **16:** 1002–1015.
31. PLATT, M.L. & P.W. GLIMCHER. 1999. Neural correlates of decision variables in parietal cortex. Nature **400:** 233–238.
32. DORRIS, M.C. & P.W. GLIMCHER. 2004. Activity in posterior parietal cortex is correlated with the relative subjective desirability of action. Neuron **44:** 365–378.
33. SAMEJIMA, K. *et al*. 2004. Estimating internal variables and parameters of a learning agent by a particular filter. *In* Advances in Neural Information Processing Systems. **16:** 1335–1342. The MIT Press.
34. HERRNSTEIN, R.J. 1961. Relative and absolute strength of response as a function of frequency of reinforcement. J. Exp. Anal. Behav. **4:** 267–272.
35. SCHULTZ, W., L. TREMBLAY & J.R. HOLLERMAN. 2000. Reward processing in primate orbitofrontal cortex and basal ganglia. Cereb. Cortex **10:** 272–284.
36. ROESCH, M.R., A.R. TAYLOR & G. SCHOENBAUM. 2006. Encoding of time-discounted rewards in orbitofrontal cortex is independent of value representation. Neuron **51:** 509–520.
37. BAYER, H.M. & P.W. GLIMCHER. 2005. Midbrain dopamine neurons encode a quantitative reward prediction error signal. Neuron **47:** 129–141.
38. FIORILLO, C.D., P.N. TOBLER & W. SCHULTZ. 2003. Discrete coding of reward probability and uncertainty by dopamine neurons. Science **299:** 1898–1902.
39. NAKAHARA, H. *et al*. 2004. Dopamine neurons can represent context-dependent prediction error. Neuron **41:** 269–280.
40. SATOH, T. *et al*. 2003. Correlated coding of motivation and outcome of decision by dopamine neurons. J. Neurosci. **23:** 9913–9923.
41. O'DOHERTY, J. *et al*. 2004. Dissociable roles of ventral and dorsal striatum in instrumental conditioning. Science **304:** 452–454.
42. PESSIGLIONE, M. *et al*. 2006. Dopamine-dependent prediction errors underpin reward-seeking behaviour in humans. Nature **442:** 1042–1045.
43. TANAKA, S.C. *et al*. 2004. Prediction of immediate and future rewards differentially recruits cortico-basal ganglia loops. Nat. Neurosci. **7:** 887–893.
44. TANAKA, S.C. *et al*. 2006. Brain mechanism of reward prediction under predictable and unpredictable environmental dynamics. Neural Netw. **19:** 1233–1241.
45. NAKAMURA, K. & O. HIKOSAKA. 2006. Role of dopamine in the primate caudate nucleus in reward modulation of saccades. J. Neurosci. **26:** 5360–5369.
46. NAKAMURA, K. & O. Hikosaka. 2006. Facilitation of saccadic eye movements by postsaccadic electrical stimulation in the primate caudate. J. Neurosci. **26:** 12885–12895.
47. YOSHIDA, W. & S. ISHII. 2006. Resolution of uncertainty in prefrontal cortex. Neuron **50:** 781–789.
48. PASUPATHY, A. & E.K. MILLER. 2005. Different time courses of learning-related activity in the prefrontal cortex and striatum. Nature **433:** 873–876.
49. BERNS, G.S. *et al*. 2001. Predictability modulates human brain response to reward. J. Neurosci. **21:** 2793–2798.

50. KNUTSON, B. *et al.* 2003. A region of mesial prefrontal cortex tracks monetarily re-warding outcomes: characterization with rapid event-related fMRI. Neuroimage **18:** 263–272.
51. HOLROYD, C.B. *et al.* 2004. Dorsal anterior cingulate cortex shows fMRI response to internal and external error signals. Nat. Neurosci. **7:** 497–498.
52. SHIMA, K. & J. TANJI. 1998. Role for cingulate motor area cells in voluntary movement selection based on reward. Science **282:** 1335–1338.
53. KENNERLEY, S.W. *et al.* 2006. Optimal decision making and the anterior cingulate cortex. Nat. Neurosci. **9:** 940–947.
54. MATSUMOTO, K., W. SUZUKI & K. TANAKA. 2003. Neuronal correlates of goal-based motor selection in the prefrontal cortex. Science **301:** 229–232.
55. HAMPTON, A.N., P. BOSSAERTS & J.P. O'DOHERTY. 2006. The role of the ventrome-dial prefrontal cortex in abstract state-based inference during decision making in humans. J. Neurosci. **26:** 8360–8367.
56. AMODIO, D.M. & C.D. FRITH. 2006. Meeting of minds: the medial frontal cortex and social cognition. Nat. Rev. Neurosci. **7:** 268–277.
57. HABER, S.N., J.L. FUDGE & N.R. MCFARLAND. 2000. Striatonigrostriatal pathways in primates form an ascending spiral from the shell to the dorsolateral striatum. J. Neurosci. **20:** 2369–2382.
58. HABER, S.N. 2004. The Human Nervous System. Elsevier. London.

Basal Ganglia Mechanisms
of Reward-Oriented Eye Movement

OKIHIDE HIKOSAKA

Laboratory of Sensorimotor Research, National Eye Institute, National Institute of Health, Bethesda, Maryland, USA

ABSTRACT: Expectation of reward facilitates motor behaviors that enable the animal to approach a location in space where the reward is expected. It is now known that the same expectation of reward profoundly modifies sensory, motor, and cognitive information processing in the brain. However, it is still unclear which brain regions are responsible for causing the reward-approaching behavior. One candidate is the dorsal striatum where cortical and dopaminergic inputs converge. We tested this hypothesis by injecting dopamine antagonists into the caudate nucleus (CD) while the monkey was performing a saccade task with a position-dependent asymmetric reward schedule. We previously had shown that: (1) serial GABAergic connections from the CD to the superior colliculus (SC) via the substantia nigra pars reticulata (SNr) exert powerful control over the initiation of saccadic eye movement and (2) these GABAergic neurons encode target position and are strongly influenced by expected reward, while dopaminergic neurons in the substantia nigra pars compacta (SNc) encode only reward-related information. Before injections of dopamine antagonists the latencies of saccades to a given target were shorter when the saccades were followed by a large reward than when they were followed by a small reward. After injections of dopamine D1 receptor antagonist the reward-dependent latency bias became smaller. This was due to an increase in saccade latency on large-reward trials. After injections of D2 antagonist the latency bias became larger, largely due to an increase in saccade latency on small-reward trials. These results indicate that: (1) dopamine-dependent information processing in the CD is necessary for the reward-dependent modulation of saccadic eye movement and (2) D1 and D2 receptors play differential roles depending on the positive and negative reward outcomes.

KEYWORDS: saccadic eye movement; caudate nucleus; substantia nigra pars reticulata; substantia nigra pars compacta; superior colliculus; dopamine; D1 receptor; D2 receptor; GABA

Address for correspondence: Okihide Hikosaka, Laboratory of Sensorimotor Research, National Eye Institute, National Institute of Health, 49 Convent Drive, Bldg. 49, Rm. 2A50, Bethesda, MD 20892-4435, USA. Voice: 301-402-7959; fax: 301-402-0511.
oh@lsr.nei.nih.gov

Ann. N.Y. Acad. Sci. 1104: 229–249 (2007). © 2007 New York Academy of Sciences.
doi: 10.1196/annals.1390.012

INTRODUCTION

The initiation of body movements can be influenced by expected rewards. Recent studies using trained monkeys revealed that many neurons in what are usually called cognitive or sensorimotor areas are modulated by reward, including the dorsolateral prefrontal cortex,[1–6] posterior parietal cortex,[7–9] premotor cortex,[10,11] and dorsal striatum.[12–19] Furthermore, other factors that influence behavior, such as learning,[20] memory,[21] and attention,[22] seem dependent on reward. Similar findings have been reported in functional imaging studies using human subjects.[23–25] These data appear to provide a strong background for understanding the neural mechanism of reward-modulated behavior. However, it is still difficult to connect these findings to produce coherent stories. There are at least two problems that prevent our understanding. First, the input–output relationship of the recorded neurons is often unclear. It is unclear whether the neurons really contribute to the reward-dependent modulation of behavior or how they receive reward-related information. Second, the reward-modulated behavior is often not well defined. When the behavior is changed by reward, it is possible that different body movements are used.

Considering the first problem, the basal ganglia seem an ideal place to study the reward-dependent motivational control of behavior. It has been well known that basal ganglia receive substantial reward-related information and strongly influence body movements. First, the motor function of the basal ganglia is illustrated by various kinds of movement disorders (e.g., inability to initiate or suppress movements) in basal ganglia dysfunctions (e.g., Parkinson's disease).[26] This *motor* function is achieved by the outputs of the basal ganglia to the brainstem motor areas (e.g., superior colliculus) and the *movement-related* areas in the cerebral cortex through the thalamus.[27,28] Second, the reward-related information to the basal ganglia is likely to be derived from substantial inputs from the limbic system (e.g., amygdala) to the ventral striatum (e.g., nucleus accumbens)[29,30] and dorsal striatum (i.e., CD nucleus and putamen),[31] and to dopamine (DA) neurons in and around the substantia nigra (SN).[32] In particular, DA neurons, which appear to carry an essential signal for reward-based learning,[33] are an important part of the basal ganglia system and project most heavily within the basal ganglia. Third, sequential and parallel inhibitory connections in the basal ganglia are thought to be suitable for the selection and learning of optimal behavior.[34,35] Fourth, the basal ganglia are thought to play an important role in learning of sensorimotor procedures or habits.[36–38] Finally, sensorimotor-cognitive signals originating from the cerebral cortex are funneled through the basal ganglia and are returned to the cerebral cortex.[39] In short, the basal ganglia are located in a perfect position to control motor behaviors based on reward information.

The basal ganglia are also ideal for solving the second problem mentioned above, that is, to define the kind of body movement as a behavioral measure. It has been well documented that the neural circuits involving the CD and the

SNr exert a powerful control over the generation of saccadic eye movement.[40] Another advantage of using saccadic eye movement is that its mechanism in the brainstem has been studied perhaps more extensively than any other motor behavior.[41] Furthermore, an important component of reward-modulated behavior is orienting movement[31] and an important component of orienting movement is saccadic eye movement.[42] The animal must orient its eye, head, and body to the location where reward is available before procuring it.[43] Saccadic eye movement is particularly important for humans[44,45] and monkeys.[46] As will be shown in this article, the initiation of saccadic eye movement is clearly facilitated or suppressed depending on the reward outcome.

In the present review, I first summarize the neuronal mechanism in the basal ganglia for the control of saccadic eye movement. I will then describe the reward-dependent changes of neuronal activity along the saccade-related neuronal circuit in the basal ganglia and related structures. Finally, I will present evidence that the reward-dependent modulation basal ganglia neuronal activity is caused by dopaminergic inputs to the CD.

OCULOMOTOR CONTROL BY THE BASAL GANGLIA

A fixed, vacant facial expression of patients with Parkinson's disease, which is often called the *Parkinson's mask*, is due to the paucity of movements in the face, including the paucity of eye movements. Similar symptoms were observed in MPTP-induced DA-deficient monkeys (FIG. 1A).[47] Most affected among various kinds of eye movements are smooth pursuit and saccades, which require voluntary control.[40] Parkinsonian patients are often impaired in shifting their gaze from one position in space to another (deficit in saccades). Compared to age-matched control subjects, saccades of parkinsonian patients tend to be small in amplitude (i.e., hypometric), slow, and delayed (i.e., long latency). Curiously, the deficit in saccades is often more severe if there is no visible object and the saccades must rely on memory. Selective deficits in memory-guided saccades are observed in other basal ganglia disorders, including Huntington's disease. The similarity between Parkinson's and Huntington's diseases is noteworthy because they are caused by different mechanisms, the former by a loss of neurons in the SNc and the latter by a loss of neurons in the CD. This suggests that the SN and the CD work together for the control of saccadic eye movements.

How the basal ganglia might control eye movements has been studied by recording single cell activity in animals trained on the visually guided and memory-guided saccade tasks. Electrical activity of single neurons was recorded with microelectrodes and was correlated with saccadic eye movements. Saccade-related activity has been found in various nuclei in the basal ganglia, including the SN,[48–50] CD,[51] subthalamic nucleus (STN),[52] and globus pallidus (GP).[53] Anatomical studies have shown that these saccade-related

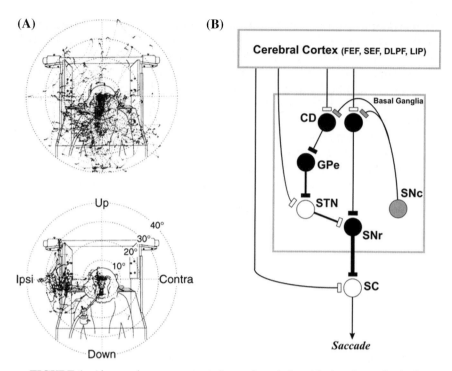

FIGURE 1. Abnormal eye movements in monkeys induced by basal ganglia dysfunctions. (**A**) Eye movements of a monkey while he was looking in a mirror before (*top*) and 31 days after (*bottom*) dopaminergic denervation of the left caudate nucleus by MPTP. Shown are the trajectories of eye movements superimposed on the view seen in the mirror by the monkey.[47] (**B**) Basal ganglia neural network involved in the control of saccadic eye movement. CD = caudate nucleus; SNr = substantia nigra pars reticulata; SC = superior colliculus; SNc = substantia nigra pars compacta; Gpe = globus pallidus external segment; STN = subthalamic nucleus; FEF = frontal eye field; SEF = supplementary eye field; DLPF = dorsolateral prefrontal cortex; LIP = area LIP in parietal cortex. Excitatory and inhibitory neurons and synapses are indicated by open and filled symbols, respectively. Gray symbol indicates dopaminergic neuron, which exerts modulatory effects on CD neurons. The thickness of the line (axon) roughly indicates the level of spontaneous activity. The direct excitation of SC neurons by inputs from the cerebral cortex is gated by the inhibitory input from the SNr.

parts are connected within the basal ganglia and with saccade-related regions outside the basal ganglia[40] (FIG. 1B). For example, the saccade-related part of the CD receives inputs from the frontal eye field and the supplementary eye field in the frontal cerebral cortex[54] while the saccade-related part of the SNr projects their axons to the superior colliculus (SC).[55]

An important feature of the basal ganglia circuits is that they use inhibitory connections as a primary means to convey signals[56] (FIG. 1B). Each area contains projection neurons and interneurons. While cortical inputs to the CD are excitatory and use glutamate as a transmitter, projection neurons in all areas in the basal ganglia, except the STN, are thought to be GABAergic and inhibitory. This means that the polarity of a signal is reversed each time it passes through projection neurons in one area. There are at least three parallel pathways in which saccade-related signals can be processed in the basal ganglia (FIG. 1B)[57]: (1) direct pathway from the CD to the SNr; (2) indirect pathway from the CD to the SNr through the GPe and/or the STN; (3) hyper-direct pathway from the STN to the SNr. Since the direct pathway consists of a series of two inhibitory connections (CD and SNr), the net effect is facilitatory. Since the indirect pathway and the hyper-direct pathways consist of three and one inhibitory connection, respectively, the net effect is inhibitory. The basal ganglia thus could facilitate or inhibit motor processes by selectively using these pathways. The target area for saccadic eye movement is the SC.

With respect to saccadic motor control in general, the basal ganglia system is situated as a side path, which has been added to the direct effect of the cerebral cortex on the SC (FIG. 1B). Probably the most important question is: How unique is the function of the basal ganglia, compared to the direct cortico-SC effect? The answers to this question should be found in the content of information carried by neurons in the CD and SNr.

The neural circuit in the basal ganglia related to eye movements originates in the CD and converges on the SNr, which then projects to the SC[40] (FIGS. 1B and 2B). The CD is a large structure in the basal ganglia and, together with the putamen, is called the striatum or the dorsal striatum. A majority of inputs to the basal ganglia are destined for the striatum, which thus acts as the input station of the basal ganglia. In addition to the cortical inputs, the entire CD (together with the putamen and the ventral striatum) receives diffuse inputs from dopaminergic neurons in the SNc and its surrounding regions (FIGS. 1B and 2B).[58] It is likely that particular combinations of these inputs create signals unique to CD neurons.

Neurons in the SNr are characterized by their rapid and tonic firing (FIG. 2A). Their firing frequency is usually between 40 and 100 spikes/sec in monkeys.[40] Furthermore, virtually all of them are GABAergic. These facts indicate that neurons in the SC should be kept inhibited by them. This inhibition can be reduced or eliminated by injecting a small amount of a GABA agonist (muscimol) into the SNr. After muscimol injection in the SNr, animals can no longer maintain stable eye position and make saccades continually and probably involuntarily (FIG. 2C). This happens to all animals tested: monkeys,[59] cats,[60] and rats.[61] Rats, compared to monkeys, exhibit a wider range of involuntary movements in addition to eye movements. These findings may be relevant to the fact that patients with basal ganglia dysfunction usually exhibit some type of involuntary movements, such as tremor, dyskinesia, dystonia, ballism, and chorea. These involuntary movements are likely to be caused by a reduction of the basal ganglia-induced inhibition.

(A)   **(B)**

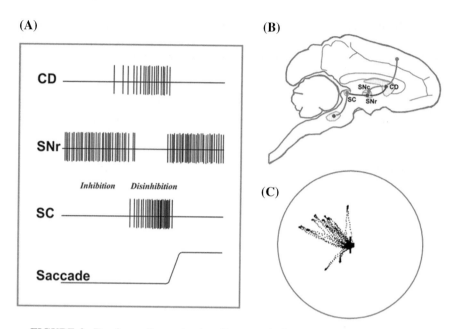

FIGURE 2. Basal ganglia mechanism for control of saccadic eye movement. **(A)** A cardinal saccade mechanism in the basal ganglia. The SC is normally inhibited by rapid firing of GABAergic neurons in the SNr. The tonic inhibition can be interrupted by GABAergic inputs from the CD. This disinhibition, together with excitatory cortical inputs, allows SC neurons to fire in burst, which leads to a saccade to a contralateral location. **(B)** Simplified neural circuits in the basal ganglia for control of saccadic eye movement in a parasagittal view of the macaque brain. **(C)** Involuntary eye movement of a monkey after muscimol injection into right SNr, shown as trajectories of saccades during 2-sec fixation periods. The monkey was unable to keep fixating at the central spot of light.[59]

Neurons in the SNr change (usually decrease) their firing rates in preparation for saccadic eye movement[62] (FIG. 2A). Many SNr neurons stop firing in response to a visual stimulus if the animal is ready to make a saccade to it. Other neurons do so just before the saccade. The inhibition of SNr neurons is caused, at least partly, by the GABAergic input from the CD. Electrical stimulation in the CD induces inhibitions and occasionally facilitations in SNr neurons.[63] The former is likely to be mediated by the direct CD-SNr inhibitory connection, while the latter is likely to be mediated by the indirect pathway through the GP (FIG. 1B).

Many of these SNr neurons project to the intermediate layer of the SC[40] and have inhibitory synaptic contacts with saccadic burst neurons.[55,64] This means that the SNr-induced tonic inhibition on SC neurons is removed or reduced before saccades (FIG. 2A). Note that saccadic neurons in the SC receive excitatory inputs from many brain areas, especially saccade-related cortical areas: the frontal eye field (FEF), supplementary eye field (SEF), and lateral intraparietal area (LIP)[65-67] (FIG. 1B). These excitatory cortical inputs, together with the SNr-induced disinhibition, would make SC neurons fire in a burst and the signal is sent to the brainstem saccade generators. Note, however, SNr neurons may increase their activity before saccade. In such a case, the SC would be less likely to generate a signal to induce saccades.

To summarize, the SNr-induced inhibition on SC neurons acts as a gate for saccade generation (FIG. 2A). SC neurons are constantly bombarded by excitatory inputs from many brain areas because there are so many objects that can attract our attention and gaze. However, these inputs are often incapable of inducing a burst of spikes in SC neurons due to the SNr-induced tonic inhibition. Only when the SNr-induced inhibition is reduced, SC neurons would exhibit a burst of spikes reliably. This is probably a very efficient mechanism to select an appropriate action in a particular context.

NEURONAL ACTIVITY IN THE BASAL GANGLIA IS MODULATED BY EXPECTED REWARD

A feature common to CD and SNr neurons is that their activity is often strikingly dependent on the behavioral context. Another feature is that many CD neurons fire tonically in an anticipatory manner before an expected task-related event occurs, such as before the onset of an expected target or the delivery of an expected reward.[14,68] And *reward* turned out to be a key factor that characterizes the information processing in the basal ganglia, as described below.

To examine the effect of reward on saccadic eye movement, we have used saccade tasks in which the amount of reward is unequal among possible target positions. We chose position as a cue for reward for two reasons. First, the goal of saccadic eye movement is to localize an object in space.[42,44] Second, when an animal forages for food, the most crucial behavior is to localize a place where the food is available.[31] Positional cues have widely been used in learning tasks, such as conditioned place preference task.[69]

In our saccade tasks the target was presented randomly at one out of two or four directions, but only one direction was associated with a big reward while the others were associated with a small or no reward (FIG. 3A). The big-reward direction was fixed in a block of 20–60 trials and is changed in the next block (FIG. 3B). Let us call this 1DR (one-direction rewarded) saccade task. We used visual and memory versions. In the visual-1DR task, the monkey makes a saccade to the target immediately after its onset (FIG. 3A, *top*).[17] In the memory-1DR task, the target position was cued and the monkey has to make saccade to the cued position after a time delay based on memory (FIG. 3A, *bottom*).[12]

(A)

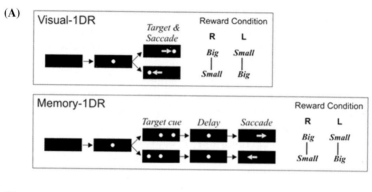

(B)

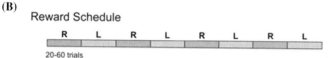

(C)

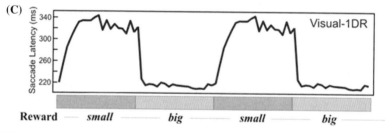

FIGURE 3. Saccade tasks with positional bias of reward, which we call one direction-rewarded (1DR) saccade tasks. (**A**) It has two versions: Visual-1DR task (*top*) and Memory-1DR task (*bottom*). In both tasks the monkey first fixates at the central spot of light and then makes a saccade to the target after the fixation point goes off. The target is chosen pseudo-randomly out of two directions (as shown here) or four directions. In the visual-1DR task (*top*) the target comes on at the same time as the fixation point goes off and hence the saccade is made to the visible target. In the memory-1DR task (*center*) the target is illuminated briefly (target cue) while the monkey is fixating and the monkey must withhold saccade until the fixating point goes off; hence, the saccade is made to the remembered target. Within a block of 20–60 trials of 1DR tasks, the saccade to one particular direction is followed by a big amount of reward, whereas the saccade to any of the other directions is followed by a small amount of reward or no reward. Even for the small or no reward direction, the monkey must make a saccade correctly to the target; otherwise, the trial is repeated until the saccade is made correctly. (**B**) One set of experiments consists of several blocks of trials during which the big-reward direction is alternated (in the two-direction condition, as shown here as R and L, which are indicated in (A) or randomized (in the four-direction condition). (**C**) Changes in saccade latency with the changes in the reward condition in the visual-1DR task.[70] The mean saccade latencies in one monkey are plotted across trials in two blocks in which saccades were followed by small and big rewards, respectively. Two cycles are shown repeated to facilitate visual impression.

Saccadic parameters changed dramatically in 1DR tasks. In the visual-1DR task, latencies were much shorter when saccades were followed by a big reward than when they were followed by a small reward.[17,70] In the schematic example shown in FIGURE 3A *top*, the latencies of rightward saccades are shorter in the reward condition R (rightward saccades associated with bigger rewards than leftward saccades) than in the condition L (leftward saccades associated with bigger rewards). Conversely, the latencies of leftward saccades are shorter in the condition L than in the condition R. During the experiment the reward conditions R and L were alternated usually every 20 trials with no external instruction (FIG. 3B), but the saccade latency changed reliably (FIG. 3C). One interesting finding common to all monkeys tested was that saccade latency decreased quickly during a small-to-big reward transition and increased more slowly during a big-to-small reward transition (FIG. 3C).[70] Thus, saccadic eye movement provides a solid and reliable behavioral measure to study the neural mechanisms of reward-oriented behavior. A series of experiments, reviewed below, suggest that the basal ganglia play a key role in the reward-dependent modulation of saccades.

The memory-1DR task allowed us to study the effects of expected reward on the preparatory process of saccadic eye movements. Kawagoe *et al.*[12] found that visual responses of CD projection neurons were modulated strongly by the expected reward. It has been known that many CD neurons respond to the target cue in the memory-guided saccade task, especially when it is presented in the contralateral hemifield.[71] Such visual responses are greatly enhanced and diminished if the saccade to the visual stimulus is expected to be followed by a bigger and smaller reward, respectively (FIG. 4 *bottom*).[12] The reward modulation was often so strong that the neuron's original direction selectivity was shifted or reversed. Other CD neurons maintained their direction preference, but their selectivity was enhanced or depressed depending on the expected reward. A minority of CD neurons showed the opposite pattern: they responded to the visual cue only or preferentially when it indicated no reward.[12] Overall, statistically significant modulation by the expected reward was observed in about 80% of visually responsive CD neurons. Similar reward-dependent modulation was found among SNr neurons.[72]

How have CD and consequently SNr neurons come to acquire activity dependent on expected reward? One possibility is that CD neurons receive signals from other brain areas that have already been modulated by expected reward. In fact, neurons exhibiting visuomotor activities that are modulated by expected reward are found in the FEF,[10] SEF,[73,74] dorsolateral prefrontal cortex,[2-4] and LIP,[7] all of which project to the saccade-related region of the CD. A second possibility is that the reward-modulation first occurs in the basal ganglia, not in the cerebral cortex. In this case, the cerebral cortex would receive reward-modulated signals from the basal ganglia through the thalamus.[75] A recent study by Pasupathy s Miller[76] seems consistent with this idea. In the following

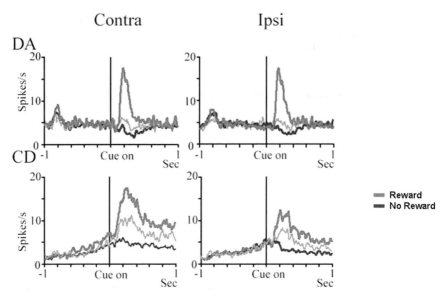

FIGURE 4. Comparison of cue responses between DA neurons and CD neurons. Population cue responses in 1DR-rewarded (*red*), 1DR-nonrewarded (*blue*), and ADR (*gray*) conditions are shown separately for contralateral (*left*) and ipsilateral (*right*) cues.[84]

section we will examine the hypothesis that the reward-modulation occurs, at least partly, in the basal ganglia depending on the inputs from DA neurons.

DOPAMINERGIC MODULATION OF INFORMATION PROCESSING IN THE CAUDATE NUCLEUS (CD)

Studies from our laboratory, including those described above, indicated that there are mechanisms in the basal ganglia that modify saccadic eye movements depending on whether or not the saccades are followed by reward. An important question was: Where does the reward-related information originate from? A candidate for the reward information carrier would be dopaminergic neurons that are located in and around the SNc and project to the CD in addition to other brain areas. To test this hypothesis we carried out two types of experiments: (a) comparison of information carried by CD neurons and DA neurons, and (b) test of DA causality.

Direct evidence for the relationship of DA to reward is based on recent single unit studies on midbrain DA neurons in trained animals. Schultz and colleagues demonstrated that DA neurons respond to the delivery of reward.[77] A key finding was that the response was correlated with the difference between the expected reward and the actual reward. Thus, the response to a reward is stronger if it is not expected.[78] The response is negative (i.e., a decrease in firing) if the expected reward is not given.[79] In short, DA neurons encode *reward prediction error.*[33] This signal corresponds nicely to a principal factor in learning theories that account for classical conditioning[80] as well as in reinforcement learning theory that was developed more recently.[20,81–83]

DA neuronal activity in the memory-1DR task followed this principle.[84] In an animal trained extensively in 1DR, DA neurons responded to the cue positively (with a phasic increase in firing) if the cue indicated an upcoming reward (FIG. 4 *top*, red); they responded to the cue negatively (with a phasic decrease in firing) if the cue indicated no reward (FIG. 4 *top*, blue). DA neurons showed no spatial selectivity: they responded to the cue at any position equally as long as it indicates an upcoming reward or as long as it indicates no reward. DA neurons exhibited no response to the cue when all positions were equally rewarded (FIG. 4 *top*, gray). DA neurons now represented the difference between the expected value of reward cue and the actual value of reward cue, consistent with an extension of the *reward prediction error* theory.[85] DA neurons' *positive* and *negative* responses are correlated respectively with the increase and decrease in reward prediction. If all positions are equally rewarded, the likelihood of reward is 100% before the cue and this is not changed by the presentation of the cue; hence, DA neurons show no response.

Here we see an intriguing relationship between CD projection neurons and DA neurons (FIG. 4).[84] In both CD and DA neurons, the visual responses in 1DR task are strongly modulated by the expected reward, although CD projection neurons exhibit only a positive response while DA neurons exhibit positive and negative responses. They are different in that CD projection neurons, not DA neurons, show spatial selectivity. The reward sensitivity of DA and CD neurons developed similarly (1) in a short time course after the reward-position contingency was reversed and (2) in a long time course while the animal learned this task.[86] These results suggest that DA neurons, with their connection to CD neurons, modulate the spatially selective signals in CD neurons in the reward-predicting manner and CD neurons in turn modulate saccade parameters with their polysynaptic connections to the oculomotor brainstem.

DOPAMINERGIC MODULATION OF INFORMATION PROCESSING IN THE CD

To test the causal role of DA in reward-dependent signals in the basal ganglia, we blocked dopaminergic synaptic transmission in the CD by injecting DA antagonists locally.[87] According to our model, if the DA input is blocked, the reward-dependent neuronal activity in the CD and consequently the reward modulation of saccade behavior should be diminished. Among at least five types of DA receptors, mainly D1 and D2, receptors are expressed in CD projection neurons. We thus injected a D1 antagonist and a D2 antagonist into the region of the CD where saccade-related neurons are clustered while the

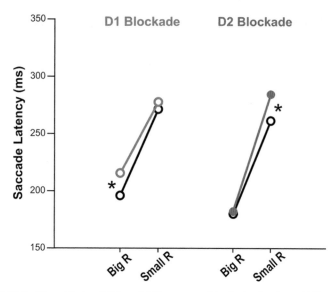

FIGURE 5. The effects of D1 and D2 receptor blockades in the middle part of the CD on the latencies of saccades followed by big and small rewards. Indicated are the mean saccade latencies before blockade (*black*), during D1 blockade (*red*), and during D2 blockade. Asterisk indicates a statistically significant difference (Mann–Whitney U test, $P < 0.01$). The numbers of experiments were 8 for D1 blockade and 13 for D2 blockade.[87]

monkey performed a reward-biased saccade task. We found that D1 antagonist attenuates the reward modulation of saccade behavior (FIG. 5, *left*). In contrast, injecting D2 antagonist into the same region enhanced the reward-dependent changes (FIG. 5, *right*).

I will now discuss possible mechanisms underlying the D1 and D2 blockade effects. The attenuation of the reaction time bias by D1 blockade was due to the prolongation of reaction times on large-reward trials and unchanged reaction times on small-reward trials (FIG. 5, *left*). The prolongation of saccade reaction times can be explained by a change in the function of the basal ganglia circuits, specifically the direct pathway (FIG. 6, *left*). First, in the CD, D1 receptors are preferentially expressed by neurons that belong to the direct pathway (i.e., projecting to the SN directly).[88,89] Second, in the anesthetized animals, DA increases the excitability of CD neurons and this effect was reduced by D1 antagonist.[90,91] These observations suggest that D1-antagonist injection into the CD would attenuate the responses of SN-projecting CD neurons, which leads to a weaker disinhibition of neurons in the SC and consequently the prolongation of saccade reaction time (indicated by *green arrows* in FIG. 6, *left*).

In contrast, the enhancement of the reaction time bias by D2 blockade was due to the prolongation of reaction times on small-reward trials and unchanged

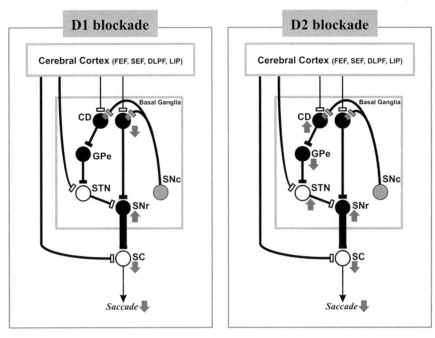

FIGURE 6. A hypothetical scheme showing how the blockade of DA receptor activation might work, assuming that D1 receptors (*red*) and D2 receptors (*blue*) are differentially expressed by CD projection neurons belonging to the direct pathway and those belonging to the indirect pathway. *Left*: The hypothetical effects of the blockade of D1 receptors are shown by green arrows, upward and downward indicating increase and decrease of neuronal activity. The primary effect would be a decrease in CD neuronal activity because D1 receptor-mediated effect is thought to be facilitatory. *Right*: The hypothetical effects of the blockade of D2 receptors. The primary effect would be an increase in CD neuronal activity because D2 receptor-mediated effect is thought to be inhibitory.

reaction times on large-reward trials (FIG. 5, *right*). The prolongation of saccade reaction times can also be explained by a change in the function of the basal ganglia circuits, but in this case the indirect pathway (FIG. 6, *right*). It has been shown that D2 receptors are preferentially expressed by neurons that belong to the indirect pathway[88,89] (FIG. 6, *right*) and that the D2-mediated effect on CD neurons is inhibitory.[91] As indicated by green arrows in FIGURE 6, *right*, D2 blockade would remove the inhibition of CD neurons, increase the CD-induced inhibition of neurons in the globus pallidus external segment (GPe), decrease the GPe-induced inhibition of SNr neurons (directly or indirectly through the STN), increase the SNr-induced inhibition of SC neurons, and consequently the prolongation of saccade reaction times.

But why should the effects of D1 blockade occur selectively for large-reward trials while the effects of D2 blockade occur selectively for small-reward trials?

To solve this puzzle we need to understand that DA neurons exhibit a short burst of spikes in response to unexpected reward[78] or a reward-indicating sensory stimulus[77,84,86] and a pause of firing in response to unexpected omission of reward[79] or a no-reward-indicating stimulus.[84] Let us suppose that D1 receptor activation has a higher threshold than D2 receptor activation. Some support for this postulate can be found in a paper by Richfield *et al.*,[92] which indicates that the sensitivity to DA is higher for D2 receptor than D1 receptor. The differential effects of D1 and D2 blockades could then be explained by the model shown in FIGURE 7. We now represent sensitivity of D1- or D2-mediated effects by *threshold*; the modulatory effects by DA become significant only when available DA is higher than the threshold for each receptor. During the control experiment (FIG. 7, *left*) D1 receptor activation is effective when the phasic increase of DA after the presentation of big-reward-indicating cue exceeds D1 threshold while D2 receptor activation is effective tonically except for a brief dip after the small-reward-indicating cue. By D1 blockade (FIG. 7, *center*), D1 threshold is elevated and therefore the phasic D1 effect for big-reward condition is reduced, leading to prolongation of SRT in the

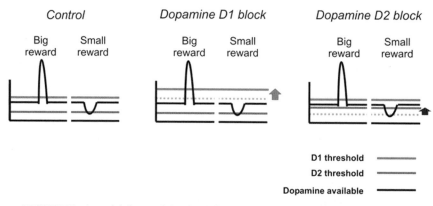

FIGURE 7. A model that explains the different effects by D1 and D2 receptor blockades on the reward-dependent modulation of saccades. In each of the three conditions (Control, D1 block, D2 block), *black lines* indicate the changes in the level of DA in the caudate during the biased-reward saccade task after presentation of the target indicating a large reward (*left*) and a small reward (*right*). The sensitivities of D1 and D2 receptors to DA are expressed as *thresholds,* which are indicated by red and blue lines, respectively (see text). If the level of DA is higher than the threshold of a given receptor type, DA influences the output of caudate projection neurons through the receptor. *Left column:* Before injection of D1 or D2 antagonist, D1 receptor activation occurs phasically after the target indicating a large reward, while D2 receptor activation is sustained except for a brief decrease after the target indicating a small reward. *Middle column:* After D1 antagonist injection, D1 threshold is elevated, which leads to a decrease in the phasic D1 effect in the large-reward condition. *Right column:* After D2 antagonist injection, D2 threshold is elevated, which leads to a further weakening of the D2 effect in the small-reward condition.

big-reward condition (FIG. 6, *left*). By D2 blockade (FIG. 7, *right*), D2 threshold is elevated and therefore the D2 effect is further reduced in the small-reward condition, leading to prolongation of SRT in the small-reward condition (FIG. 6, *right*).

The above interpretation assumes that DA neurons act quickly on CD neurons to change saccade reaction time on a single trial. However, this may not be realistic, considering the modulatory nature of DA actions. An alternative mechanism may be DA-dependent plasticity in corticostriatal synapses.[93–96] A conjunction of presynaptic activity in corticostriatal inputs and postsynaptic activity in CD neurons leads to long-term potentiation only if a large phasic increase in D1 receptor activation occurs simultaneously.[94] In our paradigm, if a particular target is repeatedly associated with a large reward that would cause DA neuron activation,[84] the corticostriatal synapses carrying the target signal should undergo long-term potentiation and therefore CD neurons respond to the target progressively more strongly, leading to shorter saccade reaction times. D1 antagonist should suppress such changes, as we observed as the longer saccade reaction times on large-reward trials. On the other hand, the D2-mediated inhibitory effect on caudate neurons is necessary to keep minimum facilitatory effects on the SC for saccades to be generated even on small-reward trials.

These results suggest that: (1) the reward modulation of saccadic eye movement, at least partly, originates from the CD; (2) the DA input to CD projection neurons is responsible for the reward modulation; (3) the DA effect is mediated by D1 and D2 receptors in differential manners.

FUTURE DIRECTIONS OF RESEARCH ON REWARD-DEPENDENT MOTIVATIONAL BEHAVIOR

We have shown that reward-dependent modulation of saccadic eye movements occurs, at least partly, in the CD where cortical inputs carrying predominantly spatial information and dopaminergic inputs carrying exclusively reward-related information converge. A causal role of DA in the reward modulation was also demonstrated in our study in which DA antagonists injected locally in the CD changed the reward modulation of saccadic eye movements.

The importance of DA neurons becomes more evident as we understand the complexity of information they carry. DA neurons respond to rewards or sensory stimuli that reliably predict the rewards. The response is positive (an increase in activity) or negative (a decrease in activity) if the value of the reward is higher or lower, respectively, than predicted. This kind of information, which is called reward prediction error, is thought to act as a teaching signal for appetitive learning to maximize the gain of reward. However, it is not a trivial task to predict the upcoming reward. Theoretically, it requires the knowledge or memory of the entire past reward history, which is assumed to be extrapolated to future. Furthermore, the prediction needs to take account of sequential or temporal patterns of reward outcome, if any.[97]

A big problem is that it is virtually unknown which parts of the brain provide DA neurons with such complex reward-related signals. It has been known anatomically that midbrain DA neurons receive inputs from many brain areas, including the ventral and dorsal striatum, subthalamic nucleus, pedunculopontine nucleus, amygdala, lateral hypothalamus, SC, and lateral habenula,[32, 98–102] but none of these areas has been proved to be a critical determinant of reward-predictive DA signals. We need to solve this issue to gain a broader perspective on motivational behavior.

REFERENCES

1. INOUE, M. *et al.* 1985. Reward related neuronal activity in monkey dorsolateral prefrontal cortex during feeding behavior. Brain Res. **326:** 307–312.
2. WATANABE, M. 1996. Reward expectancy in primate prefrontal neurons. Nature **382:** 629–632.
3. LEON, M.I. & M.N. SHADLEN. 1999. Effect of expected reward magnitude on the response of neurons in the dorsolateral prefrontal cortex of the macaque. Neuron **24:** 415–425.
4. KOBAYASHI, S. *et al.* 2002. Influence of reward expectation on visuospatial processing in macaque lateral prefrontal cortex. J. Neurophysiol. **87:** 1488–1498.
5. WATANABE, M. *et al.* 2002. Coding and monitoring of motivational context in the primate prefrontal cortex. J. Neurosci. **22:** 2391–2400.
6. BARRACLOUGH, D.J., M.L. CONROY & D. LEE. 2004. Prefrontal cortex and decision making in a mixed-strategy game. Nat. Neurosci. **7:** 404–410.
7. PLATT, M.L. & P.W. GLIMCHER. 1999. Neural correlates of decision variables in parietal cortex. Nature **400:** 233–238.
8. GLIMCHER, P.W. 2001. Making choices: the neurophysiology of visual-saccadic decision making. Trends Neurosci. **4:** 654–659.
9. SUGRUE, L.P., G.S. CORRADO & W.T. NEWSOME. 2004. Matching behavior and the representation of value in the parietal cortex. Science **304:** 1782–1787.
10. ROESCH, M.R. & C.R. OLSON. 2003. Impact of expected reward on neuronal activity in prefrontal cortex, frontal and supplementary eye fields and premotor cortex. J. Neurophysiol. **90:** 1766–1789.
11. ROESCH, M.R. & C.R. OLSON. 2004. Neuronal activity related to reward value and motivation in primate frontal cortex. Science **304:** 307–310.
12. KAWAGOE, R., Y. TAKIKAWA & O. HIKOSAKA. 1998. Expectation of reward modulates cognitive signals in the basal ganglia. Nat. Neurosci. **1:** 411–416.
13. TREMBLAY, L., J.R. HOLLERMAN & W. SCHULTZ. 1998. Modifications of reward expectation-related neuronal activity during learning in primate striatum. J. Neurophysiol. **80:** 964–977.
14. HOLLERMAN, J.R., L. TREMBLAY & W. SCHULTZ. 1998. Influence of reward expectation on behavior-related neuronal activity in primate striatum. J. Neurophysiol. **80:** 947–963.

15. Takikawa, Y., R. Kawagoe & O. Hikosaka. 2002. Reward-dependent spatial selectivity of anticipatory activity in monkey caudate neurons. J. Neurophysiol. **87:** 508–515.
16. Lauwereyns, J. et al. 2002. Feature-based anticipation of cues that predict reward in monkey caudate nucleus. Neuron **33:** 463–473.
17. Lauwereyns, J. et al. 2002. A neural correlate of response bias in monkey caudate nucleus. Nature **418:** 413–417.
18. Watanabe, K., J. Lauwereyns & O. Hikosaka. 2003. Neural correlates of re-warded and unrewarded eye movements in the primate caudate nucleus. J. Neurosci. **23:** 10052–10057.
19. Cromwell, H.C. & W. Schultz. 2003. Effects of expectations for different reward magnitudes on neuronal activity in primate striatum. J. Neurophysiol. **89:** 2823–2838.
20. Dayan, P. & B. Balleine. 2002. Reward, motivation, and reinforcement learning. Neuron **36:** 285–298.
21. Baxter, M.G. & E.A. Murray. 2002. The amygdala and reward. Nat. Rev. Neurosci. **3:** 563–573.
22. Maunsell, J.H. 2004. Neuronal representations of cognitive state: reward or attention? Trends Cogn. Sci. **8:** 261–265.
23. Delgado, M.R. et al. 2000. Tracking the hemodynamic responses to reward and punishment in the striatum. J. Neurophysiol. **84:** 3072–3077.
24. Knutson, B. et al. 2000. FMRI visualization of brain activity during a monetary incentive delay task. Neuroimage **12:** 20–27.
25. O'Doherty, J. et al. 2001. Representation of pleasant and aversive taste in the human brain. J. Neurophysiol. **85:** 1315–1321.
26. Denny-Brown, D. 1962. The Basal Ganglia and Their Relation to Disorders of Movement. Oxford University Press. London.
27. Graybiel, A.M. & C.W. Ragsdale. 1979. Fiber connections of the basal ganglia. In: Development of Chemical Specificity of Neurons. M. Cuenod, G.W. Kreutzberg & F.E. Bloom, Eds.: 239–283. Elsevier. Amsterdam.
28. DeLong, M.R. & A.P. Georgopoulos. 1981. Motor functions of the basal ganglia. In: The Nervous System. Vol. sect.1, part 2, vol. II, chapt. 19. V.B. Brooks, Ed.: 1017–1061. American Physiological Society. Bethesda, MD.
29. Mogenson, G.J., D.L. Jones & C.Y. Yim. 1980. From motivation to action: functional interface between the limbic system and the motor system. Prog. Neurobiol. **14:** 69–97.
30. Haber, S.N. & N.R. McFarland. 1999. The concept of the ventral striatum in nonhuman primates. Ann. N. Y. Acad. Sci. **877:** 33–48.
31. Swanson, L.W. 2000. Cerebral hemisphere regulation of motivated behavior. Brain Res. **886:** 113–164.
32. Fudge, J.L. & S.N. Haber. 2000. The central nucleus of the amygdala projection to dopamine subpopulations in primates. Neuroscience **97:** 479–494.
33. Schultz, W. 1998. Predictive reward signal of dopamine neurons. J. Neurophysiol. **80:** 1–27.
34. Hikosaka, O. et al. 1993. Role of basal ganglia in initiation and suppression of saccadic eye movements. In: Role of the Cerebellum and Basal Ganglia in Voluntary Movement. N. Mano, I. Hamada & M.R. Delong, Eds.: 213–219. Elsevier. Amsterdam.

35. MINK, J.W. 1996. The basal ganglia: focused selection and inhibition of competing motor programs. Prog. Neurobiol. **50:** 381–425.
36. SALMON, D.P. & N. BUTTERS. 1995. Neurobiology of skill and habit learning. Curr. Opin. Neurobiol. **5:** 184–190.
37. GRAYBIEL, A.M. 1998. The basal ganglia and chunking of action repertoires. Neurobiol. Learn. Mem. **70:** 119–136.
38. PACKARD, M.G. & B.J. KNOWLTON. 2002. Learning and memory functions of the Basal Ganglia. Annu. Rev. Neurosci. **25:** 563–593.
39. ALEXANDER, G.E., M.R. DELONG & P.L. STRICK. 1986. Parallel organization of functionally segregated circuits linking basal ganglia and cortex. Annu. Rev. Neurosci. **9:** 357–381.
40. HIKOSAKA, O., Y. TAKIKAWA & R. KAWAGOE. 2000. Role of the basal ganglia in the control of purposive saccadic eye movements. Physiol. Rev. **80:** 953–978.
41. SPARKS, D.L. 2002. The brainstem control of saccadic eye movements. Nat. Rev. Neurosci. **3:** 952–964.
42. HAYHOE, M. & D. BALLARD. 2005. Eye movements in natural behavior. Trends Cogn. Sci. **9:** 188–194.
43. EWERT, J.-P. 1980. Neuroethology. Springer. Berlin.
44. JOHANSSON, R.S. et al. 2001. Eye-hand coordination in object manipulation. J. Neurosci. **21:** 6917–6932.
45. TRIESCH, J. et al. 2003. What you see is what you need. J. Vis. **3:** 86–94.
46. MIYASHITA, K. et al. 1996. Anticipatory saccades in sequential procedural learning in monkeys. J. Neurophysiol. **76:** 1361–1366.
47. MIYASHITA, N., O. HIKOSAKA & M. KATO. 1995. Visual hemineglect induced by unilateral striatal dopamine dificiency in monkeys. Neuroreport **6:** 1257–1260.
48. HIKOSAKA, O. & R.H. WURTZ. 1983. Visual and oculomotor functions of monkey substantia nigra pars reticulata. I. Relation of visual and auditory responses to saccades. J. Neurophysiol. **49:** 1230–1253.
49. JOSEPH, J.P. & D. BOUSSAOUD. 1985. Role of the cat substantia nigra pars reticulata in eye and head movements. I. Neural activity. Exp. Brain Res. **57:** 286–296.
50. HANDEL, A. & P.W. GLIMCHER. 1999. Quantitative analysis of substantia nigra pars reticulata activity during a visually guided saccade task. J. Neurophysiol. **82:** 3458–3475.
51. HIKOSAKA, O., M. SAKAMOTO & S. USUI. 1989. Functional properties of monkey caudate neurons. I. Activities related to saccadic eye movements. J. Neurophysiol. **61:** 780–798.
52. MATSUMURA, M. et al. 1992. Visual and oculomotor functions of monkey subthalamic nucleus. J. Neurophysiol. **67:** 1615–1632.
53. KATO, M. & O. HIKOSAKA. 1995. Function of the indirect pathway in the basal ganglia oculomotor system: visuo-oculomotor activities of external pallidum neurons. In: Age-Related Dopamine-Deficient Disorders. Vol. 14. M. Segawa & Y. NOMURA, Eds.: 178–187. Karger. Basal.
54. PARTHASARATHY, H.B., J.D. SCHALL & A.M. GRAYBIEL. 1992. Distributed but convergent ordering of corticostriatal projections: analysis of the frontal eye field and the supplementary eye field in the macaque monkey. J. Neurosci. **12:** 4468–4488.
55. HIKOSAKA, O. & R.H. WURTZ. 1983. Visual and oculomotor functions of monkey substantia nigra pars reticulata. IV. Relation of substantia nigra to superior colliculus. J. Neurophysiol. **49:** 1285–1301.

56. KITA, H. 1993. GABAergic circuits of the striatum. *In*: Chemical Signalling in the Basal Ganglia. G.W. Arbuthnott & P.C. Emson, Eds.: 51–72. Elsevier. Amsterdam.

57. NAMBU, A., H. TOKUNO & M. TAKADA. 2002. Functional significance of the cortico-subthalamo-pallidal 'hyperdirect' pathway. Neurosci. Res. **43:** 111–117.

58. UNGERSTEDT, U. 1971. Stereotaxic mapping of the monoamine pathways in the rat brain. Acta Physiologica Scandinavica **Suppl 367:** 1–48.

59. HIKOSAKA, O. & R.H. WURTZ. 1985. Modification of saccadic eye movements by GABA-related substances. II. Effects of muscimol in monkey substantia nigra pars reticulata. J. Neurophysiol. **53:** 292–308.

60. BOUSSAOUD, D. & J.P. JOSEPH. 1985. Role of the cat substantia nigra pars reticulata in eye and head movements. II. Effects of local pharmacological injections. Exp. Brain Res. **57:** 297–304.

61. SAKAMOTO, M. & O. HIKOSAKA. 1989. Eye movements induced by microinjection of GABA agonist in the rat substantia nigra pars reticulata. Neurosci. Res. **6:** 216–233.

62. HIKOSAKA, O. & R.H. WURTZ. 1989. The basal ganglia. *In*: The Neurobiology of Saccadic Eye Movements. Vol. 3. R.H. Wurtz & M.E. Goldberg, Eds.: 257–281. Elsevier. Amsterdam.

63. HIKOSAKA, O., M. SAKAMOTO & N. MIYASHITA. 1993. Effects of caudate nucleus stimulation on substantia nigra cell activity in monkey. Exp. Brain Res. **95:** 457–472.

64. KARABELAS, A.B. & A.K. MOSCHOVAKIS. 1985. Nigral inhibitory termination on efferent neurons of the superior colliculus: an intracellular horseradish peroxidase study in the cat. J. Comp. Neurol. **239:** 309–329.

65. SOMMER, M.A. & R.H. WURTZ. 2000. Composition and topographic organization of signals sent from the frontal eye field to the superior colliculus. J. Neurophysiol. **83:** 1979–2001.

66. TEHOVNIK, E.J. *et al.* 2000. Eye fields in the frontal lobes of primates. Brain Res. Rev. **32:** 413–448.

67. PARÉ, M. & R.H. WURTZ. 2001. Progression in neuronal processing for saccadic eye movements from parietal cortex area LIP to superior colliculus. J. Neurophysiol. **85:** 2545–2562.

68. HIKOSAKA, O., M. SAKAMOTO & S. USUI. 1989. Functional properties of monkey caudate neurons. III. Activities related to expectation of target and reward. J. Neurophysiol. **61:** 814–832.

69. CARR, G.D. & N.M. WHITE. 1983. Conditioned place preference from intra-accumbens but not intra-caudate amphetamine injections. Life Sci. **33:** 2551–2557.

70. WATANABE, K. & O. HIKOSAKA. 2005. Immediate changes in anticipatory activity of caudate neurons associated with reversal of position-reward contingency. J. Neurophysiol. **94:** 1879–1887.

71. HIKOSAKA, O., M. SAKAMOTO & S. USUI. 1989. Functional properties of monkey caudate neurons. II. Visual and auditory responses. J. Neurophysiol. **61:** 799–813.

72. SATO, M. & O. HIKOSAKA. 2002. Role of primate substantia nigra pars reticulata in reward-oriented saccadic eye movement. J. Neurosci. **22:** 2363–2373.

73. AMADOR, N., M. SCHLAG-REY & J. SCHLAG. 2000. Reward-predicting and reward-detecting neuronal activity in the primate supplementary eye field. J. Neurophysiol. **84:** 2166–2170.

74. STUPHORN, V., T.L. TAYLOR & J.D. SCHALL. 2000. Performance monitoring by the supplementary eye field. Nature **408:** 857–860.
75. MIDDLETON, F.A. & P.L. STRICK. 2000. Basal ganglia and cerebellar loops: motor and cognitive circuits. Brain Res. Rev. **31:** 236–250.
76. PASUPATHY, A. & E.K. MILLER. 2005. Different time courses of learning-related activity in the prefrontal cortex and striatum. Nature **433:** 873–876.
77. SCHULTZ, W., P. APICELLA & T. LJUNGBERG. 1993. Responses of monkey dopamine neurons to reward and conditioned stimuli during successive steps of learning a delayed response task. J. Neurosci. **13:** 900–913.
78. MIRENOWICZ, J. & W. SCHULTZ. 1994. Importance of unpredictability for reward responses in primate dopamine neurons. J. Neurophysiol. **72:** 1024–1027.
79. HOLLERMAN, J.R. & W. SCHULTZ. 1998. Dopamine neurons report an error in the temporal prediction of reward during learning. Nat. Neurosci. **1:** 304–309.
80. RESCORLA, R.A. & A.R. WAGNER. 1972. A theory of pavlovian conditioning: variations in the effectiveness of reinforcement and nonreinforcement. *In*: Classical Conditioning II: Current Research and Theory. A.H. Black & W.F. Prokasy, Eds.: 64–99. Appleton-Century-Croft. New York.
81. BARTO, A.G. 1995. Adaptive critics and the basal ganglia. *In*: Models of Information Processing in the Basal Ganglia. J.C. Houk, J.L. Davis & D.G. Beiser, Eds.: 215–232. MIT Press. Cambridge, MA.
82. HOUK, J.C., J.L. ADAMS & A. BARTO. 1995. A model of how the basal ganglia generate and use neural signals that predict reinforcement. *In*: Models of Information Processing in the Basal Ganglia. J.C. Houk, J.L. Davis & D.G. Beiser, Eds.: 249–270. MIT Press. Cambridge, MA.
83. SCHULTZ, W. & A. DICKINSON. 2000. Neuronal coding of prediction errors. Annu. Rev. Neurosci. **23:** 473–500.
84. KAWAGOE, R., Y. TAKIKAWA & O. HIKOSAKA. 2004. Reward-predicting activity of dopamine and caudate neurons – a possible mechanism of motivational control of saccadic eye movement. J. Neurophysiol. **91:** 1013–1024.
85. TOBLER, P.N., A. DICKINSON & W. SCHULTZ. 2003. Coding of predicted reward omission by dopamine neurons in a conditioned inhibition paradigm. J. Neurosci. **23:** 10402–10410.
86. TAKIKAWA, Y., R. KAWAGOE & O. HIKOSAKA. 2004. A possible role of midbrain dopamine neurons in short- and long-term adaptation of saccades to position-reward mapping. J. Neurophysiol. **92:** 2520–2529.
87. NAKAMURA, K. & O. HIKOSAKA. 2006. Role of dopamine in the primate caudate nucleus in reward modulation of saccades. J. Neurosci. **26:** 5360–5369.
88. GERFEN, C.R. *et al.* 1990. D1 and D2 dopamine receptor-regulated gene expression of striatonigral and striatopallidal neurons. Science **250:** 1429–1432.
89. SURMEIER, D.J., W.-J. SONG & Z. YAN. 1996. Coordinated expression of dopamine receptors in neostriatal medium spiny neurons. J. Neurosci. **16:** 6579–6591.
90. GONON, F. 1997. Prolonged and extrasynaptic excitatory action of dopamine mediated by D1 receptors in the rat striatum *in vivo*. J. Neurosci. **17:** 5972–5978.
91. WEST, A.R. & A.A. GRACE. 2002. Opposite influences of endogenous dopamine D1 and D2 receptor activation on activity states and electrophysiological properties of striatal neurons: studies combining *in vivo* intracellular recordings and reverse microdialysis. J. Neurosci. **22:** 294–304.
92. RICHFIELD, E.K., J.B. PENNEY & A.B. YOUNG. 1989. Anatomical and affinity state comparisons between dopamine D1 and D2 receptors in the rat central nervous system. Neuroscience **30:** 767–777.

93. CALABRESI, P. *et al.* 1996. The corticostriatal projection: from synaptic plasticity to dysfunctions of the basal ganglia. Trends Neurosci. **19:** 19–24.
94. REYNOLDS, J.N. & J.R. WICKENS. 2002. Dopamine-dependent plasticity of corticostriatal synapses. Neural Netw. **15:** 507–521.
95. LOVINGER, D.M., J.G. PARTRIDGE & K.C. TANG. 2003. Plastic control of striatal glutamatergic transmission by ensemble actions of several neurotransmitters and targets for drugs of abuse. Ann. N. Y. Acad. Sci. **1003:** 226–240.
96. MAHON, S., J.M. DENIAU & S. CHARPIER. 2004. Corticostriatal plasticity: life after the depression. Trends Neurosci. **27:** 460–467.
97. NAKAHARA, H. *et al.* 2004. Dopamine neurons can represent context-dependent prediction error. Neuron **41:** 269–280.
98. HABER, S.N., J.L. FUDGE & N.R. MCFARLAND. 2000. Striatonigrostriatal pathways in primates form an ascending spiral from the shell to the dorsolateral striatum. J. Neurosci. **20:** 2369–2382.
99. IRIBE, Y. *et al.* 1999. Subthalamic stimulation-induced synaptic responses in substantia nigra pars compacta dopaminergic neurons *in vitro*. J. Neurophysiol. **82:** 925–933.
100. KOBAYASHI, Y. *et al.* 2002. Contribution of pedunculopontine tegmental nucleus neurons to performance of visually guided saccade tasks in monkeys. J. Neurophysiol. **88:** 715–731.
101. DOMMETT, E. *et al.* 2005. How visual stimuli activate dopaminergic neurons at short latency. Science **307:** 1476–1479.
102. SUTHERLAND, R.J. 1982. The dorsal diencephalic conduction system: a review of the anatomy and functions of the habenular complex. Neurosci. Biobehav. Rev. **6:** 1–13.

Contextual Control of Choice Performance

Behavioral, Neurobiological, and Neurochemical Influences

JOSEPHINE E. HADDON AND SIMON KILLCROSS

School of Psychology, Cardiff University, Cardiff, United Kingdom

ABSTRACT: An important aspect of decision making is the ability of responses to be controlled by different cues in different situations or contexts, especially when there is conflict between alternative responses or actions. Recently, a context-dependent biconditional task has been developed for rats that mimic some aspects of response conflict seen in human cognitive paradigms, such as the Stroop task. In this task, contextual cues are used to disambiguate conflicting response information provided by audiovisual compound stimuli. Here we review current findings that investigate some of the behavioral, neurobiological, and neurochemical mechanisms that underlie this use of contextual or task-setting information to resolve response conflict, and discuss future ways in which this research can be extended.

KEYWORDS: decision making; biconditional discrimination; response conflict; context; prefrontal; schizophrenia

CONTEXTUAL MODULATION OF DECISION-MAKING PERFORMANCE

Decision making involves the use of multiple cues to enable a choice to be made between different actions to achieve a particular goal or outcome. An important aspect of decision making is the ability for responding to be controlled by different cues in different situations. In everyday life, one has to learn to perform particular actions in the presence of specific cues. Sometimes, however, the commonly performed action has to be overridden to allow appropriate responding. To achieve this flexibility in responding, it is necessary to both learn about events in a manner specific to particular contexts, and to use this information to govern responding at a relevant point in time. The

Address for correspondence: S. Killcross, School of Psychology, Tower Building, Park Place, Cardiff University, Cardiff, CF10 3AT, United Kingdom. Voice: +44-29-20-87-5393; fax: +44-29-20-87-4858. KillcrossAS@cardiff.ac.uk

Ann. N.Y. Acad. Sci. 1104: 250–269 (2007). © 2007 New York Academy of Sciences.
doi: 10.1196/annals.1390.000

contextual modulation of behavior is thought to play a role in a wide variety of situations from relapse in drug addiction[1] to the formation of episodic memories.[2-4] Moreover, this contextual control has a pervasive influence on human behavior from the use of task instructions to guide effortful, goal-directed responding to the implicit or incidental control of behavior, and is thought to be especially important when faced with conflicting or competing response information.[5,6]

An important requirement of goal-directed responding is the need to select the appropriate response in situations when there is conflict between different actions or responses. Response conflict can occur as a consequence of ambiguous stimuli, when multiple responses are possible, or when a dominant action needs to be suppressed, and is observed in a number of cognitive tasks (i.e., continuous performance task,[7] task-switching paradigms,[8] and the Stroop task[9,10]). Cohen and colleagues[6,11,12] have proposed a model of choice behavior in which context-specific or task-setting information is used to resolve cue and response conflict. Failure to use contextual information often leads to inappropriate or impulsive behaviors and an inability to coordinate one's actions to achieve a goal. This kind of impairment is classically seen with patients with disorders of the frontal lobe[6] including patients with Parkinson's disease and schizophrenia.[12,13] In particular, patients with schizophrenia and frontal lobe damage are severely impaired on a number of cognitive tasks involving response conflict, such as the continuous performance task[7] and the Stroop task.[9,10] For example, patients with frontal damage have difficulty performing the Stroop task and display increased reaction times when selecting the subordinate but task-appropriate response (color naming) over the dominant tendency to read the word.[12,14]

Contextual Control of Response Conflict

To facilitate our understanding of the processes involved in the contextual modulation of response conflict, such as that observed in a number of human cognitive paradigms, and the neurobiological mechanisms underpinning these processes, we sought to examine similar behaviors in laboratory rats. This review will summarize previous and unpublished results from our current research, which address some of the mechanisms underlying contextual control of instrumental performance in rats. Although not the focus of this review, it should be noted that other researchers[15-17] have contributed to our understanding of knowledge in this area.

The Stroop Task

The Stroop task is often used to assess the use of task-setting instructions on goal-directed behavior in humans. The classic task[9,10] requires participants to read the word or name the color of the ink the word is written in. Color–word compounds comprise either congruent (i.e., the word GREEN in green ink) or incongruent (i.e., the word GREEN in red ink) word and color ink combinations. For successful performance, participants must use the task-relevant attribute of the compound to control responding while ignoring the alternative, task-irrelevant, attribute. This ability to use the task-relevant attribute becomes especially important when the aim of the task is to name the color of an incongruent color–word compound as this requires participants to inhibit the strong tendency to read the word, resulting in a greater number of errors and longer latencies to respond.[9,10]

Context-Dependent Biconditional Discrimination Task in Rats

One way in which the Stroop task can be conceptualized is as an instructed conditional discrimination task in which correct responding to a compound stimulus is dictated by the current task instruction; that is, if given task instruction T1, respond according to the color of a stimulus, whereas if given task instruction T2, respond by reading the word. To study this type of process in animals, we developed a novel task that led to context-dependent learning of stimulus–response associations and the subsequent use of these contextual cues to govern choice responding when presented with conflicting cue information.

In this context-dependent biconditional task, contextual cues are used to govern choice responding when animals are presented with audiovisual compounds that provide conflicting response information.[18] Rats are trained on two instrumental biconditional discrimination tasks, one auditory and one visual, in two distinct contexts (TABLE 1). In one context, C1, production of lever press LP1, but not LP2, in the presence of auditory cue A1 produces reinforcement; production of lever press LP2, but not LP1, produces reinforcement during the presence of auditory cue A2. In the second context, C2, reinforcement follows lever press LP1 during visual cue V1, and lever press LP2 during visual cue V2. At test, rats are then presented with audiovisual compounds of the training stimuli in each of the training contexts, C1 and C2. These audiovisual compounds are composed of the stimulus elements that dictated the same

TABLE 1. Experimental design

| | Biconditional Training | TEST Sessions | |
		Congruent	Incongruent
C1	A1:LP1 → R1, A2:LP2 → R1	A1V1, A2V2	A1V2, A2V1
C2	V1:LP1 → R2, V2:LP2 → R2	A1V1, A2V2	A1V2, A2V1

C1/C2, R1/R2, LP1/LP2, A1/A2, and V1/V2 were different experimental chambers, reinforcers, lever presses, auditory and visual stimuli.

lever press responses during training (i.e., using the example above, A1V1 and A2V2) or stimulus elements that were associated with different lever press responses during training (i.e., A1V2 and A2V1), termed congruent and incongruent stimulus compounds, respectively. Responses during incongruent stimulus compounds are defined as correct or incorrect according to whether they are appropriate to the test context. For example, if presented with incongruent stimulus compound, A1V2, in context C2, then the context-appropriate response would be to produce LP1, as in context C1 production of LP1 during auditory cue A1 had been followed by reinforcement.

Using this procedure,[18] we demonstrated that the contextual cues came to control responding, such that in normal animals the context was able to control performance during incongruent stimulus compounds (FIG. 1). Successful performance during incongruent stimulus compounds revealed that rats were able to respond according to the context-appropriate stimulus element, thus, indicating that rats had learned about a particular discrimination (stimulus–response pairings) within a specific context, and that this contextual information was

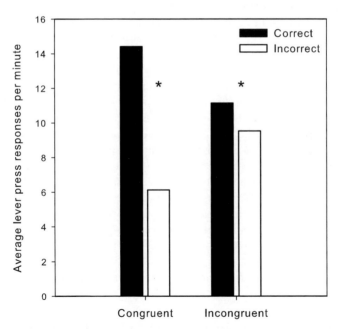

FIGURE 1. Context-appropriate responding to incongruent test compounds. Congruent stimulus compounds are composed of stimulus elements that required the same responses during training, whereas incongruent stimulus compounds required different responses during training. Correct responding to incongruent stimulus compounds indicates that the rats are using contextual cues to disambiguate the conflicting information from stimulus elements. *$P < 0.05$ significantly more correct than incorrect lever press responses. Adapted from Haddon, George & Killcross (in press).

able to govern responding when faced with the conflicting response information provided by novel incongruent compounds. This finding is consistent with human performance on the Stroop task in which participants are able to correctly respond to ambiguous visual cues according to the task instruction or "context" provided.

Relation to the Stroop Task

The context-dependent biconditional discrimination task can be viewed as reflecting some aspects of the response conflict seen in cognitive paradigms, such as the Stroop task in humans. The classic Stroop task requires the use of task instructions to disambiguate conflicting response information provided by incongruent color–word compounds, paralleling the way in which contextual cues disambiguate conflicting response information provided by audiovisual compounds in the rat task outlined above. In the Stroop task, participants have already learned the stimulus–response pairings associated with the individual stimulus elements of the color–word compounds, that is, to respond "red" when presented with the word or the color red, and to respond "green" to the word or color green. The lifetime acquisition of these stimulus–response associations may be seen as analogous to learning the basic auditory and visual stimulus–response pairings in solving the biconditional discriminations in particular contexts. At test, human participants may be presented with compounds of these previously acquired stimulus–response pairings that are either congruent (e.g., the word RED in the color red or the word GREEN in the color green) or incongruent (the word RED in the color green or the word GREEN in the color red), such that the stimulus elements require either the same or different responses, respectively. Hence the test compounds in the Stroop task are analogous to the audiovisual compounds in the rat task. In the two tasks participants are required to disambiguate conflicting response information provided by incongruent stimulus compounds by using task instructions or contextual cues to indicate the task-relevant stimulus element.

Behavioral Influences on the Contextual Control of Response Conflict

The successful development of this novel paradigm has allowed for a direct, and more detailed, examination of the underlying neurobiology and neurochemistry involved in the behavioral control of response competition in Stroop-like tasks than is currently possible with human neuroimaging and neuropsychological studies. Consequently, the use of this procedure may contribute to our understanding of the disruptions in cognitive function seen with a number of neuropsychological disorders, including schizophrenia.

Differential Training

An essential aspect of the response conflict seen in the Stroop task is that participants have greater experience reading words than they have naming colors. This difference in practice results in the classic asymmetry of Stroop interference; participants find it more difficult to name the color of incongruent stimulus compounds because of the greater interference produced by the strong tendency to read the word.[9,10] In further experiments,[18] we demonstrated that following asymmetric training on the two biconditional tasks (in which one discrimination was "overtrained" relative to the other "undertrained" discrimination), similar response competition and interference effects were observed in rats compared to those observed in human participants performing the Stroop task (FIG. 2). Rats demonstrated increased interference from the overtrained stimulus elements when required to produce the task-appropriate but undertrained response to incongruent stimulus compounds. This increased

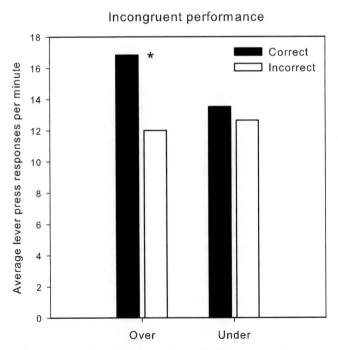

FIGURE 2. Context-appropriate responding to incongruent stimulus compounds in the overtrained but not the undertrained context. Context-appropriate responding to incongruent stimulus compounds is only observed in the overtrained context. The dominant response-set (overtrained discrimination) interferes with responding to incongruent stimulus compounds in the undertrained context. *$P < 0.05$ significantly more correct than incorrect lever press responses. Adapted from Haddon, George & Killcross (in press).

interference from the overtrained cues manifested as increased incorrect responses and decreased correct responses, and is consistent with observations of increased difficulty in naming the color of incongruent color–word combinations in humans.

These results (and those observed following equivalent training on the discriminations) can be explained in terms of the model of contextual control in which task-setting information is used to resolve cue and response conflict.[6,11,12] According to this view, the contextual stimuli serve as task-setting cues that boost the activation of the context- or task-relevant stimulus–response pathway, allowing the animals to respond according to the stimulus element that was previously trained in the test context. In the case of differential training, interference may manifest itself because the stimulus–response pathways associated with the overtrained stimulus elements are stronger than those associated with the undertrained stimulus element. When rats are presented with an incongruent stimulus compound in the undertrained context, despite the boosted activity of the undertrained stimulus–response pathway by the task-setting cues, the strong activation of the overtrained stimulus–response pathways remains and so context-appropriate responding is not observed.

Motivational Factors

As it is difficult to test explicitly the influence of motivational factors on decision-making processes in humans, it is of interest to investigate the role of motivational factors in rats in situations of response conflict similar to those used in humans. Consequently, in further behavioral studies, we adapted the context-dependent biconditional discrimination task to investigate the role of motivational factors on performance. Consider the situation in which different outcomes are consistently presented in the two different contexts, and thus consistently associated with one of the two biconditional discrimination tasks (auditory and visual). When the training stimuli are combined at test to produce the audiovisual compounds, not only are the two types of discrimination in competition, but the stimulus elements are also associated with different outcomes. Therefore, by manipulating the value of the different outcomes by employing a specific satiety reinforcer devaluation procedure, we investigated the way in which motivational variables related to the desirability of a particular outcome may influence the decision to respond when an animal is faced with conflicting response information.[19]

Reinforcer devaluation was found to result in a specific disruption of context-appropriate responding to incongruent stimulus compounds, suggesting that the use of contextual information to disambiguate conflicting response information is modulated by the value of the expected reward. Context-appropriate responding was not observed when animals were tested in a context previously associated with the devalued outcome. In models of response conflict, such as those proposed by Cohen and colleagues,[6,11,12] the context or task instructions are seen as providing abstract goal information, and the motivational influences related to the expected value of genuine outcomes are rarely, if ever, considered. As such, these models are essentially stimulus–response in nature. The performance following reinforcer devaluation suggests that in contrast to this stimulus–response approach, the abstract representation of goals provided by contextual or task-setting cues does not capture those aspects of choice performance found to come under the control of motivational or incentive systems. Thus, the use of task-setting cues to govern behavior and the influence of motivationally salient outcomes on choice behavior are, at least, partially dissociable. We have proposed an extension to current models of choice behavior, such as that proposed by Cohen and colleagues. This extension includes a representation of the outcome of an action that is distinct from contextual or goal information, and as such may allow such models to account for the influence of reinforcer devaluation on the contextual control of response conflict.[19]

NEUROBIOLOGICAL INFLUENCES

Prefrontal Cortex

The contextual control of response conflict has been shown to be dependent upon the integrity of the rat prefrontal cortex.[20–23] Large excitotoxic lesions to the prefrontal cortex (incorporating both the prelimbic and anterior cingulate cortices) were found selectively to abolish the ability to produce the context-appropriate response to incongruent stimulus compounds, while sparing congruent, and hence biconditional, performance (FIG. 3). This finding is consistent with both neuropsychological and neuroimaging research, which has implicated the human prefrontal cortex in task-appropriate responding in the Stroop task and similar assays of cognitive control.[12,14,24–26] For example, patients with frontal damage have difficulty performing the Stroop task and display increased reaction times when selecting the subordinate but task-appropriate response (color naming) over the dominant tendency to read the word.[12,14] This finding is also in agreement with models proposing a role for the prefrontal cortex in the ability to maintain and use goals to direct behavior, especially in situations of response conflict.[6,11,12]

Some researchers[27] have argued that there is little strong evidence for functional specialization within the prefrontal cortex. In contrast to this suggestion, our research has revealed that different subregions of the rat prefrontal cortex (specifically the prelimbic, infralimbic, and anterior cingulate cortices) have a profoundly different influence on the use of contextual cues to disambiguate conflicting response information provided by incongruent stimulus compounds.

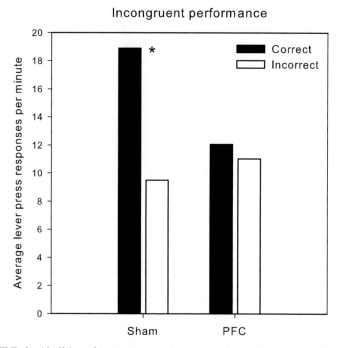

FIGURE 3. Abolition of context-appropriate responding to incongruent stimulus compounds in animals with PFC lesions. Context-appropriate responding is only observed in sham-operated animals. Prefrontal lesioned animals were unable to use the contextual cues to disambiguate the conflicting response information provided by incongruent stimulus compounds. *$P < 0.05$ significantly more correct than incorrect lever press responses. Adapted from Haddon & Killcross.[20]

Prelimbic Cortex

Further reversible inactivation studies demonstrated that the complete abolition of behavioral control by task-setting cues observed following large prefrontal cortex lesions appears to be a consequence of the extensive damage to the prelimbic region of the medial prefrontal cortex produced by these lesions. Temporary inactivation with the $GABA_A$ agonist muscimol of the prelimbic, but not the infralimbic, cortex produced a profound deficit in the use of task-setting contextual cues.[22,23] A specific role for the prelimbic cortex in goal-directed behavior is consistent with previous findings.[28–30] Damage to the prelimbic cortex leads to performance that is essentially habit based; responding is governed by stimulus–response associations and animals are insensitive to the current value of the goal of their actions.[31]

Moreover, the finding of impaired contextual control is also consistent with alternative characterizations of aspects of frontal function, such as behavioral flexibility in response to novel or confusing situations, in the active maintenance of a task or goal (working memory), and in situations in which novel distracters have to be ignored.[11,32–36] For example, it has been proposed that this frontal-dependent enhancement of task-relevant information by task-setting cues may necessitate the maintenance of relevant information over time[5] and may also allow appropriate responding in the face of more compelling (i.e., as a consequence of novelty) alternative behavioral choices.

Anterior Cingulate Cortex

Pretraining excitotoxic lesions to the anterior cingulate subregion of the prefrontal cortex resulted in a transient impairment in task-appropriate responding to contextual cues during the first 10 sec of stimulus presentation (FIG. 4). Context-appropriate responding emerged during the final 50 sec of stimulus presentation.[20] This finding is in accordance with theories implicating the anterior cingulate cortex in response-related processes, such as the detection of response conflict[24,25,37] or in "selection for action" processes.[38] These response-related theories suggest that the anterior cingulate cortex promotes

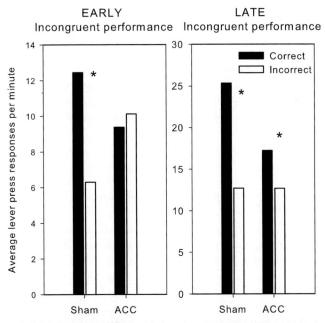

FIGURE 4. Context-appropriate responding abolished early in incongruent stimulus presentation in animals with anterior cingulate lesions. Context-appropriate responding emerged in the last 50 s of stimulus presentation. *$P < 0.05$ significantly more correct than incorrect lever press responses. Adapted from Haddon & Killcross.[20]

the exertion of greater cognitive control in situations when increased control is required.[39] Although further experimentation is required to address the specific function of the anterior cingulate cortex, the results are in agreement with this general property, demonstrating decreased correct responding early in incongruent stimulus presentation as a result of a failure to rapidly enhance activation of task-relevant pathways after the detection of error or conflict.

Infralimbic Cortex

In contrast to the results from manipulations of the prelimbic and anterior cingulate cortices, temporary inactivation of the infralimbic region of the prefrontal cortex was not found to influence the ability to use contextual cues to disambiguate the conflicting response information provided by incongruent stimulus compounds.[22,23] Previous research is consistent with this finding; lesions to the infralimbic cortex have been found to lead to performance that remains sensitive to the current value of the goal, even in situations in which normal animals no longer show this sensitivity, that is, following extensive training.[30] Moreover, temporary inactivation of this region has been reported to reinstate goal-directed responding in extensively trained animals where behavior would normally be controlled by stimulus–response habits.[29] Thus, the failure to observe an influence of infralimbic inactivation is entirely consistent with this viewpoint. If responding is goal-directed, infralimbic inactivation would preserve the animals' ability to use contextual cues to direct task-appropriate responding. Further in line with this suggestion are preliminary studies demonstrating that infralimbic inactivation ameliorates the interference from overtrained stimulus–response pairings when animals are required to produce the task-appropriate, but undertrained, response to incongruent stimulus compounds (unpublished data).

Hippocampal Formation

Damage to the hippocampal formation has often been reported to result in a failure to demonstrate context-specific learning, while preserving other behaviors or knowledge that are distinct from contextual information. For example, hippocampal damage has been found to result in impaired context-specific latent inhibition[40] and impaired background contextual conditioning.[41–43] In contrast to these results, excitotoxic lesions to the hippocampal formation failed to influence the contextual control of biconditional discrimination performance, despite the contextual information being incidental to the solution of the discrimination tasks during training (FIG. 5). Moreover, animals with hippocampal damage actually demonstrated responding to congruent and incongruent stimulus compounds that was better than the performance of

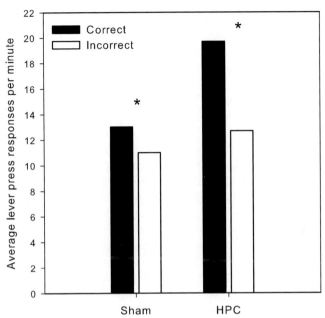

FIGURE 5. Contextual control of incongruent performance remains intact following excitotoxic lesions to the hippocampal formation. Both sham-operated and hippocampal-lesioned animals were able to use the contextual cues to disambiguate the conflicting information from stimulus elements. *$P < 0.05$ significantly more correct than incorrect lever press responses.

sham-operated controls (unpublished data). In further tests, animals with hippocampal lesions failed to show an influence of reinforcer devaluation on Pavlovian (magazine approach) performance when the test context was associated with a devalued reinforcer. In contrast, the same animals were able to show an influence of reinforcer devaluation when using an instrumental (lever press) measure of performance. This finding is in contrast with previous research that has suggested a role for the hippocampal formation in the formation of context–outcome associations,[44,45] and provides some questions for theories suggesting a role for the hippocampal formation in incidental learning about contextual cues.[43]

The result of the hippocampal lesion study suggests that the contextual control of responding to incongruent stimulus compounds is unlike the contextual modulation of associations thought to be related to the hippocampal formation, and perhaps episodic memory. Rather, this prefrontal- dependent control appears to be best characterized as an explicit goal-directed process in which contextual or task-setting cues are used as the basis of higher order rules to enable the appropriate control of choice responses.

DOPAMINERGIC INFLUENCES

Researchers have suggested that the use of contextual or task-setting cues to govern responses is dependent upon dopamine function within the prefrontal cortex, and dysfunction of dopamine modulation has also been implicated in the pathology of schizophrenia.[12,46] Although recent research has suggested a role for dopamine in the use of conditional information to guide behavior[47] little research has examined the locus of action of these systemic drug effects. With this in mind, we have recently used the context-dependent biconditional task to examine the modulation of response conflict by prefrontal dopamine. Following training on the auditory and visual tasks, rats received infusions of either a control substance (artificial cerebrospinal fluid, aCSF) or the D1 dopamine receptor agonist SKF-38393 (SKF) into the prelimbic region of the prefrontal cortex and responses to congruent and incongruent compounds were recorded. Overall, boosting prefrontal dopamine receptor activation by directly infusing SKF resulted in improved performance during incongruent compounds (in which response conflict is high) but impaired performance during congruent compounds (in which response conflict is low). Moreover, the performance of individual animals following SKF infusion was found to be related to their baseline performance following aCSF infusion. More specifically, negative linear correlations were observed between baseline performance and the change in performance following SKF infusion during both congruent and incongruent compounds. Animals demonstrating poor baseline performance showed improved performance following SKF infusion; in contrast, animals demonstrating good baseline performance showed poorer performance following infusion of SKF. Although a correlation between baseline performance and performance change following SKF infusion was evident during both congruent and incongruent compounds (after all, both involve some level of conflict), the effect was more marked during incongruent performance (FIG. 6).

In addition to supporting previous work indicating a role for prefrontal dopamine in the use of contextual cues to resolve response conflict, the discovery of a relationship between baseline performance and subsequent performance following SKF infusion is also consistent with the inverted-U hypothesis that relates efficiency of cognitive performance to the level of D1 receptor stimulation.[48,49] According to this hypothesis, in rats with lower baseline performance, dopamine activity within the prefrontal cortex is suboptimal and so can be boosted by infusion of the D1 agonist to near optimal levels, resulting in improved performance. In contrast, in rats with higher baseline performance, dopamine activity within the prefrontal cortex is already optimal, and so increasing dopamine activity by D1 agonist infusion produces too much activity and leads to a deterioration in performance. The finding of an influence of D1 receptor activity on this task, however, does not rule out a role for other types of dopamine receptor in the contextual control of response conflict. Indeed, recent research has demonstrated that the principle of the inverted-U function

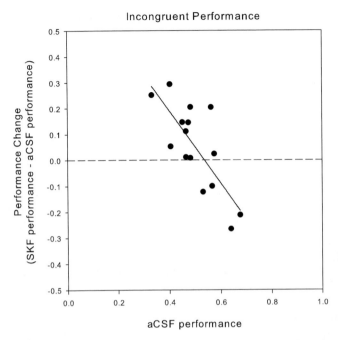

FIGURE 6. Incongruent performance change following infusion of the dopamine receptor agonist SKF-38393 into the prefrontal cortex. Animals demonstrating poor baseline performance showed improved performance following SKF infusion; in contrast, animals demonstrating good baseline performance showed poorer performance following infusion of SKF. ($R^2 = 0.59$, $P < 0.001$).

of D1 receptor does not necessarily apply to other prefrontal functions.[50] For example, psychopharmacological studies have shown that only D1 receptor activity is necessary for working memory performance.[51,52] In contrast, D1 and D2 receptor function has been reported to interact to mediate behavioral flexibility,[53] and blockade of all three dopamine receptor subtypes has been shown to impair conditioned punishment.[50]

SUMMARY AND FUTURE CONSIDERATIONS

The studies discussed in this review represent only a first step toward a more complete understanding of the behavioral, neurobiological, and neuro-chemical mechanisms that underlie decision making in the face of response conflict. First, although these studies have addressed some of the behavioral influences on the contextual control of incongruent responding including the effects of differential training and outcome devaluation on performance, they have not extensively investigated the situations in which interference from the task-irrelevant stimulus elements is increased or decreased. In particular, in what circumstances can goal-directed responding prevail? In human cognitive paradigms, such as the Stroop task, performance has been shown to be dependent upon working memory load (which can be manipulated by varying the ratio of congruent and incongruent trials); exposure to relatively few congruent trials encourages goal-directed responding as participants are continually required to use task-setting cues to govern appropriate performance.[54] Moreover, MacLeod[55] demonstrated that the amount of practice on, and separation of, the two dimensions (color and word) both strongly influenced the degree of interference observed. As a consequence, further studies could examine whether cognitive control in rats, like humans, is influenced by manipulations that vary the degree to which a task set or goal is activated.

Second, this review has emphasized a role for the prefrontal cortex in the use of task-setting cues to govern responses, consistent with findings from human cognitive paradigms. However, it is likely that other brain regions may also have an influence on performance of the context-dependent biconditional discrimination task. For example, Cohen and colleagues[56] recently proposed roles for the ventral tegmental area and locus coeruleus in the adaptive updating of task representations, and the allocation of cognitive control, respectively. Thus, damage to either of these regions may influence context-appropriate responding to incongruent stimulus compounds, especially in situations where the task instructions or contexts are rapidly changing and so response conflict is high. In addition, we have shown that context-appropriate responding on this task is dependent upon the current value of the outcome.[19] Consequently, one might expect that damage to regions of the brain thought to be important in representing the motivational value of outcomes (i.e., basolateral amygdala[57,58]) or response–outcome associations (i.e., posterior dorsomedial striatum[59]) may have a discrete influence on the control of task performance by outcome devaluation, but not performance governed by task-setting cues.

Third, further studies have highlighted the importance of prefrontal dopamine function in the use of task-setting cues to govern responding in the face of competing response information. Both too much and too little prefrontal dopaminergic activity was found to result in impaired incongruent performance. However, this is not to say that other neurotransmitters do not play a role in performance of this task. Indeed, other neurotransmitters are known to modulate prefrontal function (i.e., norepinephrine,[48] perhaps via the locus coeruleus[56]), or have a modulatory influence on dopamine transmission (i.e., N-Methyl-D-Aspartate (NMDA)[60]). As such, one might also expect that modulation of these neurotransmitters might influence the contextual control of response conflict.

Finally, the studies discussed in this review provide support for the view of conflict control proposed by Cohen and colleagues[6,11,12] in which the activation of contextual or task-setting units bias responses along task-appropriate stimulus–response pathways. This review has also highlighted conditions in which this model fails to make any predictions, for example, the influence of reinforcer devaluation on performance. We have proposed an extension to this model, which includes a representation of the current value of an outcome,[19] however, the influence of an outcome (and hence reinforcer devaluation) on choice control could occur in two ways. Changes in responding following reinforcer devaluation could be a result of an influence of the outcome representations via the stimulus–response pathways (between a biconditional cue and its associated response) or via an effect of these representations at the level of the contextual cues, in both cases leading to a reduction in the ability of these pathways to modulate responding. Hence, further experimentation will be needed to further differentiate between these two potential influences of reinforcer devaluation.

CONCLUSIONS

To conclude, we have developed a novel context-dependent biconditional discrimination paradigm, which closely reflects some aspects of the response competition seen in human cognitive paradigms, such as the Stroop task. As such this procedure provides an important tool for understanding the behavioral and neurobiological processes underlying decision making in such tasks. Moreover, we have shown that this paradigm is highly sensitive to both prefrontal cortex manipulations and prefrontal dopaminergic function, both of which are thought to be dysfunctional in a number of neuropsychological disorders, such as schizophrenia,[12,61,62] and so this paradigm may be of importance in the assessment of preclinical agents for the treatment of the cognitive impairments observed in such disorders.

ACKNOWLEDGMENTS

This work was supported by an Independent Investigator award (Southwest Florida named award), from the National Alliance for Research into Schizophrenia and Depression, to SK, and a BBSRC CASE award in conjunction with Merck, Sharp and Dohme, to JH.

REFERENCES

1. SHAHAM, Y., U. SHALEV, L. LU, et al. 2003. The reinstatement model of drug relapse: history, methodology and major findings. Psychopharmacology. **168:** 3–20.
2. BURGESS, N. 2002. The hippocampus, space, and viewpoints in episodic memory. Quart. J. Exp. Psychol. A. **55:** 1057–1080.

3. BUTLER, J. & C. ROVEE-COLLIER. 1989. Contextual gating of memory retrieval. Dev. Psychobiol., **22:** 533–552.
4. CHUN, M., & E. PHELPS. 1999. Memory deficits for implicit contextual information in amnesic subjects with hippocampal damage. Nat. Neurosci., **2:** 775–776.
5. COHEN, J.D., T.S. BRAVER & R.C. O'REILLY. 1998. A computational approach to prefrontal cortex, cognitive control and schizophrenia: recent developments and current challenges. *In* The Prefrontal Cortex: Executive and Cognitive Functions. A.C. Roberts, T.W. Robbins & L. Weiskrantz, Eds.: 195–220. Oxford University Press. UK.
6. MILLER, E.K., & J.D. COHEN. 2001. An integrative theory of prefrontal cortex function. Annu. Rev. Neurosci. **24:** 167–202.
7. ROSVOLD, H.E., A.F. MIRSKY, I. SARANSON, *et al*. 1956. A continuous performance test of brain damage. J. Consult. Psychol. **20:** 343–350.
8. WYLIE, G. & A. ALLPORT. 2000. Task switching and the measurement of "switch costs." Psychol. Res. **63:** 212–233.
9. STROOP, J.R. 1935. Studies of interference in serial verbal reactions. J. Exp. Psychol. **18:** 643–662.
10. MACLEOD, C.M. 1991. Half a century of research on the Stroop effect: an integrative review. Psychol. Bull. **109:** 163–203.
11. COHEN, J.D., K. DUNBAR & J.L. MCCLELLAND. 1990. On the control of automatic processes: a parallel distributed processing account if the Stroop effect. Psychol. Rev. **97:** 332–361.
12. COHEN, J.D. & D. SERVAN-SCHREIBER. 1992. Context, cortex, and dopamine: a connectionist approach to behavior and biology in schizophrenia. Psychol. Rev. **99:** 45–77.
13. BRAVER, T.S., D.M. BARCH & J.D. COHEN. 1999. Cognition and control in schizophrenia: a computational model of dopamine and prefrontal function. Biol. Psychiatry **46:** 312–328.
14. PERRET, E. 1974. The left frontal lobe of man and the suppression of habitual responses in verbal categorical behavior. Neuropsychologia **12:** 323–330.
15. BRAVER, T.S. & D.M. BARCH. 2002. A theory of cognitive control, aging cognition, and neuromodulation. Neurosci. Biobehav. Rev. **26:** 809–817.
16. DALLEY, J.W., R.N. CARDINAL & T.W. ROBBINS. 2004. Prefrontal executive and cognitive functions in rodents: neural and neurochemical substrates. Neurosci. Biobehav. Rev. **28:** 771–784.
17. MILLER, E.K. 2000. The prefrontal cortex and cognitive control. Nat. Rev. Neurosci. **1:** 59–65.
18. HADDON, J.E., D.N. GEORGE & A.S. KILLCROSS. Contextual control of biconditional task performance in rats: an analogue of the Stroop task. Quart. J. Exp. Psychol. (In press.)
19. HADDON, J.E. & A.S. KILLCROSS. 2006a. Both motivational and training factors affect response conflict choice performance in rats. Neural Networks. Special Issue on the Neurobiology of Decision Making **19:** 1192–1202.
20. HADDON, J.E. & A.S. KILLCROSS. 2006b. Prefrontal cortex lesions disrupt the contextual control of response conflict. J. Neurosci. **26:** 2933–2940.
21. HADDON, J.E. & A.S. KILLCROSS. 2005. Medial prefrontal cortex lesions abolish contextual control of competing responses. J. Exp. Anal. Behav. **84:** 485–504.
22. KILLCROSS, A.S., J-P. MARQUIS & J.E. HADDON. 2005. GABAA agonist injection in prelimbic cortex, but not in infralimbic cortex, impairs contextual control of response conflict in rats. Behav. Pharmacol. **16:** S27.

23. Marquis, J-P, A.S. Killcross & J.E. Haddon. 2007. Inactivation of the prelimbic, but not infralimbic, prefrontal cortex, impairs the contextual control of response conflict in rats. Eur. J. Neurosci. **25:** 559–566.

24. Botvinick, M., L.E. Nystrom, K. Fissell, *et al*. 1999. Conflict monitoring versus selection-for-action in anterior cingulate cortex. Nature **402:** 179–181.

25. Carter, C.S., T.S. Braver, D.M. Barch, *et al*. 1998. Anterior cingulate cortex, error detection, and the online monitoring of performance. Science **280:** 747–749.

26. MacDonald, A.W., J.D. Cohen, V.A. Stenger, *et al*. 2000. Dissociating the role of dorsolateral prefrontal cortex and anterior cingulate cortex in cognitive control. Science **288:** 1835–1838.

27. Duncan, J. & A.M. Owen. 2000. Common regions of the human frontal lobe recruited by diverse cognitive demands. Trends Neurosci. **23:** 475–483.

28. Corbit, L.H. & B.W. Balleine. 2003. The role of the prelimbic cortex in instrumental conditioning. Behav. Brain Res. **146:** 145–157.

29. Coutureau, E. & S. Killcross. 2003. Inactivation of the infralimbic cortex reinstates goal-directed responding in overtrained rats. Behav. Brain Res. **146:** 167–174.

30. Killcross, S. & E. Coutureau. 2003. Coordination of actions and habits in the medial prefrontal cortex of rats. Cereb. Cortex. **13:** 400–408.

31. Balleine, B. & A. Dickinson. 1998. Goal-directed instrumental action: contingency and incentive learning and their cortical substrates. Neuropharmacology. **37:** 407–419.

32. Ragozzino, M.E., S. Detrick & R.P. Kesner. 1999a. Involvement of the prelimbic cortex in behavioural flexibility for place and response learning. J. Neurosci. **19:** 4585–4594.

33. Ragozzino, M.E., J. Kim, D. Hassert, *et al*. 2003. The contribution of the rat prelimbic-infralimbic areas to different forms of task-switching. Behav. Neurosci. **117:** 1054–1065.

34. Ragozzino, M.E., C. Wilcox, M. Raso, *et al*. 1999b. Involvement of rodent prefrontal cortex subregions in strategy switching. Behav. Neurosci. **113:** 32–41.

35. Delatour, B. & P. Gisquet-Verrier. 2000. Functional role of rat prelimbic-infralimbic cortices in spatial memory: evidence for their involvement in attention and behavioral flexibility. Behav. Brain Res. **109:** 113–128.

36. Rougier, N.P., D. Noelle, T.S. Braver, *et al*. 2005. Prefrontal cortex and the flexibility of cognitive control: rules without symbols. Proc. Natl. Acad. Sci. **102:** 7338–7343.

37. De Wit, S., Y. Kosaki, B.W. Balleine, *et al*. 2006. Dorsomedial prefrontal cortex resolves response conflict in rats. J. Neurosci. **26:** 5224–5229.

38. Frith, C.D., K.J. Friston, P.F. Liddle, *et al*. 1991. Willed action and the prefrontal cortex in man: A study with PET. Proc. Royal Soc. London B **244:** 241–246.

39. Rushworth, M.F., M.E. Walton, S.W. Kennerley, *et al*. 2004. Action sets and decisions in the medial prefrontal cortex. Trends Cogn. Sci. **8:** 410–417.

40. Honey, R. & M. Good. 1993. Selective hippocampal lesions abolish the contextual specificity of latent inhibition and conditioning. Behav. Neurosci. **107:** 23–33.

41. Phillips, R. & J. LeDoux. 1994. Lesions of the dorsal hippocampal formation interfere with background but not foreground contextual fear conditioning. Learn. Memory **1:** 34–44.

42. RUDY, J. & R. SUTHERLAND. 1995. Configural association theory and the hippocampal formation: an appraisal and reconfiguration. Hippocampus **5:** 375–389.

43. O'REILLY, R.C. & J.W. RUDY. 2001. Conjunctive representations in learning and memory: principles of cortical and hippocampal function. Psychol.Rev., **108:** 311–345.

44. FROHARDT, R.J., F.A. GUARRACI, & M.E. BOUTON. 2000. The effects of neurotoxic hippocampal lesions on two effects of context after fear extinction. Behav. Neurosci. **114:** 227–240.

45. WHITE, N.M. 1996. Addictive drugs as reinforcers: multiple partial actions on memory systems. Addiction **91:** 921–949.

46. WEINBERGER, D.R. 1988. Premorbid neuropathology in schizophrenia. Lancet **2:** 445.

47. DUNN, M., D. FUTTER, C. BONARDI, et al. 2004. Attenuation of d-amphetamine-induced disruption of conditional discrimination performance by α-flupenthixol. Psychopharmacology **177:** 296–306.

48. ARNSTEN, A.F.T. 1998. Catecholamine modulation of prefrontal cortical cognitive function. Trends Cogn. Sci., **2:** 436–447.

49. ZAHRT, J., J.R. TAYLOR & A.F.T. ARNSTEN. 1997. Supranormal stimulation of D1 dopamine receptors in the rodent prefrontal cortex impairs working memory performance. J. Neurosci. **17:** 8528–8535.

50. FLORESCO, S.B. & O. MAGYAR. 2006. Mesocortical dopamine modulation of executive functions: beyond working memory. Psychopharmacology, E-publication, May 3.

51. SAWAGUCHI, T. & P.S. GOLDMAN-RAKIC. 1991. D1 dopamine receptors in prefrontal cortex: involvement in working memory. Science **251:** 947–950.

52. SEAMANS, J.K., S.B. FLORESCO & A.G. PHILLIPS. 1998. D1 receptor modulation of hippocampal prefrontal cortical circuits integrating spatial memory with executive functions in the rat. Behav. Neurosci. **109:** 1063–1073.

53. FLORESCO, S.B., O. MAGYAR, S. GHODS-SHARAFI, et al. 2006. Multiple dopamine receptor subtypes in the medial prefrontal cortex of the rat regulate set-shifting. Neuropsychopharmacology **31:** 297–309.

54. KANE, M.J. & R.W. ENGLE. 2003. Working-memory capacity and the control of attention: the contributions of goal neglect, response competition, and task set to Stroop interference. J. Exp. Psychol. General **132:** 47–70.

55. MACLEOD, C.M. 1998. Training on integrated versus separated Stroop tasks: the progression of interference and facilitation. Memory Cognition **26:** 201–211.

56. COHEN, J.D., G. ASTON-JONES & M.S. GILZENRAT. 2004. A systems-level perspective on attention and cognitive control: guided activation, adaptive gating, conflict monitoring, and exploitation vs. exploration. In Cognitive Neuroscience of Attention. M.I. Posner Ed.: 71–90. Guilford Press. New York.

57. BALLEINE, B., A.S. KILLCROSS & A. DICKINSON. 2003. The effect of lesions of the basolateral amygdala on instrumental conditioning. J. Neurosci. **23:** 666–675.

58. BLUNDELL, P., G. HALL & S. KILLCROSS. 2003. Preserved sensitivity to outcome value after lesions to the basolateral amygdala. J. Neurosci. **23:** 7702–7709.

59. YIN, H.H., S.B. OSTLUND, B.J. KNOWLTON, et al. 2005. The role of the dorsomedial striatum in instrumental conditioning. Eur. J.Neurosci. **22:** 513–523.

60. SESACK, S.R., A.Y. DEUTCH, R.H. ROTH, *et al.* 1989. Topographical organisation of the efferent projections of the medial prefrontal cortex in the rat: an antero-grade tract-tracing study with Phaseolus vulgaris leucoagglutinin. J. Comparative Neurol. **290:** 213–242.
61. KRAEPLIN, E. 1950. *Dementia Praecox and Paraphrenia.* International Universities Press. New York.
62. LEVIN, S. 1984. Frontal lobe dysfunctions in schizophrenia. II. Impairments of psychological and brain function. J. Psychiatry Res. **18:** 57–72.

A "Good Parent" Function of Dopamine

Transient Modulation of Learning and Performance during Early Stages of Training

JON C. HORVITZ,[a] WON YUNG CHOI,[b] CECILE MORVAN,[a] YANIV EYNY,[b] AND PETER D. BALSAM[c]

[a]Boston College, Department of Psychology, Chestnut Hill, Massachusetts, USA

[b]Columbia University, Department of Psychology, New York, USA

[c]Department of Psychology, Barnard College, Columbia University, and New York State Psychiatric Institute, New York, USA

ABSTRACT: While extracellular dopamine (DA) concentrations are increased by a wide category of salient stimuli, there is evidence to suggest that DA responses to primary and conditioned rewards may be distinct from those elicited by other types of salient events. A reward-specific mode of neuronal responding would be necessary if DA acts to strengthen behavioral response tendencies under particular environmental conditions or to set current environmental inputs as goals that direct approach responses. As described in this review, DA critically mediates both the acquisition and expression of learned behaviors during early stages of training, however, during later stages, at least some forms of learned behavior become independent of (or less dependent upon) DA transmission for their expression.

KEYWORDS: learning; reinforcement; D1; D2; Parkinson; habit; electrophysiology; single unit; VTA; SN; LTP; glutamate; SCH23390; raclopride

INTRODUCTION

Dopamine (DA) neurons in the substantia nigra (SN) and ventral tegmental area (VTA) respond to unexpected rewards as well as some other classes of salient events, with phasic burst-mode activity lasting several hundreds of milliseconds.[1,2] Increases in midbrain DA discharge rate lead to elevations in DA release within forebrain target sites that are disproportionally large during burst-mode firing compared to release during firing of individual action

Address for correspondence: Jon C. Horvitz, Department of Psychology, Boston College, 301 McGuinn Hall, Chestnut Hill, MA 02467. Voice: 617-552-2999; fax: 617-552-0523.
jon.horvitz@bc.edu

Ann. N.Y. Acad. Sci. 1104: 270–288 (2007). © 2007 New York Academy of Sciences.
doi: 10.1196/annals.1390.017

potentials.[3] Increases in DA transmission modulate both the *throughput* of corticostriatal glutamate (GLU) signals, selectively amplifying strong relative to weak GLU-mediated excitatory postsynaptic potentials,[4–6] and the *plasticity* of these synapses,[7,8] presumably corresponding to DA's roles in response expression and acquisition, respectively.[5] D1 receptor blockade reduces while D2 blockade increases striatal membrane excitability,[9] long-term potentiation (LTP),[8] and at least some forms of conditioned response acquisition.[10]

The types of learning likely to be promoted by phasic increases in DA activity depend upon whether the DA response is selective to reward, or whether DA neurons respond to salient events more generally. Imagine, for example, that DA-modulated plasticity increases the future likelihood that current sensory inputs receive downstream processing (beyond the level of the striatal synapse) and that motivational and behavioral relevance of the inputs is assigned at a later stage of signal processing. DA responses to the onset of a wide category of salient unexpected events could promote such a function. Similarly, such a DA response could signal to target regions that current models of response–outcome relationships must be modified (since an unexpected event occurred), without specifying the nature of the necessary modification. On the other hand, if the DA signal sets current environmental inputs as goals that direct approach responses, updates the value of goals, and/or directly increases the likelihood that the just-emitted behavior is repeated,[11–13] then DA responses would be expected to uniquely code for rewards.

Questions regarding the types of associations that underlie the acquisition and expression of operant and Pavlovian learning, and which of these are likely to be modulated by DA activity within specific forebrain target regions, have been the topic of careful analyses (see, e.g., Ref.14). However, for the purposes of the discussion here, we will bracket these questions, and consider a simplified model (FIG. 1) in which a GLU input (A) can evoke one of a number of outputs (B), and the DA response is necessary for an increase in the strength of the currently active input–output connection through promotion of synaptic plasticity. For the sake of this discussion, we will assume that inputs represent sensory events and output elements represent behavioral response tendencies.[15] It should be noted that because GLU inputs to striatal target sites carry information beyond simply sensory data, including information regarding expected outcomes,[5] we do not believe stimulus–response (S–R) strengthening is the sole content of DA-modulated learning. However, regardless of the content of the DA-modulated learning, if one assumes that the expression of the learned behavior involves information flow through the same set of DA-modulated GLU synapses that are strengthened as a result of learning, it becomes clear why DA's role in acquisition and expression has been so difficult to disentangle.[11,16]

As an organism seeks rewards in its environment, it will occasionally produce a response that leads to unexpected primary or conditioned reward, and undergo a strengthening of currently active synapses (FIG. 1). Under a low

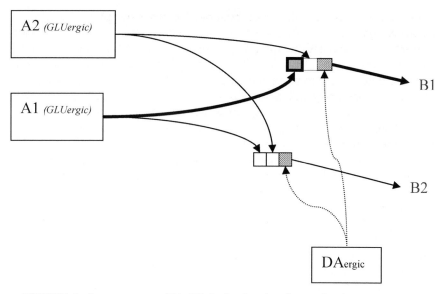

FIGURE 1. Output neurons (B1, B2) in the dorsal and ventral striatum receive GLU inputs (A1, A2) originating in cortical and limbic regions. The informational nature of these GLU inputs varies according to the striatal target region.[5] For the sake of the illustration, we imagine that the inputs (**A**) carry information about the sensory environment, and the outputs (**B**) represent behavioral response tendencies. DA activity influences information flow through,[5,76] and plasticity (LTP and LTD)[8] of the currently active A→B GLU synapse (*bold square*). DA binding to receptors on the output cells (*stippled squares*) does not promote strengthening of nonactive GLU synapses (*other squares*). The diagram is simplified regarding the nature of the input–output connectivity. There is a large (up to 10,000 to 1) convergence of information from the cortex to striatum; that is, a given striatal neuron receives a large number of cortical and/or limbic GLU inputs (for an examination of corticostriatal mapping, see review; Ref. 77). In addition, while dorsolateral striatal cells receive GLU input primarily from sensory-motor cortical regions, many striatal regions receive GLU inputs that carry information regarding sensory inputs, anticipated movements, expected outcomes, as well as information regarding appetitive and aversive valences of current inputs. Often, a single striatal neuron receives a convergence of such information, and responds to a conjunction of sensory, motor, and outcome–expectation conditions (see Ref. 5; review).

synaptic DA state, the likelihood of learning would be expected to be reduced for at least two reasons, corresponding to DA's role in GLU transmission and in plasticity of GLU synapses. First, reduced response output makes it less likely that the organism will encounter reward; if transmission across A→B synapses is less likely to occur, it is less likely that this connection will be strengthened. Second because DA promotes plasticity of active striatal GLU synapses, if A→B transmission were to occur and lead to reward under a low DA state, LTP would be disrupted,[8] reducing the likelihood that the active synapses (e.g., A1→B1) will undergo strengthening. The idea that reductions in DA transmission affect learning both by disrupting *synaptic plasticity* and *response expression* is of particular importance for those modeling DA's role in learning on the basis of experimental data on animal learning under conditions of pharmacologically altered DA transmission. Under such conditions, it is difficult to determine which of these two factors accounts for disruptions in response acquisition. Interestingly, one would assume that if a low synaptic DA state decreases GLU transmission across *previously strengthened* A→B connections, the expression of these connections and the corresponding learned behavior should be reduced. This appears to be true during early stages of learning, but *not* if the connection underwent extensive training prior to the reduction in DA transmission.[17] Because experimental disruptions in DA transmission affect both plasticity and response expression at least during early stages of learning, phenomena observed in slice preparations, such as opposing effects of D1 and D2 receptor antagonists on LTP, may not show correlates in behavioral learning since normal (or enhanced) learning often depends upon normal behavioral expression during the learning trials. This article will focus on (1) the conditions under which DA cells are activated, (2) the role of D1 and D2 receptor transmission in the acquisition of a conditioned response to a sensory cue, and (3) the role of DA in the *expression* of conditioned responding during early versus late stages of training. For reasons related to experimental methodology, we will discuss DA's role in response expression prior to its role in acquisition.

REWARD VERSUS NONREWARD DA RESPONSES

The magnitude of the phasic DA response to reward is inversely related to the animal's reward *expectation* at the time that it is delivered.[2,18] Phasic responses of DA neurons can thus be said to code the discrepancy between predicted and actual reward, that is, a reward prediction error.[2,19] Like the midbrain DA response, associative learning is greatest when presentation of the unconditioned stimulus (US) is unexpected.[20,21] The phasic DA excitatory response to unexpected primary and conditioned rewards has therefore been suggested to provide a learning signal, strengthening striatal input–output connections.[2,22,23]

Like unexpected rewards, some salient nonreward events produce a phasic DA activation. VTA DA neurons show burst firing in response to nonreward auditory and visual events (light flashes and loud clicks of 1 msec duration) that are salient by virtue of their intensity and rapid onset, and for which behavioral and phasic DA responses undergo little habituation over hundreds of stimulus presentations (FIG. 2).[24] The approximately 70 msec latency to peak DA response for these stimuli is similar to those reported for unexpected primary and conditioned rewards.[2] However, the VTA DA response to salient nonreward stimuli differs from that typically observed following the

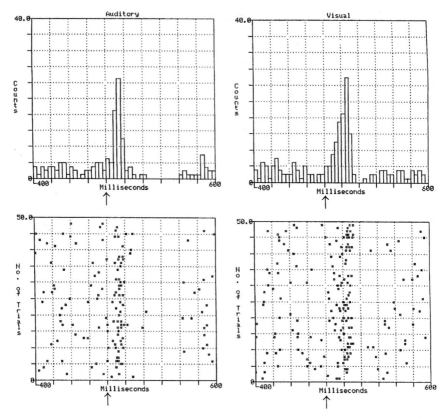

FIGURE 2. The responses of an individual VTA dopamine neuron to an auditory click (*left panel*) and a light flash (*right panel*). X axis ranges from 400 msec before to 600 msec after the auditory or visual stimulus (*arrow*) of 1 msec duration (time 0).[24]

presentation of primary and conditioned rewards in that the initial excitation, lasting approximately 100 msec, is typically followed by a period of inhibition lasting several hundreds of milliseconds. Salient nonreward stimuli also produce excitation followed by inhibition in DA cells of the SN.[25] Less intense stimuli that are salient by virtue of their novelty can also produce these types of excitatory–inhibitory responses.[26] In contrast, the phasic DA response to unexpected primary and conditioned rewards is characterized by an excitation for several hundreds of milliseconds, which is not followed by an inhibitory phase.[2]

Why might midbrain DA cells respond with excitation to both rewards and nonrewards in approximately 70 msec, and subsequently show an inhibition for nonrewards? Redgrave[27] noted that the 70 msec phasic DA response latency is likely to be too rapid for the reward versus nonreward status of a stimulus to be evaluated by the nervous system under natural conditions in which, for example, an animal is required to produce a saccade to foveate a visual stimulus, a response that requires longer than 70 msec. It is possible that DA cells respond to salient stimuli (e.g., stimuli that possess rapid onset, high intensity, or novelty) with an initial excitation before the reward or nonreward status of the event has been processed, and that the postexcitatory inhibition cancels out the preceding reward signal if the event is determined to be a nonreward.[28] One might speculate that the rapid DA excitatory response provides a more precise time-stamping of the event compared to that which would be possible if the DA neuron responded only after the reward/nonreward status of the event had been determined.[29]

The question of how DA cells respond to aversive events is more difficult to answer. A single-unit study found that only 11% of DA neurons respond to aversive stimuli, such as an air puff to the arm or quinine-adulterated liquid delivery[30] (but see Refs. 31 and 32). Microdialysis studies, however, have shown that DA concentrations in forebrain target sites are increased by aversive events, such as foot shock as well as rewarding stimuli, such as food (see Refs. 1 and 33; reviews). Some have reconciled these results by suggesting that microdialysis with its (typically) 10-min sampling period is measuring *gradual* rather than *phasic* changes in DA, and that while gradual increases in DA concentrations reflect both appetitive and aversive motivational arousal, *phasic* DA responses uniquely code a positive reward prediction error.[2,34] Alternatively, the differing results from dialysis and single-unit studies may be reconciled by assuming that DA is increased by strong aversive stimuli (such as footshock) that have been employed in dialysis studies[35–37] and not by weak aversive stimuli (such as air puffs to the arm and quinine-adulterated saline) that have typically been employed in single-unit studies.[30] It is important to note that while DA unit responses can be shown to be phasic versus gradual, dialysis measurements may reflect either gradual changes in DA concentration or phasic increases in DA concentration, which accumulate over the course of the dialysis sampling period. It is therefore possible that the onset (or offset) of strong aversive stimuli produce phasic increases in extracellular DA concentration that accumulate over the course of the DA sampling period.

However, in a recent single-unit study that did employ strong aversive stimuli (foot pinch), midbrain DA cells did not show phasic excitatory responses, but rather showed inhibition of firing.[38] In this study, the neurochemical identity of neurons presumed to be DAergic on the basis of electrophysiological criteria was confirmed on the basis of immunopositive responses to tyrosine hydroxylase. These data lend support to the idea that phasic DA excitation is not observed for even strongly aversive events, although the fact that animals were anesthetized during the single-unit recordings calls for some caution in comparing these results to the dialysis results in which similar types of aversive stimuli increase DA release in nonanesthetized animals. The hypothesis that motivationally arousing nonreward events elicit gradual rather than phasic increases in DA activity is therefore a hypothesis that has not yet been explicitly tested. Such a test requires subsecond measurements of DA activity (e.g., fast-scan cyclic voltammetry), which can distinguish phasic from gradual changes, while exposing nonanesthetized animals to the kinds of strong aversive stimuli that have been shown to elevate DA concentrations in dialysis experiments. Nevertheless, the fact that DA cells show a phasic excitatory response to a conditioned stimulus (CS) signaling reward, a phasic excitation followed by inhibition in response to salient nonrewards, yet most DA cells do not show a phasic DA excitation to aversive stimuli (and at least in anesthetized animals show pure inhibitory responses), suggests that aversive events may fall out of the category of stimuli that generate phasic DA excitation. If so, the nonreward status of some stimuli (such as foot pinch) must be transmitted to the midbrain DA cells prior to the 70 msec onset latency typically observed for DA excitatory responses. This assumption is reasonable, for a visual stimulus that signals the absence of reward under conditions in which reward is expected (when rewarded and nonrewarded trials are intermixed) can produce DA inhibition without a preceding excitatory phase when animals are required to fixate on a central screen position prior to presentation of the visual stimulus, and when screen position itself is the property that signals reward or nonreward.[39] Thus, under conditions in which environmental constraints do not slow the processing of the nonreward status of an event, midbrain DA cells appear capable of suppressing the early excitatory response and of showing inhibition alone.

DA AND RESPONSE EXPRESSION

While the precise manner in which midbrain DA cells respond to reward and some classes of nonreward stimuli remain to be precisely characterized, it is clear that a loss of, or strong disruption in, DA transmission leads to reductions in goal-directed movement for humans[40,41] and other animals.[42–44] In FIGURE 1, this may be represented as a failure of B1 to become activated by A1 (or a failure of B1 to be preferentially activated compared to B outputs that are antagonist to B1, see Ref. 5). Yet Parkinson's (PD) patients suffering nigrostriatal DA loss, show what is known as "paradoxical kinesia," that is, under certain environmental conditions, motor-impaired patients show normal or relatively normal movements. PD patients with locomotor impairment have been reported to walk quickly or even run out of a hospital room in response to a fire alarm, or to locomote normally when permitted to step over salient lines drawn on the ground.[45,46] Parkinsonian motor deficits are most pronounced when the movement requires internal guidance, and are greatly reduced when the movement is cued by a salient external stimulus.[47,48]

In accordance with the observation that DA loss in humans leaves behavior unimpaired in the presence of strong stimulus elicitors, we observe that rats

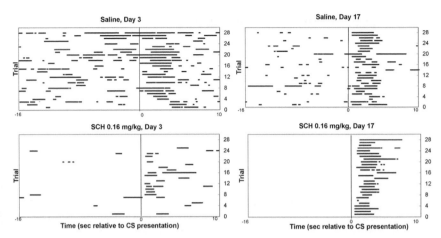

FIGURE 3. (**A**) Left panels: Raster plot of head entries (horizontal lines) from −16 sec before to 10 sec after CS presentation (x axis). Successive trials 1–28 are represented from the bottom to the top of the y axis. On day 3 of training (after approximately 60 trials) systemic 0.16 mg/kg SCH23390 strongly suppresses head entries during the intertrial interval and in response to the CS (time 0). (**B**) Right panels: Separate groups of rats receive VEH or SCH23390 on day 17 of training. In animals receiving extended training prior to D1 antagonist challenge, the drug continues to suppress spontaneous head entries emitted during the ITI (−16 to 0), but does not affect the latency to respond to the CS.[17]

under conditions of D1 receptor blockade are strongly impaired in generating approach responses to a food compartment in the absence of a salient response-eliciting cue, but can normally initiate the same behavior when it is cued by a well-trained CS. The top panels of FIGURE 3 show the response of representative drug-free rats that have learned that a brief auditory CS signals the immediate delivery of a food pellet into a food compartment. The x axis of the figure represents the period 16 sec before to 10 sec after CS presentation, and each head entry occurring during this period is depicted as a horizontal line. The y axis represents successive trials (1–28) from the bottom to the top of the y axis. The bottom panels show behavior during test sessions in which the selective D1 antagonist SCH23390[49] was administered prior to the session. FIGURE 3A shows that after 3 days of training, the drug disrupts both cued and noncued responding. However, after 16 daily sessions of 28-trial/session (approximately 450 total) CS–food pairings, the D1 antagonist (a) continues to reduce the frequency of noncued head entries during the intertrial interval, but (b) produces no impairment in the latency to perform the same head entry behavior in response to the CS (FIG. 3B). The DA-independent initiation of the approach response requires that the CS be well acquired, because approach responses to the CS during early stages of training are highly vulnerable to D1 receptor blockade.[17]

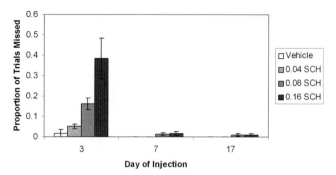

FIGURE 4. Proportion of missed trials (i.e., trials for which latency to respond to the CS was > 10 sec) as a function of SCH23390 dose on test day 3, 7, and 17. *Bars* represent the standard error of the mean. On day 3, SCH23390 produced a dose-dependent increase in the proportion of missed trials ($P < 0.0005$). In contrast, animals that received SCH23390 on either day 7 or 17 of training showed no increase in missed trials. Note that animals in each group received only a single injection of SCH23390.[17]

During this early stage of training, D1 receptor blockade does not simply produce a slowing of the approach response, but rather, appears to increase the likelihood that the animals will not emit a cue-elicited approach response. Trials on which the animal shows a latency of greater than 10 sec to emit a head entry following cue presentation are designated "missed trials." As can be seen in the bottom panel of FIGURE 3A, SCH-treated animals show a large number of "missed" trials (rows with no horizontal lines for 0 to 10 sec after CS presentation). FIGURE 4 shows that during a day 3 test session, the D1 antagonist produces a dose-dependent increase in the proportion of test trials that animals "miss." However, the D1 antagonist does not increase misses (or latencies) on days 7 or 17 of training. Because animals receive only a single drug injection, the change in response vulnerability to the D1 antagonist cannot be attributed to repeated drug administration. Further, the reduced vulnerability of the conditioned response to D1 antagonist challenge is not due to a reduced overall behavioral effectiveness of the drug in well-trained animals, for the drug suppresses locomotion (FIG. 5), as well as noncued head entries, during all phases of learning. The reduced vulnerability to D1 antagonist challenge is specific to the conditioned cue-elicited response. The cued approach is specifically mediated by D1 rather than D2 receptor activity, for during early phases of learning, the cued approach response is disrupted by D1 and not by D2 receptor antagonist treatment[50] and during later stages, it is disrupted by neither treatment.[17,51]

This raises the question of what type of change is occurring within DA target regions such that a behavioral response that was previously D1-dependent becomes independent of (or less dependent upon) DA transmission. At least two general kinds of change may be possible. First, it is possible that with extended

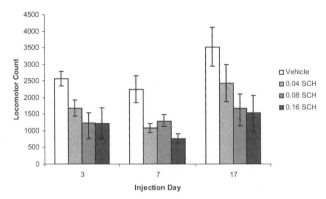

FIGURE 5. Two-way (Drug X Day) analysis of variance (ANOVA) conducted on days 3, 7, and 17 locomotor scores revealed a main effect of drug ($P < 0.000001$), a main effect of day ($P < 0.05$), and no interaction, showing that the drug produced similar locomotor suppression across different stages of learning.[17]

training, the behavioral response shifts to mediation by non-DA target areas, and therefore becomes less subject to DA modulation. It has been suggested that over the course of habit learning, learned sensory-motor representations may shift from corticostriatal–basal ganglia circuits to direct corticocortical mediation.[52,53] Alternatively, it is possible that as the conditioned behavioral response becomes well acquired, its expression continues to be mediated by the same neurons that originally mediated response expression, but DA plays a declining role in modulating that expression. For example, cortical GLU input to striatal cells may depend on D1 transmission to amplify the strength of task-relevant input signals (A→B1 in Fig. 1)[5,54] during early stages of learning. During later stages of learning, these GLU synapses may become so efficient that DA facilitation of GLU transmission is no longer necessary for normal responding.

To precisely characterize the nature of the changes that accompany the shift to DA-independent performance during *later* stages of training, it will be necessary to anatomically localize the site of DA's (D1-dependent) mediation of performance during *earlier* stages. Work in our laboratory by Won Yung Choi has been directed toward identifying the central site(s) of D1-mediated conditioned approach performance, an effort made difficult by the fact that DA forebrain target sites within which D1 receptor blockade reduces rates of lever pressing during early stages of learning (the nucleus accumbens core and prefrontal cortex[55,56]) are not substrates for D1-mediated performance of the conditioned approach.[57,58] Related to the goal of identifying the central site(s) of D1-mediated response expression during early stages of learning is the question of whether the shift to DA-independent performance occurs not only for the conditioned approach described above, but extends to other forms of learning (e.g., the operant lever press). If so, it is clear that the time course for the shift in the more complex lever press response is much longer than that for the simple cued approach.[59]

D1 PROMOTION OF APPETITIVE LEARNING

D1 and D2 receptor transmission has been shown to produce opposite effects on DA-modulated LTP. D1 antagonism reduces[7,60] while D2 antagonism or D2 receptor knockout[61,62] increases the magnitude of LTP. These opposite effects of D1 and D2 receptor blockade on LTP appear to result from their opposing effects on those intracellular cascades within striatal neurons, which lead to dopamine and cAMP-regulated phosphoprotein of 32 kDa (DARPP-32) phosphorylation. D1 receptor binding causes a G protein-mediated increase in the activity of adenylyl cyclase, increased cAMP formation, stimulation of protein kinase A, and phosphorylation (i.e., activation) of DARPP-32.[63,64] D2 binding produces the opposite effect on DARPP-32 activation, both by reducing adenylyl cyclase activity and through a mechanism that involves calcineurin activation.[64,65] It is known that the activation and deactivation state of DARPP-32 critically mediates the facilitative and inhibitory effects of D1 and D2 binding on LTP induction, respectively.[7,8] Thus, via opposing effects on the activation state of DARPP-32, D1 activity promotes and D2 activity restricts LTP.

However, the effect of D1 and D2 receptor blockade on behavioral learning is difficult to examine experimentally because D1 and D2 antagonist-induced disruptions in behavioral performance may indirectly lead to poor behavioral acquisition. If one were to administer a D1 or D2 antagonist on day 1 of cued approach training, many animals would fail to approach the food compartment at all during the session, and almost all animals would show very long latencies to respond to the CS (unpublished observations). It would be impossible to know whether these behavioral reductions were due to impaired learning or performance. If one were to test these animals during a drug-free test session on day 2, one would still be unable to determine whether increased latencies to respond to the CS on day 2 were the result of a day 1 learning deficit or the long CS–US intervals and fewer CS–US trials that the neuroleptic-treated rats experienced on day 1. (In FIG. 1, failure to express A1→B1 on day 1 precludes the strengthening of the connection.)

Even the place preference paradigm, which is one of the least problematic for assessing the effect of neuroleptic treatment on reward conditioning, suffers from the possibility of performance disruptions indirectly producing interference with learning. If DA antagonist treatment reduces exploratory behavior[17,58,66] during the sessions in which a distinctive environment is paired with reward, it will be difficult to determine whether apparent neuroleptic-induced disruptions of conditioning results from a disruption of learning or from a reduction in the animal's sampling of the CS, that is, the contextual stimuli of the environment.

One way to avoid this motor confound is to administer D1 or D2 antagonists *after* a learning session, since a disruption in learning (assessed during a later session when the drug is no longer behaviorally active) cannot be accounted for by behavioral disruptions during the learning session. Postsession D1 antagonist administration to the nucleus accumbens does disrupt some forms of appetitive learning[67] (but see Ref. 68). However, posttraining disruptions in DA transmission would only be expected to disrupt acquisition if DA facilitation of learning is due to tonic levels of DA transmission, which promote consolidation of information after the learning trials have terminated. To the extent that DA's role in learning is due to phasic DA responses at the time that the unexpected reward is presented (see above), DA antagonist administration after the conditioning session would not be expected to affect learning.

To ask whether D1 or D2 receptor blockade produces a disruption in the acquisition of appetitive learning, we took advantage of the fact that animals receiving 17 days of CS–food pairings in a nondrugged state show no reduction in the number of food pellets retrieved and no increase in the CS–food interval even when tested under relatively high doses of D1 or D2 antagonist drugs.[51,58] We therefore exposed animals to 17 days of training, in which the sound of the feeder signaled food delivery (28 trials per day, with a variable time 70 sec-inter-trial interval). On day 18, a tone was presented 3 sec prior to (and terminated 3 sec after) food delivery for animals pretreated with vehicle, D1 or D2, antagonist drugs. Another group received unpaired presentations of tone and food. On the following tone-alone drug-free test day (day 19), we examined approach responses elicited by the tone. As can be seen in FIGURE 6 (top panel), the high (0.16 mg/kg) dose of D1 antagonist SCH23390, administered during the day 18 conditioning session, reduced the animals' approach responses to the tone during the day 19 drug-free test session compared to animals that were not drugged during the conditioning session. The D2 antagonist raclopride during the day 18 conditioning session did not reduce approach responses to the tone on day 19, but rather produced a dose-dependent *increase* in responses to the CS. Importantly, neither the SCH23390 nor the raclopride groups showed increased latencies to retrieve the food pellet during the day 18 conditioning session.[10]

This D1 antagonist-induced disruption, and D2 antagonist-induced facilitation, of conditioned responding mirrors the effects of D1 and D2 antagonist drugs on striatal LTP (see Ref. 8) under conditions in which key parameters of conditioning (number of CS–US pairings and latency to respond to the CS) were themselves unaffected by the DA antagonists during the conditioning session. The finding that D1 receptor blockade reduced conditioning is consistent with a number of other studies showing that D1 receptor blockade reduces LTP[7,8] and disrupts operant and Pavlovian conditioning.[56,67,69,70] The finding

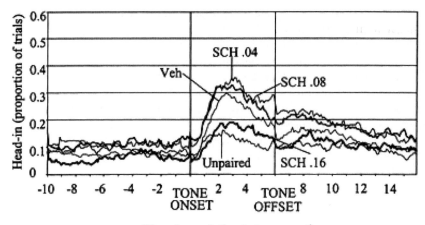

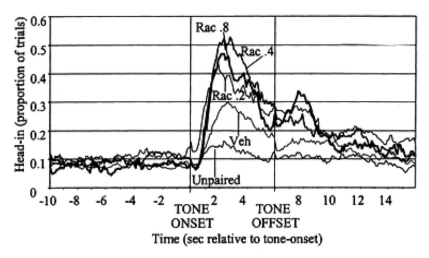

FIGURE 6. Performance of each treatment group during day 19, the drug-free test session. The y axis shows mean proportion of trials during which rats' heads were in the food compartment for each successive 100 msec bin during the 10 sec before tone onset (−10 to 0), the presentation of the tone (0–6), and the 10 sec after tone offset (6–16). As can be seen, rats that had received UNPAIRED presentations of CS and food showed reduced CS period head-in durations compared to paired CS–food controls ($P < 0.01$). Animals that were under the influence of the highest SCH23390 dose during tone–food pairings also showed a reduced CS head-in probability ($P < 0.05$). Animals that were under the influence of RAC during the tone–food pairings showed increased CS head-in probabilities compared to vehicle controls on test day ($P < 0.01$).[10]

that D2 receptor blockade increases conditioning mirrors the effects of D2 antagonism on LTP, suggesting that these opposite effects of D1 and D2 receptor blockade on learning may reflect their opposing effects on LTP.

However, there are two other plausible explanations for the observed D2 antagonist-induced promotion of conditioning. First, as noted above, to ensure that neuroleptic treatment did not disrupt approach responses during the conditioning session, animals received magazine training sessions for 17 days prior to the day 18 tone-food conditioning session. Conditioning of the context is strong after this sort of training.[71,72] Additionally, this procedure amounts to overtraining on CS1 (magazine) → food, and then conditioning CS2 (tone) → CS1(magazine) → food. It is possible that conditioning to the tone was to some degree attenuated by blocking, that is, because the context and the sound of the magazine already signaled food, the occurrence of the food was less surprising and less likely to strengthen the associative value of the tone than it would have been in the absence of overtraining.[73] A D2 antagonist-induced reduction in animals' *expectation* of food in the experimental context or following the sound of the magazine during the day 18 conditioning session could have reduced blocking and thereby produced increased conditioning to the tone compared to nondrugged animals. A second alternative explanation for the D2 antagonist-induced increase in conditioning is that while the D2 antagonist blocks postsynaptic D2 receptors, it also blocks D2 autoreceptors, leading to an increase in DA release at forebrain target sites.[74,75] Under conditions of D2 receptor blockade, D1 receptors, which remain available for DA binding, may be exposed to an increased amount of DA compared to the D1 occupancy for nondrugged animals. If one assumes that D1 transmission facilitates appetitive learning, as the present SCH23390 results and other studies cited above suggest, both D1 antagonist and D2 antagonist results can be accounted for by a D1-mediated promotion of learning.

SUMMARY

We have highlighted three issues that we believe merit careful experimental examination to understand more precisely the relationship between DA neuronal firing, response acquisition, and response expression: (1) While extracellular DA concentrations are increased by a wide category of salient events,[1] there is evidence to suggest that DA responses to primary and conditioned rewards may be distinct from those elicited by other types of salient events. This reward-specific mode of responding is necessary if DA functions to strengthen response tendencies adaptively under particular environmental conditions. (2) There is recent evidence that for some forms of neural plasticity (LTP) and learning (conditioned approach), interruption of D1 receptor transmission reduces while blockade of D2 receptors promotes plasticity and learning, suggesting that the two DA receptor families may play opposing roles in at least some forms of learning. Because data on opposing D1 and D2 roles in these functions come from a relatively small and recent set of findings, we view the claim as an early-stage hypothesis that merits further experimental examination, but one of potential importance. (3) After a large number of conditioning trials, at least some behaviors become less dependent upon DA transmission for their expression. It will be important to determine whether this reduced dependence upon DA transmission reflects a strengthening of GLU synapses during learning, such that DA is no longer needed to modulate synaptic transmission, or whether, as some have suggested, behaviors that were once represented in DA-innervated regions come to be represented in anatomically distinct regions that are not subject to DA's modulatory influence.

It is known that extended training of an operant response leads to a shift from "outcome-mediation" to S–R (automatized) performance.[76] After a small number of training sessions, animals for whom the outcome (e.g., food) value is degraded (e.g., when food is paired with LiCl-induced illness) show reduced vigor of operant responding (compared to control animals that received unpaired exposure to illness and food). In these procedures, devaluation of the outcome takes place when the animal does not have the opportunity to emit the learned behavioral response; that is, the response itself is not "punished." Reductions in the rate of responding during later sessions therefore reflect a reduction in the animal's representation of the outcome value, and demonstrate that the outcome representation modulates expression of the learned behavior. After extended training, at least some learned behaviors become insensitive to outcome devaluation, pointing to a shift from outcome-mediated to automatized responding.[76] It has recently been shown that amphetamine-induced sensitization of DA receptors prior to operant lever press training causes animals to undergo an abnormally rapid shift from outcome-mediated to S–R responding,[77] suggesting that DA transmission plays a role in the shift to automatized behavior. Current studies in our laboratory by Cecile Morvan suggest that the shift from outcome-mediated to S–R responding coincides temporally with the shift to DA-independent behavioral performance. Based upon the data above, it seems plausible that DA's role in response acquisition and expression resembles that of a "good parent" who, during early stages of learning, promotes behavioral acquisition and expression, permitting the behavior to finally be performed without DA's involvement.

REFERENCES

1. HORVITZ, J.C. 2000. Mesolimbocortical and nigrostriatal dopamine responses to salient non-reward events. Neuroscience 96: 651–656.
2. SCHULTZ, W. 1998. Predictive reward signal of dopamine neurons. J. Neurophysiol. 80: 1–27.
3. SUAUD-CHAGNY, M.F. et al. 1992. Relationship between dopamine release in the rat nucleus accumbens and the discharge activity of dopaminergic neurons during local in vivo application of amino acids in the ventral tegmental area. Neuroscience 49: 63–72.

4. CEPEDA, C. *et al.* 1998. Dopaminergic modulation of NMDA-induced whole cell currents in neostriatal neurons in slices: contribution of calcium conductances. J. Neurophysiol. **79:** 82–94.

5. HORVITZ, J.C. 2002. Dopamine gating of glutamatergic sensorimotor and incentive motivational input signals to the striatum. Behav. Brain Res. **137:** 65–74.

6. KIYATKIN, E.A. & G.V. REBEC. 1996. Dopaminergic modulation of glutamate-induced excitations of neurons in the neostriatum and nucleus accumbens of awake, unrestrained rats. J. Neurophysiol. **75:** 142–153.

7. CALABRESI, P. *et al.* 2000. Dopamine and cAMP-regulated phosphoprotein 32 kDa controls both striatal long-term depression and long-term potentiation, opposing forms of synaptic plasticity. J. Neurosci. **20:** 8443–8451.

8. CENTONZE, D. *et al.* 2001. Dopaminergic control of synaptic plasticity in the dorsal striatum. Eur. J. Neurosci. **13:** 1071–1077.

9. WEST, A.R. & A.A. GRACE. 2002. Opposite influences of endogenous dopamine D1 and D2 receptor activation on activity states and electrophysiological properties of striatal neurons: studies combining *in vivo* intracellular recordings and reverse microdialysis. J. Neurosci. **22:** 294–304.

10. EYNY, Y.S. & J.C. HORVITZ. 2003. Opposing roles of D1 and D2 receptors in appetitive conditioning. J. Neurosci. **23:** 1584–1587.

11. ETTENBERG, A. 1989. Dopamine, neuroleptics and reinforced behavior. Neurosci. Biobehav. Rev. **13:** 105–111.

12. WISE, R. 1982. Neuroleptics and operant behavior: the anhedonia hypothesis. Behav. Brain Sci. **5:** 39–87.

13. WISE, R.A. & P.P. ROMPRE. 1989. Brain dopamine and reward. Annu. Rev. Psychol. **40:** 191–225.

14. BALLEINE, B.W. 2005. Neural bases of food-seeking: affect, arousal and reward in corticostriatolimbic circuits. Physiol. Behav. **86:** 717–730.

15. WICKENS, J.R., J.N. REYNOLDS & B.I. HYLAND. 2003. Neural mechanisms of reward-related motor learning. Curr. Opin. Neurobiol. **13:** 685–690.

16. HORVITZ, J.C. & A. ETTENBERG. 1988. Haloperidol blocks the response-reinstating effects of food reward: a methodology for separating neuroleptic effects on reinforcement and motor processes. Pharmacol. Biochem. Behav. **31:** 861–865.

17. CHOI, W.Y., P.D. BALSAM & J.C. HORVITZ. 2005. Extended habit training reduces dopamine mediation of appetitive response expression. J. Neurosci. **25:** 6729–6733.

18. FIORILLO, C.D., P.N. TOBLER & W. SCHULTZ. 2003. Discrete coding of reward probability and uncertainty by dopamine neurons. Science **299:** 1898–1902.

19. MONTAGUE, P.R., P. DAYAN & T.J. SEJNOWSKI. 1996. A framework for mesencephalic dopamine systems based on predictive Hebbian learning. J. Neurosci. **16:** 1936–1947.

20. KAMIN, L.J. 1969. Selective Association and Conditioning. Fundamental Issues in Associative Learning. Dalhousie University Press. Halifax, Nova Scotia.

21. MACKINTOSH, N.J. 1975. A theory of attention: variations in the associability of stimulus with reinforcement. Psychol. Rev. **82:** 276–298.

22. BENINGER, R.J. 1983. The role of dopamine in locomotor activity and learning. Brain Res. **287:** 173–196.

23. WICKENS, J. 1990. Striatal dopamine in motor activation and reward-mediated learning: steps towards a unifying model. J. Neural Transm. Gen. Sect. **80:** 9–31.

24. HORVITZ, J.C., T. STEWART & B.L. JACOBS. 1997. Burst activity of ventral tegmental dopamine neurons is elicited by sensory stimuli in the awake cat. Brain Res. **759:** 251–258.
25. STEINFELS, G.F. *et al.* 1983. Response of dopaminergic neurons in cat to auditory stimuli presented across the sleep-waking cycle. Brain Res. **277:** 150–154.
26. SCHULTZ, W. & R. ROMO. 1990. Dopamine neurons of the monkey midbrain: contingencies of responses to stimuli eliciting immediate behavioral reactions. J. Neurophysiol. **63:** 607–624.
27. REDGRAVE, P., T.J. PRESCOTT & K. GURNEY. 1999. Is the short-latency dopamine response too short to signal reward error? Trends Neurosci. **22:** 146–151.
28. KAKADE, S. & P. DAYAN. 2002. Dopamine: generalization and bonuses. Neural Netw. **15:** 549–559.
29. REDGRAVE, P. & K. GURNEY. 2006. The short-latency dopamine signal: a role in discovering novel actions? Nat. Rev. Neurosci. **7:** 967–975.
30. MIRENOWICZ, J. & W. SCHULTZ. 1996. Preferential activation of midbrain dopamine neurons by appetitive rather than aversive stimuli. Nature **379:** 449–451.
31. GUARRACI, F.A. & B.S. KAPP. 1999. An electrophysiological characterization of ventral tegmental area dopaminergic neurons during differential Pavlovian fear conditioning in the awake rabbit. Behav. Brain Res. **99:** 169–179.
32. KIYATKIN, E.A. 1988. Functional properties of presumed dopamine-containing and other ventral tegmental area neurons in conscious rats. Int. J. Neurosci. **42:** 21–43.
33. SALAMONE, J.D. 1994. The involvement of nucleus accumbens dopamine in appetitive and aversive motivation. Behav. Brain Res. **61:** 117–133.
34. DAW, N.D., S. KAKADE & P. DAYAN. 2002. Opponent interactions between serotonin and dopamine. Neural Netw. **15:** 603–616.
35. ABERCROMBIE, E.D. *et al.* 1989. Differential effect of stress on *in vivo* dopamine release in striatum, nucleus accumbens, and medial frontal cortex. J. Neurochem. **52:** 1655–1658.
36. SORG, B.A. & P.W. KALIVAS. 1991. Effects of cocaine and footshock stress on extracellular dopamine levels in the ventral striatum. Brain Res. **559:** 29–36.
37. YOUNG, A.M., M.H. JOSEPH & J.A. GRAY. 1993. Latent inhibition of conditioned dopamine release in rat nucleus accumbens. Neuroscience **54:** 5–9.
38. UNGLESS, M.A., P.J. MAGILL & J.P. BOLAM. 2004. Uniform inhibition of dopamine neurons in the ventral tegmental area by aversive stimuli. Science **303:** 2040–2042.
39. TAKIKAWA, Y., R. KAWAGOE & O. HIKOSAKA. 2004. A possible role of midbrain dopamine neurons in short- and long-term adaptation of saccades to position-reward mapping. J. Neurophysiol. **92:** 2520–2529.
40. MARSDEN, C.D. 1984. Which motor disorder in Parkinson's disease indicates the true motor function of the basal ganglia? Ciba Found. Symp. **107:** 225–241.
41. STERN, E.R. *et al.* 2005. Maintenance of response readiness in patients with Parkinson's disease: evidence from a simple reaction time task. Neuropsychology **19:** 54–65.
42. CARLI, M., J.L. EVENDEN & T.W. ROBBINS. 1985. Depletion of unilateral striatal dopamine impairs initiation of contralateral actions and not sensory attention. Nature **313:** 679–682.
43. FOWLER, S.C. & J.R. LIOU. 1998. Haloperidol, raclopride, and eticlopride induce microcatalepsy during operant performance in rats, but clozapine and SCH 23390 do not. Psychopharmacology (Berl.) **140:** 81–90.

44. SALAMONE, J.D. & M. CORREA. 2002. Motivational views of reinforcement: implications for understanding the behavioral functions of nucleus accumbens dopamine. Behav. Brain Res. **137:** 3–25.
45. JAHANSHAHI, M. 1998. Willed action and its impairments. Cogn. Neuropsychol. **15:** 483.
46. MARTIN, J.P. 1967. The Basal Ganglia and Posture. Pitman Medical. London.
47. FRISCHER, M. 1989. Voluntary vs autonomous control of repetitive finger tapping in a patient with Parkinson's disease. Neuropsychologia **27:** 1261–1266.
48. SCHETTINO, L.F. *et al.* 2004. Deficits in the evolution of hand preshaping in Parkinson's disease. Neuropsychologia **42:** 82–94.
49. IORIO, L.C. *et al.* 1983. SCH 23390, a potential benzazepine antipsychotic with unique interactions on dopaminergic systems. J. Pharmacol. Exp. Ther. **226:** 462–468.
50. CHOI, W. & J.C. HORVITZ. 2003. D1 but not D2 Receptor mediation of the expression of a Pavlovian approach response. Society for Neuroscience Abstracts. **29:** 716.1.
51. HORVITZ, J.C. & Y.S. EYNY. 2000. Dopamine D2 receptor blockade reduces response likelihood but does not affect latency to emit a learned sensory-motor response: implications for Parkinson's disease. Behav. Neurosci. **114:** 934–939.
52. ASHBY, F.G., J.M. ENNIS & B.J. SPIERING. 2007. A neurobiological theory of automaticity in perceptual categorization. Psychol. Rev. In press.
53. CARELLI, R.M., M. WOLSKE & M.O. WEST. 1997. Loss of lever press-related firing of rat striatal forelimb neurons after repeated sessions in a lever pressing task. J. Neurosci. **17:** 1804–1814.
54. O'DONNELL, P. 2003. Dopamine gating of forebrain neural ensembles. Eur. J. Neurosci. **17:** 429–435.
55. BALDWIN, A.E., K. SADEGHIAN & A.E. KELLEY. 2002. Appetitive instrumental learning requires coincident activation of NMDA and Dopamine D1 Receptors within the medial prefrontal cortex. J. Neurosci. **22:** 1063–1071.
56. SMITH-ROE, S.L. & A.E. KELLEY. 2000. Coincident activation of NMDA and dopamine D1 receptors within the nucleus accumbens core is required for appetitive instrumental learning. J. Neurosci. **20:** 7737–7742.
57. CHOI, W. 2005. The role of D1 and D2 dopamine receptors in the expression of a simple appetitive response at different stages of learning. Dissertation, Columbia University. New York.
58. CHOI, W. & J.C. HORVITZ. 2002. Functional dissociations produced by D1 antagonist SCH 23390 infusion to the nucleus accumbens core versus dorsal striatum under an appetitive approach paradigm. Society for Neuroscience Abstracts. **28.**
59. MORVAN, C.I. *et al.* 2006. The shift to dopamine-independent expression of an overtrained pavlovian approach response coincides with the shift to S-R performance. Society for Neuroscience Abstracts. **463.10.**
60. KERR, J.N. & J.R. WICKENS. 2001. Dopamine D-1/D-5 receptor activation is required for long-term potentiation in the rat neostriatum *in vitro*. J. Neurophysiol. **85:** 117–124.
61. CALABRESI, P. *et al.* 1997. Abnormal synaptic plasticity in the striatum of mice lacking dopamine D2 receptors. J. Neurosci. **17:** 4536–4544.
62. YAMAMOTO, Y. *et al.* 1999. Expression of N-methyl-D-aspartate receptor-dependent long-term potentiation in the neostriatal neurons in an *in vitro* slice after ethanol withdrawal of the rat. Neuroscience **91:** 59–68.

63. HEMMINGS, H.C., Jr., A.C. NAIRN & P. GREENGARD. 1984. DARPP-32, a dopamine- and adenosine 3':5'-monophosphate-regulated neuronal phosphoprotein. II. Comparison of the kinetics of phosphorylation of DARPP-32 and phosphatase inhibitor 1. J. Biol. Chem. **259:** 14491–14497.

64. NISHI, A., G.L. SNYDER & P. GREENGARD. 1997. Bidirectional regulation of DARPP-32 phosphorylation by dopamine. J. Neurosci. **17:** 8147–8155.

65. GREENGARD, P., P.B. ALLEN & A.C. NAIRN. 1999. Beyond the dopamine receptor: the DARPP-32/protein phosphatase-1 cascade. Neuron **23:** 435–447.

66. AHLENIUS, S. *et al.* 1987. Suppression of exploratory locomotor activity and increase in dopamine turnover following the local application of cis-flupenthixol into limbic projection areas of the rat striatum. Brain Res. **402:** 131–138.

67. DALLEY, J.W. *et al.* 2005. Time-limited modulation of appetitive Pavlovian memory by D1 and NMDA receptors in the nucleus accumbens. Proc. Natl. Acad. Sci. USA. **102:** 6189–6194.

68. HERNANDEZ, P.J. *et al.* 2005. AMPA/kainate, NMDA, and dopamine D1 receptor function in the nucleus accumbens core: a context-limited role in the encoding and consolidation of instrumental memory. Learn. Mem. **12:** 285–295.

69. AZZARA, A.V. *et al.* 2001. D1 but not D2 dopamine receptor antagonism blocks the acquisition of a flavor preference conditioned by intragastric carbohydrate infusions. Pharmacol. Biochem. Behav. **68:** 709–720.

70. BENINGER, R.J. & R. MILLER. 1998. Dopamine D1-like receptors and reward-related incentive learning. Neurosci. Biobehav. Rev. **22:** 335–345.

71. BALSAM, P.D. & J. GIBBON. 1988. Formation of tone-US associations does not interfere with the formation of context-US associations in pigeons. J. Exp. Psychol. Anim. Behav. Process **14:** 401–412.

72. BALSAM, P.D. & A.L. SCHWARTZ. 1981. Rapid contextual conditioning in autoshaping. J. Exp. Psychol. Anim. Behav. Process **7:** 382–393.

73. KAMIN, L.J. 1969. Fundamental Issues in Instrumental Learning. N.J. Mackintosh & W.K. HONIG, Eds.: 42–64. Dalhousie University Press. Halifax, Nova Scotia.

74. ALTAR, C.A. *et al.* 1987. Dopamine autoreceptors modulate the *in vivo* release of dopamine in the frontal, cingulate and entorhinal cortices. J. Pharmacol. Exp. Ther. **242:** 115–120.

75. UNGERSTEDT, U. *et al.* 1985. Functional classification of different dopamine receptors. Psychopharmacology Suppl. **2:** 19–30.

76. DICKINSON, A. 1985. Actions and habits: the development of behavioral autonomy. Phil. Trans. R. Soc. London (biol.) **308:** 67–78.

77. NELSON, A. & S. KILLCROSS. 2006. Amphetamine exposure enhances habit formation. J. Neurosci. **26:** 3805–3812.

Serotonin and the Evaluation of Future Rewards

Theory, Experiments, and Possible Neural Mechanisms

NICOLAS SCHWEIGHOFER,[a] SAORI C. TANAKA,[b] AND KENJI DOYA[b,c]

[a]Department of Biokinesiology and Physical Therapy, University of Southern California, Los Angeles, California, USA

[b]Computational Neuroscience Laboratories, Advanced Telecommunications Research Institute, Kyoto, Japan

[c]Initial Research Project, Okinawa Institute of Science and Technology, Okinawa, Japan

ABSTRACT: The ability to select an action by considering both delays and amount of reward outcome is critical for survival and well-being of animals and humans. Previous animal experiments suggest a role of serotonin in action choice by modulating the evaluation of delayed rewards. It remains unclear, however, through which neural circuits, and through what receptors and intracellular mechanisms, serotonin affects the evaluation of delayed rewards. Here, we review experimental studies and computational theory of decisions under delayed rewards, and propose that serotonin controls the timescale of reward prediction by regulating neural activity in the basal ganglia.

KEYWORDS: discounting; impulsivity; reinforcement learning; discount rate; basal ganglia

INTRODUCTION

A neuromodulator, such as serotonin, is a neurotransmitter that has spatially distributed and temporally extended effects on the recipient neurons and circuits.[1-3] Neuromodulators have traditionally been assumed to be involved in the control of general arousal.[2,4] Recent studies in molecular biology and neuroscience, however, have provided a more complex picture, with sometimes hard-to-reconcile data on the spatial localization and physiological effects of

Address for correspondence : Nicolas Schweighofer, Department of Biokinesiology and Physical Therapy, University of Southern California, 1450 E. Alcazar Street, Los Angeles, CA, 90089. Voice: 323-442-1838; fax: 323-442-1515.
schweigh@usc.edu

Ann. N.Y. Acad. Sci. 1104: 289–300 (2007). © 2007 New York Academy of Sciences.
doi: 10.1196/annals.1390.011

different neuromodulators and their receptors. In particular, despite numerous physiological and pharmacological studies, the role of serotonin is unclear. Such an understanding of the role of serotonin is all the more needed that serotonin dysfunction is thought to be linked to a variety of common mood and behavior disorders, such as depression[5] and impulsivity.[6]

Here, we focus on one important (though definitely *not* exclusive) aspect of serotonin, suggested by lesion and pharmacological data, in choice behaviors with delayed rewards. Specifically, we propose a theory on the role of serotonin from the viewpoint that it is a medium for signaling specific global parameters, which controls the timescale of the evaluation of delayed rewards. The article is organized as follows. We begin by discussing the concepts of reward values and reward discounting, first in animals and humans, and then in light of the framework of reinforcement learning theory developed in artificial intelligence. We then review the role of the serotonergic system in impulsive behavior, and show that existing data point to a role of serotonin in regulating the rate of discounting of delayed rewards. Finally, we suggest functional–anatomical models of the role of serotonin in the evaluation of future rewards.

REWARDS DISCOUNTING IN ANIMALS AND HUMANS

When choosing between a larger but delayed reward, and a smaller but more immediate reward, we compare the "values" associated with each reward, and often choose the reward associated with the larger value.[7] Critical to these choices are the *shape* and the *steepness* of the reward values, which monotonically decrease as a function of the delay: the rewards are said to be discounted as a function of the delays (FIG. 1).

Two models that characterize the *shape* of reward discounting have been proposed: exponential[8–10] and hyperbolic.[11–19] The exponential discounting model leads to maximal gain under the assumption of a constant probability of reward loss per unit time and exact estimate of the time of the future reward delivery. The reward value V is then an exponential function of the delay:

$$V = R \, \exp(-k_e \, D), \tag{1}$$

where $k_e \geq 0$ is the decay rate, or equivalently

$$V = R \gamma^D, \tag{2}$$

where γ the discount factor ($0 \leq \gamma < 1$). A small discount factor, γ, which is equivalent to a large decay rate k_e as $\gamma = \exp(-k_e)$, results in steeper discounting.

In repeated reward choice animal experiments, assuming a constant intertrial interval, if the animal consistently makes a choice that gives the same reward R after the same delay D, the average reward rate is the hyperbolic function of the delay[20]:

$$V = R/(T + D), \quad \text{with } T > 0, \tag{3}$$

(A) **(B)**

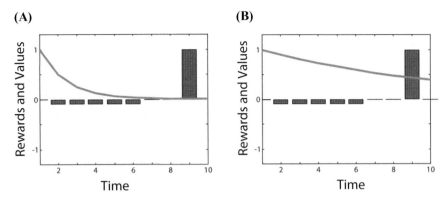

FIGURE 1. The effect of the discount factor γ in decision making. The lines show the discounted value and the bars the potential rewards. In a scenario where repeated small negative rewards (costs) are expected before receiving a large positive reward, the cumulative future reward V becomes slightly negative if the discount factor is small (**A**, at Time 10) and positive if the discount factor is large enough (**B**, at Time 10). In the situation in A, if there is baseline behavior in which the expected rewards and costs are zero, the agent will take no action.

where T is the sum of all times except the delay in each trial (which is often equal to the intertrial interval). When T is large, discounting is steep. In human experiments, where subjects make a number of one off choices, the hyperbolic function is often given by

$$V = R/(1 + k_y D), \text{with} k_y > 0, \tag{4}$$

where $k_y \geq 0$ is the discounting parameter. When k_y is large, discounting is steep. A remarkable feature of hyperbolic discounting is that the rate of growth of the value, that is, $V(D)/V(D + 1) = (1 + k_y (D + 1))/(1 + k_y D) = 1 + 1/(D + 1/k_y)$, increase as the delay D becomes close to zero, while the growth rate is constant in exponential model $(V(D)/V(D + 1) = 1/\gamma)$.[21] As a consequence, hyperbolic discounting (but not exponential discounting) can result in an "irrational" preference reversal. For instance, a person may prefer one apple today to two apples tomorrow, but at the same time prefer two apples in 51 days to one apple in 50 days.[22] Thus, hyperbolic discounting is often presented as a struggle between oneself and one's alter ego in the future, or similarly, between a myopic doer and a far-sighted planner.[23,24]

Most behavioral studies that have directly compared the two types of discounting in animals or humans have concluded that hyperbolic discounting better fits delayed reward choice data than does exponential discounting, for example.[12–14,25–27] We have, however, recently questioned the notion of hyperbolic reward discounting as a universal principle in humans.[28] In a reward decision task with temporal constraints in which each choice affects the time remaining for later trials, and in which the delays vary at each trial, we demonstrated that most subjects adopted exponential discounting, and by doing so maximized their total gain.

The *steepness* of discounting specifies how far in the future delayed rewards should be considered. A large discount rate (the k_y parameter in equations 4, or similarly the k_e parameter in equation 1, which corresponds to small discount factor γ in equation 2), which results in steep discounting, biases individuals to acquire small and more immediate rewards, as delayed rewards have very small values. Individuals with impulse-control disorders, as well as heroin-, alcohol-, cigarette-, and cocaine- addicted individuals, have steeper discounting functions than controls.[16,29–32] To maximize future gains, the discount rate should be carefully adjusted, as we have shown in,[28] and as we will further discuss below.

REWARDS DISCOUNTING IN ARTIFICIAL AGENTS

The theory of reinforcement learning, developed in artificial intelligence but initially loosely based on psychology,[10] is a particularly attractive model of animal learning, as it provides a normative model of how an adaptive agent should update its evaluation of values and behavioral policy. Furthermore, dopamine neurons have the formal characteristics of the teaching signal known as the temporal difference (TD) error.[33] Thus, reinforcement learning has both desirable theoretical properties and offers a good model of the basal ganglia and dopaminergic system.[34–36] The main issue in the theory of reinforcement learning is to maximize the long-term cumulative reward. Thus, central to reinforcement learning is the estimation of the value function

$$V(s(t)) = E\left[\sum_{k=0}^{\infty} \gamma^k r(t+k)\right], \tag{5}$$

where $r(t)$, $r(t+1)$, $r(t+2)\ldots$ are the rewards acquired by following a certain action policy $P(a\}s)$ starting from the state $s(t)$, and γ is a discount factor such that $0 \leq \gamma < 1$. We note here that this formulation of the value function reduces to that of the exponential value of a single reward in animal and human experiments (see above). The value function for the states before and after the transition should satisfy the consistency equation

$$V(s(t-1)) = E\left[r(t) + \gamma V(s(t))\right]. \tag{6}$$

Therefore, any deviation from the consistency equation, expressed as

$$\delta(t) = r(t) + \gamma V(s(t)) - V(s(t-1)), \tag{7}$$

should be zero on average. This signal is the TD error and is used as the teaching signal to learn the value function

$$\Delta V(s(t-1)) = \alpha \delta(t), \tag{8}$$

where α is a learning rate.

The policy is usually defined via the action value function $Q(s(t),a)$, which represents how much future rewards the agent would get by taking the action a at state $s(t)$ and following the current policy in subsequent steps. One common way for stochastic action selection that encourages exploitation is to compute the probability to take an action by the soft-max function

$$P(a_i \mid s(t)) = \frac{e^{\beta \, Q(s(t),a_i)}}{\sum_{j=1}^{M} e^{\beta \, Q(s(t),a_j)}},\qquad(9)$$

where the meta-parameter β is called the inverse temperature.

Crucial to successful reinforcement learning is the careful setting of the three meta-parameters α, β, and γ.

- The learning rate α controls the speed of learning, as small learning rates induce slow learning, and large learning rates induce instability of memory update.
- The inverse temperature β controls the exploitation–exploration trade-off. Ideally, β should initially be low to allow large exploration, when the agent does not have a good mapping of which actions will be rewarding, and gradually increase as the agent reaps higher and higher rewards.
- The discount factor γ determines how far into the future the agent should consider reward prediction and action selection. The setting of the discount factor is particularly important when there is a conflict between the immediate and long-term outcomes (FIG. 1). In real life, it is often the case that one would have to pay some immediate cost (negative reward) to achieve a larger future reward, for example, long travels in foraging, or daily cultivation for harvest. It is also the case that one should avoid positive immediate reward if it is associated with a large negative reward in the future. If γ is small, the agent learns to behave only for short-term rewards. Although a large γ (close to 1) promotes the agent to learn to act for long-term rewards, there are at least three reasons why γ should not be too large. First, any real learning agent, either artificial or biological, has a limited lifetime. Thus, a discounted value function is equivalent to a nondiscounted value function for an agent with a constant death rate of $1-\gamma$. Second, an agent has to acquire some rewards in time; for instance, an animal must find food before it starves; a robot must recharge its battery before it is exhausted. Third, if the environmental dynamics is highly stochastic or the dynamics is nonstationary, long-term prediction is unreliable. The complexity of learning a value function has been shown to increase with the increase of $1/(1-\gamma)$.[37]

SEROTONIN AND THE TIMESCALE OF REWARD PREDICTION

In a seminal paper, Soubrie[6] suggested that serotonergic neurons are brought into play whenever behavioral inhibition is required. He pointed out that reduced serotonin was linked to impulsive behavior: the animal is less able to "wait." Low levels of serotonin are often associated with behaviors regarded as impulsive, such as aggression,[38,39] or failures to not respond in response to a stimulus in a no-go trial.[40] "Impulsivity" is a multidimensional phenomenon, however; Evenden described three varieties of impulsivity[41] (1) unreliable sensory discrimination (attention/preparation), (2) making premature responses in situations that require the postponement of actions (execution), and (3) choosing a smaller immediate reinforcer rather than a larger delayed reinforcer (outcome).

Serotonin dysfunction seems to specifically lead to the third type of impulsivity, as shown by delayed reward choice experiments. Using Mazur's[42] adjusting delay paradigm, Wogar et al.[43] showed that serotonin is involved in maintaining effectiveness of delayed reinforcers: rats with lesioned ascending serotonergic system did not wait for the large reinforcer when offered a choice between immediate small and large delayed reinforcers. Further, it has been shown that serotonin depletion results in failure of delayed rewards to motivate behavior.[44] On the other hand, increased serotonin levels decrease impulsive choice.[45,46] Two variables can possibly lead to this effect of serotonin on delayed reward choices: the reinforcer magnitude or the delay to the reward. Mobini and coworkers showed that serotonin depletion in the forebrain steepens hyperbolic discounting, that is, lower serotonin resulted in higher value of the parameter k_y in equation 4.[47] These authors found no modulation of the magnitude of the reward, however. Although serotonergic fibers arise from both dorsal and median raphe nuclei, the dorsal raphe nucleus seems to be the source of serotonergic neurons involved in impulsivity.[48]

These experimental data are consistent with the hypothesis that serotonin neurons in the dorsal raphe nucleus control the timescale of reward evaluation. In this hypothesis, serotonin controls the discount factor (larger γ in the exponential model of equation 2 and smaller k_y in the hyperbolic model of equation 4 with the higher level of serotonin), which controls the evaluation of future reward. With low serotonin levels, because delayed rewards have a low value, agents choose the small immediate reward over the large delayed reward, characteristics of impulsivity.

MODULATION OF THE DISCOUNT RATE BY SEROTONIN

Serotonin Regulation of the Discount Rate

A body of experimental evidence strongly suggests that learning of the value functions occurs in the central nervous system, presumably in the basal ganglia.

In particular, it has been found that dopamine neuron activity resembles closely the temporal difference (TD) error (e.g., Refs. 33,35). Further, recent experimental studies suggest that the striatum computes value functions,[49–53] and it has been suggested that action selection occurs in the globus pallidus.[54,55] Dopamine induces long-lasting plasticity in corticostriatal synapses,[56,57] and thus could allow learning of the value functions. Serotonergic neurons project to the basal ganglia,[58–60] and control dopamine release in the striatal,[61,62] and thus could modulate the computation of value function (see role of the discount factor γ in equation 5), and the dopaminergic activity (see role of the discount factor γ in equation 7).

The control of the timescale of reward prediction could be achieved by activating or deactivating multiple reward prediction pathways in the basal ganglia. A parallel learning mechanism in the corticobasal ganglia loops used for reward prediction at a variety of timescales would have the merit of enabling flexible selection of a relevant timescale appropriate for the task and the environment at the time of decision making.

This view is supported by our previous brain imaging study,[50] in which we developed a "Markov decision task" to probe decision making in a dynamic context, with small losses followed by a large positive reward (as in FIG. 1). By analyzing subjects' performance data using a reward value model with different discount factors (as in equation 2), we found a gradient of activation within the striatum for prediction error of rewards at different timescales. The graded maps are consistent with the topographic corticostriatal organization,[63] and suggest that areas that project to the more dorsoposterior part of the striatum are involved in reward prediction at a longer timescale. These results are also consistent with the observations that localized damages within the limbic and cognitive corticobasal ganglia loops manifest as deficits in evaluation of future rewards[64–68] and learning of multistep behaviors.[69]

A possible mechanism underlying these observations is that these different corticobasal ganglia subloops are differentially activated by the ascending serotonergic system from the dorsal raphe nucleus. Although serotonergic projections are relatively diffuse and global, differential expression of serotonergic receptors in the cortical areas and in the ventral and dorsal striatum[58,60] could result in differential modulation. The distribution of serotonin receptor subtypes is not uniform within the striatum, as various subtypes receptors subtypes, with different affinities and intracellular effects, are differentially distributed in the ventral and dorsal parts of the striatum.[58,59] Such differential distributions could allow differential striatal modulation of activities under different serotonin levels. A positron emission tomography (PET) experiment using particular receptor radioligands may shed light on these mechanisms.[70,71]

Regulation of Serotonin Levels

How could serotonergic neurons themselves be regulated? In a previous computational study, we proposed a simple, yet robust and biologically plausible algorithm that regulates the reinforcement learning meta-parameters, which include the discount factor.[72] The algorithm is based on Stochastic Real Value Units (SRV) algorithm.[73] An SRV unit output is produced by adding to the weighted sum of its input pattern a small random perturbation that provides the unit with the variability necessary to explore its activity space. When a perturbation results in increased probability of receiving extra rewards, the unit's input synaptic efficacies are adjusted such that the output moves in the direction in which it was perturbed. We expanded the idea of SRV units and take it on to a level higher—that is, we proposed that neuromodulator neurons are themselves SRV-like units. According to this hypothesis, serotonergic, or "gamma" neuron, would be governed by both a slowly varying mean activity term and a noise term. The noise term corresponds to a spontaneous change in the tonic firing of the gamma neuron. We proposed that mean activity of the serotonergic neuron is updated by Hebbian-like learning rule that correlates with the random perturbation and the difference between a short-term and a long-term running reward average. If a positive perturbation in the neurons' firing rate yields a state of affair slightly superior to that the animal expects, then the discount rate γ is increased, and vice versa.

Although untested, this simple algorithm is biologically plausible. Spontaneous fluctuations of the tonic firing of the neuromodulator neuron may arise naturally with the wake–sleep cycle and/or the level of activity of the animal.[74] The difference between a short-term and a long-term running average of the reward could be carried by dopaminergic neuron activity.[75] As dopaminergic neurons send projections to serotonergic neurons,[76] we predicted that dopamine-dependent plasticity is present in these neurons.

CONCLUDING REMARKS

Serotonin seems to play a major role in depression, as selective serotonin reuptake inhibitors (SSRI) and other serotonin-enhancing drugs are known to be effective for unipolar depression and bipolar disorders. The therapeutic mechanisms of these drugs are still not well understood, however.[5] Our theory of the role of serotonin, although primarily aimed at explaining impulsive behavior, may also perhaps explain certain aspects of depressive behavior: low serotonin levels could lead to the situation shown in the left of FIGURE 1, in which the optimal policy is not to act. Future experiments using delayed reward paradigms could be designed to study impulsivity in depressed patients.

ACKNOWLEDGMENTS

This work was supported in part by CREST, and by grants NIH P20 RR020700–02 and NSF IIS 0535282 to NS.

REFERENCES

1. KATZ, P.S. 1999. Beyond Neurotransmission: Neuromodulation and Its Importance for Information Processing. Oxford University Press. Oxford, UK.
2. SAPER, C.B. 2000. Brain stem modulation of sensation, movement and consciousness. *In* Principles of Neural Science E.R. Kandel, J.H. Schwartz & T.M. Jessel, Eds.: 889–909. McGraw-Hill. New York.
3. MARDER, E. & V. THIRUMALAI. 2002. Cellular, synaptic and network effects of neuromodulation. Neural Netw. **15:** 479–493.
4. ROBBINS, T.W. 1997. Arousal systems and attentional processes. Biol. Psychol. **45:** 57–71.
5. WONG, M.L. & J. LICINIO. 2001. Research and treatment approaches to depression. Nat. Rev. Neurosci. **2:** 343–351.
6. SOUBRIE, P. 1986. Serotonergic neurons and behavior. J. Pharmacol. **17:** 107–112.
7. PLATT, M.L. 2002. Neural correlates of decisions. Curr. Opin. Neurobiol. **12:** 141–148.
8. SAMUELSON, P.A. 1937. A note on measurement of utility. Rev. Econ. Stud. **4:** 155–161.
9. KAGEL, J.H., L. GREEN & T. CARACO. 1986. When foragers discount the future: constraints or adaptation? Anim. Behav. **34:** 271–283.
10. SUTTON, R.S. & A.G. BARTO. 1998. Reinforcement Learning. The MIT Press. Cambridge, MA.
11. AINSLIE, G. 1975. Specious reward: a behavioral theory of impulsiveness and impulse control. Psychol. Bull. **82:** 463–496.
12. MAZUR, J.E. 1987. An adjusting procedure for studying delayed reinforcement. *In* Quantitative Analysis of Behavior. Vol. V: The Effect of Delay and Intervening Events. M.L. Commons, *et al.*, Eds. Erlbaum. London.
13. RODRIGUEZ, M.L. & A.W. LOGUE. 1988. Adjusting delay to reinforcement: comparing choice in pigeons and humans. J. Exp. Psychol. Anim. Behav. Process **14:** 105–117.
14. RACHLIN, H., A. RAINERI & D. CROSS. 1991. Subjective probability and delay. J. Exp. Anal. Behav. **55:** 233–244.
15. BATESON, M. & A. KACELNIK. 1996. Rate currencies and the foraging starling: the fallacy of the averages revisited. Behav. Ecol. **7:** 341–352.
16. BICKEL, W.K., A.L. ODUM & G. J. MADDEN. 1999. Impulsivity and cigarette smoking: delay discounting in current, never, and ex-smokers. Psychopharmacology (Berl). **146:** 447–454.
17. HO, M.Y. *et al.* 1999. Theory and method in the quantitative analysis of "impulsive choice" behaviour: implications for psychopharmacology. Psychopharmacology (Berl.) **146:** 362–372.
18. KIRBY, K.N., N.M. PETRY & W. K. BICKEL. 1999. Heroin addicts have higher discount rates for delayed rewards than non-drug-using controls. J. Exp. Psychol. Gen. **128:** 78–87.
19. PETRY, N.M. 2001. Pathological gamblers, with and without substance use disorders, discount delayed rewards at high rates. J. Abnorm. Psychol. **110:** 482–487.
20. KACELNIK, A. 1997. Normative and descriptive models of decision making: time discounting and risk sensitivity. *In* Characterizing Human Psychological Adaptations: 51–70. Wiley. Chichester.

21. LAIBSON, D.I. 2003. Intertemporal decision making. *In* Encyclopedia of Cognitive Science. Nature Publishing Group. London.
22. THALER, R.H. & H.M. SHEFRIN. 1981. An economic theory of self-control. J. Pol. Economy **89:** 392–410.
23. AINSLIE, G. 2005. Precis of breakdown of will. Behav. Brain Sci. **28:** 635–650.
24. THALER, R.H. 1981. Some empirical evidence on dynamic inconsistency. economic letters. **8:** 201–207.
25. KIRBY, K.N. & N.N. MARAKOVIC. 1995. Modeling myopic decisions: evidence for hyperbolic delay-discounting within subjects and amounts. Org. Behav. Human Decision Proc. **64:** 22–30.
26. MYERSON, J. & L. GREEN. 1995. Discounting of delayed rewards: models of individual choice. J. Exp. Anal. Behav. **64:** 263–276.
27. ANGELETOS, G.M. *et al.* 2001. The hyperbolic consumption model: calibration, simulation, and empirical evaluation. J. Eco. Prospect. **15:** 47–68.
28. SCHWEIGHOFER, N. *et al.* 2006. Humans can adopt optimal discounting strategy under real-time constraints. Plos Comp. Biol. **11:** 1349–1356.
29. CREAN, J.P., H. DE WIT & J. B. RICHARDS. 2000. Reward discounting as a measure of impulsive behavior in a psychiatric outpatient population. Exp. Clin. Psychopharmacol. **8:** 155–162.
30. MADDEN, G.J. *et al.* 1997. Impulsive and self-control choices in opioid-dependent patients and non-drug-using control participants: drug and monetary rewards. Exp. Clin. Psychopharmacol. **5:** 256–262.
31. VUCHINICH, R.E. & C.A. SIMPSON. 1998. Hyperbolic temporal discounting in social drinkers and problem drinkers. Exp. Clin. Psychopharmacol. **6:** 292–305.
32. COFFEY, S.F. *et al.* 2003. Impulsivity and rapid discounting of delayed hypothetical rewards in cocaine-dependent individuals. Exp. Clin. Psychopharmacol. **11:** 18–25.
33. SCHULTZ, W. 1998. Predictive reward signal of dopamine neurons. J. Neurophysiol. **80:** 1–27.
34. HOUK, J.C., J. DAVIS & D. BEISER, Eds. 1995. Models of Information Processing in the Basal Ganglia. 249–270. The MIT Press. Cambridge, MA.
35. MONTAGUE, P.R., P. DAYAN & T. J. SEJNOWSKI. 1996. A framework for mesencephalic dopamine systems based on predictive Hebbian learning. J. Neurosci. **16:** 1936–1947.
36. DOYA, K. 2000. Complementary roles of basal ganglia and cerebellum in learning and motor control. Curr. Opin. Neurobiol. **10:** 732–739.
37. LITTMAN, M.L., T.L. DEAN & L. P. KAELBLING. 1995. On the complexity of solving Markov decision problems. Eleventh International Conference on Uncertainty in Artificial Intelligence.
38. BUHOT, M.C. 1997. Serotonin receptors in cognitive behaviors. Curr. Opin. Neurobiol. **7:** 243–254.
39. ROBBINS, T.W. 2000. From arousal to cognition: the integrative position of the prefrontal cortex. Prog. Brain Res. **126:** 469–483.
40. HARRISON, A.A., B.J. EVERITT & T. W. ROBBINS. 1999. Central serotonin depletion impairs both the acquisition and performance of a symmetrically reinforced go/no-go conditional visual discrimination. Behav. Brain Res. **100:** 99–112.
41. EVENDEN, J.L. 1999. Varieties of impulsivity. Psychopharmacology (Berl.) **146:** 348–361.

42. MAZUR, J.E., M. SNYDERMAN & D. COE. 1985. Influences of delay and rate of reinforcement on discrete-trial choice. J. Exp. Psychol. Anim. Behav. Process **11:** 565–575.
43. WOGAR, M.A., C.M. BRADSHAW & E. SZABADI. 1993. Effect of lesions of the ascending 5-hydroxytryptaminergic pathways on choice between delayed reinforcers. Psychopharmacology (Berl.) **111:** 239–243.
44. RAHMAN, S. *et al.* 2001. Decision making and neuropsychiatry. Trends Cogn. Sci. **5:** 271–277.
45. BIZOT, J. *et al.* 1999. Serotonin and tolerance to delay of reward in rats. Psychopharmacology (Berl.) **146:** 400–412.
46. POULOS, C.X., J.L. PARKER & A.D. LE. 1996. Dexfenfluramine and 8-OH-DPAT modulate impulsivity in a delay-of-reward paradigm: implications for a correspondence with alcohol consumption. Behav. Pharmacol. **7:** 395–399.
47. MOBINI, S. *et al.* 2000. Effect of central 5-hydroxytryptamine depletion on intertemporal choice: a quantitative analysis. Psychopharmacology (Berl.) **149:** 313–318.
48. CARLI, M. & R. SAMANIN. 2000. The 5-HT(1A) receptor agonist 8-OH-DPAT reduces rats' accuracy of attentional performance and enhances impulsive responding in a five-choice serial reaction time task: role of presynaptic 5-HT(1A) receptors. Psychopharmacology (Berl.) **149:** 259–268.
49. O'DOHERTY, J. *et al.* 2004. Dissociable roles of ventral and dorsal striatum in instrumental conditioning. Science **304:** 452–454.
50. TANAKA, S.C. *et al.* 2004. Prediction of immediate and future rewards differentially recruits cortico-basal ganglia loops. Nat. Neurosci. **7:** 887–893.
51. SHIDARA, M., T.G. AIGNER & B.J. RICHMOND. 1998. Neuronal signals in the monkey ventral striatum related to progress through a predictable series of trials. J. Neurosci. **18:** 2613–2625.
52. TREMBLAY, L. & W. SCHULTZ. 2000. Reward-related neuronal activity during go-nogo task performance in primate orbitofrontal cortex. J. Neurophysiol. **83:** 1864–1876.
53. SAMEJIMA, K. *et al.* 2005. Representation of action-specific reward values in the striatum. Science **310:** 1337–1340.
54. BERNS, G.S. & T.J. SEJNOWSKI. 1998. A computational model of how the basal ganglia produce sequences. J. Cogn. Neurosci. **10:** 108–121.
55. DOYA, K. 2000. Metalearning and neuromodulation. Math. Sci. **38:** 19–24.
56. REYNOLDS, J.N., B.I. HYLAND & J.R. WICKENS. 2001. A cellular mechanism of reward-related learning. Nature **413:** 67–70.
57. REYNOLDS, J.N. & J.R. WICKENS. 2002. Dopamine-dependent plasticity of corticostriatal synapses. Neural Netw. **15:** 507–521.
58. COMPAN, V. *et al.* 1998. Selective increases in serotonin 5-HT1B/1D and 5-HT2A/2C binding sites in adult rat basal ganglia following lesions of serotonergic neurons. Brain Res. **793:** 103–111.
59. VARNAS, K., C. HALLDIN & H. HALL. 2004. Autoradiographic distribution of serotonin transporters and receptor subtypes in human brain. Hum. Brain Mapp. **22:** 246–260.
60. MIJNSTER, M.J. 1997. Regional and cellular distribution of serotonin 5-hydroxytryptamine2a receptor mRNA in the nucleus accumbens, olfactory tubercle, and caudate putamen of the rat. J. Comp. Neurol. **389:** 1–11.

61. SERSHEN, H., A. HASHIM & A. LAJTHA. 2000. Serotonin-mediated striatal dopamine release involves the dopamine uptake site and the serotonin receptor. Brain Res. Bull. **53:** 353–367.
62. DE DEURWAERDERE, P & U. SPAMPINATO. 1999. Role of serotonin(2A) and serotonin(2B/2C) receptor subtypes in the control of accumbal and striatal dopamine release elicited *in vivo* by dorsal raphe nucleus electrical stimulation. J. Neurochem. **73:** 1033–1042.
63. MIDDLETON, F.A. & P. L. STRICK. 2000. Basal ganglia and cerebellar loops: motor and cognitive circuits. Brain Res. Brain Res. Rev. **31:** 236–250.
64. BECHARA, A., H. DAMASIO & A.R. DAMASIO. 2000. Emotion, decision making and the orbitofrontal cortex. Cereb. Cortex **10:** 295–307.
65. CARDINAL, R.N. *et al.* 2001. Impulsive choice induced in rats by lesions of the nucleus accumbens core. Science **292:** 2499–2501.
66. ROLLS, E.T. 2000. The orbitofrontal cortex and reward. Cereb. Cortex **10:** 284–294.
67. EAGLE, D.M. *et al.* 1999. Effects of regional striatal lesions on motor, motivational, and executive aspects of progressive-ratio performance in rats. Behav. Neurosci. **113:** 718–731.
68. PEARS, A. *et al.* 2003. Lesions of the orbitofrontal but not medial prefrontal cortex disrupt conditioned reinforcement in primates. J. Neurosci. **23:** 11189–11201.
69. HIKOSAKA, O. 1999. Parallel neural networks for learning sequential procedures. Trends Neurosci. **22:** 464–471.
70. HUANG, Y. *et al.* 2005. Synthesis of potent and selective serotonin 5-HT1B receptor ligands. Bioorg. Med. Chem. Lett. **15:** 4786–4789.
71. LARISCH, R. *et al.* 2003. Influence of synaptic serotonin level on [18F]altanserin binding to 5HT2 receptors in man. Behav. Brain Res. **139:** 21–29.
72. SCHWEIGHOFER, N. & K. DOYA. 2003. Meta-learning in reinforcement learning. Neural Netw. **16:** 5–9.
73. GULLAPALLI, V. 1990. A stochastic reinforcement learning algorithm for learning real-valued functions. Neural Netw. **3:** 671–692.
74. JACOBS, B.L. & C.A. FORNAL. 1993. 5-HT and motor control: a hypothesis. Trends Neurosci. **16:** 346–352.
75. DAW, N.D., S. KAKADE & P. DAYAN. 2002. Opponent interactions between serotonin and dopamine. Neural Netw. **15:** 603–616.
76. HAJ-DAHMANE, S. 2001. D2-like dopamine receptor activation excites rat dorsal raphe 5-HT neurons *in vitro*. Eur. J. Neurosci. **14:** 125–134.

Receptor Theory and Biological Constraints on Value

GREGORY S. BERNS,[a] C. MONICA CAPRA,[b] AND CHARLES NOUSSAIR[b]

[a]Department of Psychiatry and Behavorial sciences, Emory University School of Medicine, Atlanta, Georgia, USA

[b]Department of Economics, Emory University, Atlanta, Georgia, USA

ABSTRACT: Modern economic theories of value derive from expected utility theory. Behavioral evidence points strongly toward departures from linear value weighting, which has given rise to alternative formulations that include prospect theory and rank-dependent utility theory. Many of the nonlinear forms for value assumed by these theories can be derived from the assumption that value is signaled by neurotransmitters in the brain, which obey simple laws of molecular movement. From the laws of mass action and receptor occupancy, we show how behaviorally observed forms of nonlinear value functions can arise.

KEYWORDS: expected utility theory; prospect theory; neuroeconomics; reward; receptor theory

INTRODUCTION

The last 5 years have seen a vast increase in the number of neuroimaging papers that attempt to identify the neural code for "value," whether described as a "reward," a "reinforcer," or by the economic term, "utility." The concept of value that is investigated is a "score" that is relevant to behavior. An extreme view of the relationship between value and behavior is that of classical economics, which assumes that people make optimal decisions. Optimal decisions are those decisions that maximize the individual's value, or utility, they hope to obtain. This model provides a normative description of how perfectly rational agents behave. A behavioral economist might add assumptions of errors in actions, decision biases, or slow and incomplete learning in constructing descriptive models of behavior. A psychologist, on the other hand, might view decision making as the interaction between cognitive and affective forces. It is into this fray that neuroimagers have stepped. The hope is that measurement of brain activity will resolve these long-standing debates about the

Address for correspondence: Gregory S. Berns, Department of Psychiatry and Behavorial Sciences, Emory University School of Medicine, 101 Woodruff Circle, Suite 4000, Atlanta, GA 30322. Voice: +404-727-2556; fax: +404-727-3233.
gberns@emory.edu

Ann. N.Y. Acad. Sci. 1104: 301–309 (2007). © 2007 New York Academy of Sciences.
doi: 10.1196/annals.1390.013

varied motivations that people have for the decisions they make. But, like Alice's looking-glass, the answers seem to depend on who is asking the question.

In this article, we propose a parsimonious structure to consider the biophysical constraints of how neuronal systems respond to varying levels of scalar quantities—interpreted as value or, alternatively, as the determinants of value. As far as we know, neurons act as detectors of differences. That is, they fire when a change of membrane state occurs. They adapt to levels of stimulation, making them poor signalers of absolute levels of anything, be it intensity of visual stimulation, caloric intake, or value. If we consider value as a quantity, like photic stimulation, then it is possible to extrapolate, from the behavior of analogous neural systems, how the brain perceives value.

We begin with the assumption that cognitive processing of value is influenced by the biophysical properties that govern neurotransmitter and receptor binding in the brain. Although decidedly reductionist, this assumption requires only that the brain is the physical organ that controls behavior and that the brain operates according to well-known laws of physics and chemistry. While the biophysical properties of the brain may not capture all the aspects of psychological processes, the physical structure of the brain does place specific constraints on the implementation of psychological processes.

Our derivation suggests a hypothesis that measures of brain activation are biological transformations of stimuli, which can be interpreted as biological proxies for utility functions. With minimal assumptions, one can derive properties of a value function that capture many aspects of those postulated in theories of decision making, like expected utility theory[1] and prospect theory.[2] For instance, our model implies a diminishing marginal sensitivity to value and probability, which is consistent with the available evidence from economic experiments.

RECEPTOR OCCUPANCY THEORY

According to receptor occupancy theory, a biological response results from the interaction between a neurotransmitter and a cellular receptor.[3] The theory has its basis in the law of mass action. In chemical reaction notation, we consider a drug (or neurotransmitter), A, which binds to a receptor, R, and forms a drug–receptor complex, AR, which acts as a stimulus to the cell. The result is a cellular response:

$$A + R \leftrightarrow AR \to \text{Stimulus} \to \text{Response}$$

The magnitude of this response can be described by

$$\text{Response} = \left[\varepsilon \times [R] \times \frac{[A]}{K_d + [A]} \right]^{\alpha} \qquad (1)$$

where [*A*] is the concentration of the neurotransmitter, [*R*] is the concentration of the receptor, K_d is the dissociation constant, ε is the efficacy, and α is an exponent that describes the stimulus-response relationship of the cell. If we assume that [*A*] is directly, and linearly, related to the magnitude of an exogenous entity, *z*, such as the amount of money, pain, or consumption that an individual is presented with or receives, and that the effective maximal receptor concentration, $R_{\max}$, equals $\varepsilon \times [R]$, then

$$\text{Response} = \left[\frac{R_{\max} z}{K_d + z} \right]^\alpha \qquad (2)$$

Thus, we make a critical distinction between the quantity, *z*, and the measurable *response* to *z*. By the same logic, *z* can represent any psychological or economic quantity, which is represented by a scalar quantity. There are two ways that utility can be interpreted in the context of Equation 2. The first is to interpret utility as *z*, the exogenous entity, and treat it as if it were a physical quantity that is transformed into a cellular response. The second is to interpret Equation 2 as describing a transformation of a physical quantity into utility. To the degree that mental constructs of value become physically instantiated in the brain, Equation 2 describes how this transformation might occur (with certain assumptions). Under either interpretation, the utility may be that of a certain event or the value assigned to a lottery, in which case it may take the expected utility or another functional form. The current neuroeconomic literature does not address the issue of which interpretation might be more appropriate. Although several studies have reported biological correlates of utility, it is not yet clear whether these measurements correspond to utility,[4-6] a version of instantaneous pleasure or anticipated pleasure (analogous to *z* in Eq. 2), or whether these measurements are, in fact, a *response* to a psychological construct of utility. For purposes of this article, we conjecture only that biological measurements are constrained by Equation 2, but it is worth considering the two possibilities separately.

It is possible that the notion of value (or utility) is a mentally constructed quantity without physical instantiation in a single brain location. While still contained in the brain, a distributed representation might limit the practical application of Equation 2, which is fundamentally a description of the movement of molecules at a single synapse. A variety of studies, however, have suggested that the nucleus accumbens is a location that contains a centralized measure of utility. Although the proposed explanations differ from purely hedonic (experiential) utility, to expected utility, to temporal difference prediction (i.e., a derivative of expected utility), it is reasonable to ask how Equation 2 would transform these parameters into a physical response. Once transformed, this signal could be propagated to other brain regions that require access to value information.

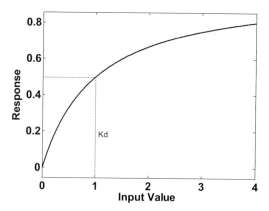

FIGURE 1. The neurotransmitter receptor response function. Assuming a linear cellular response ($\alpha = 1$ in Eq. 2), the binding of a neurotransmitter to its receptor complex follows the law of mass action and leads to an asymptotic response, regardless of the level of input stimulation (*Input Value*). The dissociation constant, K_d, is the level at which 50% of the receptor is occupied. In this plot, $R_{max} = 1$.

Consider the interpretation of Equation 2 under which z represents the objective quantity of some stimulus, but this quantity serves as an input to a specific neural system, for example, the ventral tegmental dopamine system, whose response would be subjectively experienced as *utility*. This is the strongest possible interpretation of Equation 2. In essence, this says that the response of the dopamine system corresponds to the subjective experience of utility. If utility determines decisions, then the correspondence to dopamine system activity means that dopamine release would govern decision making. If we assume that $\alpha = 1$, namely the stimulus-response relationship of the neuron is linear, then Equation 2 leads to the well-known concave utility function[7] (FIG. 1). Although the classical form for the utility function is strictly monotonically increasing and concave, which has the property of diminishing but strictly positive marginal returns, the hyperbolic form given by the law of mass action has the property of asymptotically approaching R_{max}. This corresponds to the existence of a saturation point where increasing the quantity of stimulus no longer yields greater utility.

Most neural systems detect transient changes, as in the visual system, in which neurons respond to changing levels of stimulation (as opposed to tonic, long-term levels of stimulation). A large body of data on the properties of the dopamine system also suggests that these neurons respond to deviations in reward prediction as opposed to absolute levels of reward.[8–10] Thus it seems reasonable to cast Equation 2 as a response to a change in future reward expectations:

$$\text{Response} = \left[\frac{R_{max}\frac{dz}{dt}}{K_d + \frac{dz}{dt}} \right]^{\alpha} \tag{3}$$

Thus, the greater the rate of change of expected value, the greater the response, up to the limit imposed by R_{max}. The temporal derivative, dz/dt, can be integrated over an arbitrary time interval to yield the corresponding integrated response. Equation 3 can also be applied to negative rewards, or losses. Losses are likely governed by different neural systems, although dopamine activity has been shown to decrease when expected rewards are not received. Allowing R_{max} to take on different magnitudes for gains and losses yields the well-known S-shaped value function postulated by prospect theory[2] (FIG. 2). The exponent, α, describes the transformation of receptor activation into a cellular response. A cellular system that is weakly coupled to the receptor is characterized by $\alpha > 1$, and, conversely, a strongly coupled system has $\alpha < 1$.[3] Because the degree of receptor-coupling affects the dose-response relationship, α also affects the shape of the value function (FIG. 3).

EVIDENCE

Our theory makes a strong assumption: that the response of a specific neural system is the biophysical carrier of utility. Although it is possible that "utility" is a distributed mental construct, without physical instantiation in a single location, it is worthwhile examining the evidence for the stronger prediction. The vast majority of the experimental work on the reward system has focused on the dopamine system and the brain regions to which it projects. To test Equation 3 in an experimental system, one would need to vary either the magnitude of rewards delivered to a test subject, or vary the contingencies such that the subject's expectations are changing. Both have been done. Monkeys, for example, can discriminate between the magnitudes of a liquid

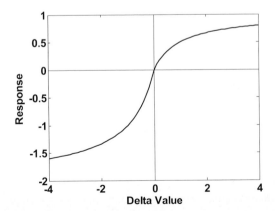

FIGURE 2. The neurotransmitter receptor response function when the input is considered as a change in value from an arbitrary baseline. Here, the R_{max} for losses was set at twice the R_{max} for gains. The function captures all the salient characteristics of the value function postulated in the prospect theory.

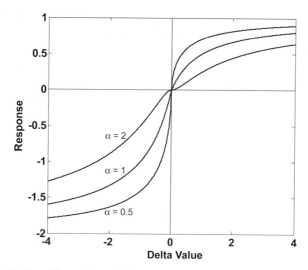

FIGURE 3. The effect of the cellular response exponent, α, on the neurotransmitter-response function. Decreasing $\alpha < 1$ implies highly efficient coupling of the cellular response to the neurotransmitter receptor complex and results in a more convexoconcave response function. Increasing $\alpha > 1$ implies inefficient coupling and results in a more complex response function with three inflection points.

reward based on the volume received. This is also reflected in corresponding levels of activity change in the midbrain dopamine neurons as well as their target sites in the striatum.[11,12] Importantly, the dopamine response displays a concave functional form, and changes in phasic dopamine activity correlate with future decision making.[13] Although there are comparatively fewer neuroimaging data in humans at a sufficient range of reward values, a recent study in which the dopamine system activity was either augmented by treatment with the dopamine precursor, L-DOPA, or blocked with the receptor antagonist, haloperidol, corresponding changes were observed in the fMRI BOLD response in the striatum.[14] Numerous fMRI studies have shown corresponding changes in striatal activity, that is, at least, monotonic in the magnitude of expected monetary reward.[15–19] These studies, however, have not offered a sufficient range of rewards to assess the curvature of the function.

A milder interpretation of Equation 3 is that value and utility are mental constructs that occur through a complex set of psychological operations, and because of the complexity, are distributed across different brain regions. Receptor theory still applies in this scenario because neurons adhere to physical laws governing molecule movement. But the one-to-one correspondence with value breaks down because the input to each disparate brain region might represent only one aspect of the process of computing value. For example, a visually presented stimulus, say in the form of a number, must first be processed by the visual system before being passed on to any putative value system. Low-level visual processing might reflect some aspects of value although it would be loose correlation at best. Animal studies use conditioning paradigms, so that the visual stimulus, in addition to its geometric properties, comes to contain value information through associative learning mechanisms. Value becomes reflected not only in dopamine neurons but parietal neurons as well.[20–22] It is not yet possible to determine which way value information flows between these two systems, but regardless of the direction, the information becomes transformed by the above equations.

The nonlinearity of value functions is generally accepted as an empirical truism. Most explanations for the nonlinearity derive from psychological or cognitive distortions, which include framing effects, loss aversion, status quo bias, etc.[23,24] Many of the same distortions, however, have been demonstrated in other animals. A well-known motivating stimulus, that has behavioral effects akin to value, is electrical stimulation of certain brain regions.[25] Brain-stimulation reward (BSR) can be titrated to replace conventional exogenous rewards like food. BSR, however, shows the same relationship to behavior as do other rewards; namely, as both the magnitude of electrical stimulation and its frequency are increased, the behavioral effect approaches an asymptote.[26] The role of dopamine in BSR is complicated by the fact that electrical stimulation of specific brain regions propagates in multiple directions, some of which eventually reach the midbrain dopamine neurons. Although BSR is more complicated than simply stimulating the dopamine system, the pure physicality of BSR eliminates the role of cognitive heuristics in the determination of the value function. And although multiple neurotransmitter systems are involved, they each obey receptor occupancy constraints.

CONCLUSIONS

To our knowledge, the physical constraints imposed by receptor occupancy theory have never been proposed as the cause of the functional forms associated with economic utility theories. The commonality across many species for concave forms of utility suggests a common biological mechanism that is independent of the cognitive heuristics used by each animal. Of course, the fact that neurotransmitters and receptors display binding kinetics with a functional form similar to that postulated by economic theory may be a coincidence. The main argument against such a causal relationship derives from the multiplicity of systems in the brain that operate in parallel to coordinate decision making. Nonlinearities in one system may be offset by nonlinearities in another system. However, the strength of our proposal is that it is testable. Manipulation of specific receptors in the context of decision making, either through pharmacological means or through genetic approaches, can isolate the contribution of each neurotransmitter receptor complex to the behaviorally derived value function.

ACKNOWLEDGMENTS

We are grateful for the contributions of Jonathan Chappelow, Sara Moore, and Giuseppe Pagnoni. This work was supported by grants from the National Institute on Drug Abuse (DA016434 and DA20116).

REFERENCES

1. VON NEUMANN, J. & O. MORGENSTERN. 1944. Theory of Games and Economic Behavior. Princeton University Press. Princeton.
2. KAHNEMAN, D. & A. TVERKSY. 1979. Prospect theory: an analysis of decision under risk. Econometrica **47:** 263–291.
3. ROSS, E.M. & T.P. KENAKIN. 2001. Pharmacodynamics. Mechanisms of drug action and the relationship between drug concentration and effect. *In* Goodman & Gilman's The Pharmacological Basis of Therapeutics, Vol. Tenth. J.G. Hardman & L.E. Limbird, Eds. McGraw-Hill. New York.
4. KNUTSON, B., J. TAYLOR, M. KAUFMAN, *et al.* 2005. Distributed neural representation of expected value. J. Neurosci. **25:** 4806–4812.
5. HSU, M., M. BHATT, R. ADOLPHS, *et al.* 2005. Neural systems responding to degrees of uncertainty in human decision-making. Science **310:** 1680–1683.
6. HUETTEL, S.A., C.J. STOWE, E.M. GORDON, *et al.* 2006. Neural signatures of economic preferences for risk and ambiguity. Neuron **49:** 765–775.
7. BERNOULLI, D. 1763/1958. Exposition of a new theory on the measurement of risk. Econometrica **22:** 23–36.
8. SCHULTZ, W., P. DAYAN & P.R. MONTAGUE. 1997. A neural substrate of prediction and reward. Science **275:** 1593–1599.
9. LJUNGBERG, T., P. APICELLA & W. SCHULTZ. 1992. Responses of monkey dopamine neurons during learning of behavioral reactions. J. Neurophys. **67:** 145–163.
10. SCHULTZ, W. & A. DICKINSON. 2000. Neuronal coding of prediction errors. Ann. Rev. Neurosci. **23:** 473–500.
11. CROMWELL, H.C. & W. SCHULTZ. 2003. Effects of expectations for different reward magnitudes on neuronal activity in primate striatum. J. Neurophys. **89:** 2823–2838.
12. TOBLER, P.N., C.D. FIORILLO & W. SCHULTZ. 2005. Adaptive coding of reward value by dopamine neurons. Science **307:** 1642–1645.
13. MORRIS, G., A. NEVET, D. ARKADIR, *et al.* 2006. Midbrain dopamine neurons encode decisions for future action. Nat. Neurosci. **9:** 1057–1063.
14. PESSIGLIONE, M., B. SEYMOUR, G. FLANDIN, *et al.* 2006. Dopamine-dependent prediction errors underpin reward-seeking behaviour in humans. Nature **442:** 1042–1045.
15. KNUTSON, B., A. WESTDORP, E. KAISER, *et al.* 2000. FMRI visualization of brain activity during a monetary incentive delay task. Neuroimage **12:** 20–27.
16. KNUTSON, B., C.M. ADAMS, G.W. FONG, *et al.* 2001. Anticipation of increasing monetary reward selectively recruits nucleus accumbens. J. Neurosci. **21:** RC159.
17. BREITER, H.C., I. AHARON, D. KAHNEMAN, *et al.* 2001. Functional imaging of neural responses to expectancy and experience of monetary gains and losses. Neuron **30:** 619–639.

18. DELGADO, M.R., L.E. NYSTROM, C. FISSEL, *et al.* 2000. Tracking the hemodynamic responses to reward and punishment in the striatum. J. Neurophys. **84:** 3072–3077.

19. O'DOHERTY, J., M.L. KRINGELBACH, E.T. ROLLS, *et al.* 2001. Abstract reward and punishment representations in the human orbitofrontal cortex. Nat. Neurosci. **4:** 95–102.

20. PLATT, M.L. & P.W. GLIMCHER. 1999. Neural correlates of decision variables in parietal cortex. Nature **400:** 233–238.

21. DORRIS, M.C. & P.W. GLIMCHER. 2004. Activity in posterior parietal cortex is correlated with the relative subjective desirability of action. Neuron **44:** 365–378.

22. SUGRUE, L.P., G.S. CORRADO & W.T. NEWSOME. 2004. Matching behavior and the representation of value in the parietal cortex. Science **304:** 1782–1787.

23. KAHNEMAN, D. & A. TVERKSY. 1984. Choices, values, and frames. Am. Psychol. **39:** 341–350.

24. TVERSKY, A. & D. KAHNEMAN. 1991. Loss aversion in riskless choice: a reference-dependent model. Quart. J. Econ. **106:** 1039–1061.

25. SHIZGAL, P. 1997. Neural basis of utility estimation. Curr. Opin. Neurobiol. **7:** 198–208.

26. SIMMONS, J.M. & C.R. GALLISTEL. 1994. Saturation of subjective reward magnitude as a function of current and pulse frequency. Behav. Neurosci. **108:** 151–160.

Reward Prediction Error Computation in the Pedunculopontine Tegmental Nucleus Neurons

YASUSHI KOBAYASHI [a,b] AND KEN-ICHI OKADA[a]

[a]*Graduate School of Frontier Biosciences, Osaka University, Machikaneyama, Toyonaka, Japan*

[b]*ATR Computational Neuroscience Laboratories, Hikaridai, Seika-cho, Soraku-gun, Kyoto, Japan*

ABSTRACT: **In this article, we address the role of neuronal activity in the pathways of the brainstem–midbrain circuit in reward and the basis for believing that this circuit provides advantages over previous reinforcement learning theory. Several lines of evidence support the reward-based learning theory proposing that midbrain dopamine (DA) neurons send a teaching signal (the reward prediction error signal) to control synaptic plasticity of the projection area. However, the underlying mechanism of where and how the reward prediction error signal is computed still remains unclear. Since the pedunculopontine tegmental nucleus (PPTN) in the brainstem is one of the strongest excitatory input sources to DA neurons, we hypothesized that the PPTN may play an important role in activating DA neurons and reinforcement learning by relaying necessary signals for reward prediction error computation to DA neurons. To investigate the involvement of the PPTN neurons in computation of reward prediction error, we used a visually guided saccade task (VGST) during recording of neuronal activity in monkeys. Here, we predict that PPTN neurons may relay the excitatory component of tonic reward prediction and phasic primary reward signals, and derive a new computational theory of the reward prediction error in DA neurons.**

KEYWORDS: **reinforcement learning; reward prediction error; brainstem; cholinergic system; saccade; monkey**

INTRODUCTION

Classical View of Pedunculopontine Tegmental Nucleus (PPTN)

In the older literature, the PPTN in the brainstem was thought to be the central part of the reticular activating system, which provides background excitation

Address for correspondence: Y. Kobayashi, Ph.D., Graduate School of Frontier Biosciences, Osaka University, 1-3 Machikaneyama, Toyonaka 560-8531, Japan. Voice/fax: +81-6-6850-6521.
yasushi@fbs.osaka-u.ac.jp

Ann. N.Y. Acad. Sci. 1104: 310–323 (2007). © 2007 New York Academy of Sciences.
doi: 10.1196/annals.1390.003

for several sensory and motor systems essential for perception and cognitive processes.[1,2] The PPTN contains both cholinergic and glutamatergic neurons,[3] and is one of the major sources of cholinergic projections in the brainstem. The cholinergic system is one of the most important modulatory neurotransmitter systems in the brain, and controls neuronal activity that depends on selective attention, and anatomical and physiological evidence supports the idea of a "cholinergic component" of conscious awareness.[4] The PPTN has reciprocal connections with the basal ganglia: the subthalamic nucleus (STN), the globus pallidus, and the substantia nigra,[5,6] and recently, it has been argued that PPTN forms a part of the basal ganglia.[7] Further, the PPTN also has reciprocal connections with catecholaminergic systems in the brainstem: the locus coeruleus (LC; noradrenergic) and the dorsal raphe nucleus (DRN; serotonergic).[8] This basal ganglia–PPTN–catecholaminergic complex has been proposed to play an important role in gating movement and controlling several forms of attentional behavior.[9] Despite these abundant anatomical findings, however, the functional importance of the PPTN is not yet fully understood.

The Role of the PPTN in Reward Processing and Learning

Recent lesion and drug administration studies using rodents have indicated that the PPTN is involved in various reinforcement processes.[10] According to a physiological study in cats, the PPTN is thought to relay either a reward signal or a salient event in a fully conditioned situation.[11] Anatomically, the PPTN receives reward input from the lateral hypothalamus and the limbic cortex.[12,13] Conversely, the PPTN abundantly projects to midbrain dopamine (DA) neurons of the substantia nigra pars compacta (SNc),[14] which encode a reward prediction error signal for reinforcement learning.[15] For DA neurons, the PPTN is one of the strongest excitatory input sources.[16] The PPTN neurons release glutamate and acetylcholine to target neurons, and glutamatergic and cholinergic inputs from the PPTN make synaptic connections with DA neurons,[17,18] and electrical stimulation of the PPTN induces a time-locked burst in DA neurons in the rat.[19] Thus, PPTN activity and acetylcholine provided by the PPTN can facilitate the DA neuron's burst firing and appear to do so via muscarinic and nicotinic receptor activation.[20,21] Furthermore, midbrain DA neurons are dysfunctional following excitotoxic lesions of the PPTN.[22] A number of studies have found impairments in learning following excitotoxic lesions of the PPTN.[23–25]

New Insights into the PPTN: Relationship with Reward Prediction Error

Humans and animals can learn to predict upcoming rewards. After learning, DA neurons respond to cues that predict reward, and also suppress responses to predicted rewards.[15,26,27] Therefore, the reward response of DA neuron corresponds to a discrepancy between the prediction of reward and the reward actually delivered. Since this activation of DA neurons also acts as a learning signal (termed the reward prediction error signal) in the striatum and other projection areas,[28] it has been suggested to play a key role in reinforcement learning. One of the most critical issues in reinforcement learning involving the basal ganglia is where and how to compute the reward prediction error signal between actual reward and the prediction of an upcoming reward.[29] However, because of the poor physiological information regarding the input signals to DA neurons, this issue has remained controversial. The PPTN is one of the strongest afferent pathways for DA neurons, and in the PPTN, acetylcholine and glutamate have strong excitatory effects on DA neurons.[22,30,31] Interestingly, one recent computational theory predicts the PPTN to be a major input source of the excitatory signal to DA neurons and possibly an important component of reinforcement learning,[32] but its functional role is still unclear. In this article, we hypothesize that the PPTN plays an important role in the computation of reward prediction error signals in DA neurons.

EXPERIMENTAL PROCEDURES

To address the PPTN contribution to the reward prediction error, we analyzed neuronal activity of the PPTN during a visually guided saccade task (VGST) for monkeys. Most data have been published previously.[33] For each trial of a VGST, initially the monkey was required to direct and maintain its gaze at a central fixation target (FT). The monkey then had to make a correct saccade to a peripheral saccade target (ST), and received a reward at the end of a trial (FIG. 1A). Here, we will show the preliminary results when we controlled motivation or reward prediction of the monkey by altering the reward size across a block of trials (FIG. 1B).[34]

All experimental procedures in this article were performed in accordance with the NIH Guidelines for the Care and Use of Laboratory Animals, and approved by the Committee for Animal Experiment at Okazaki National Institutes and Osaka University. The details of the surgical and data acquisition methods were published previously.[33]

RESULTS

Effect of Reward Prediction on Behavior and Neuronal Activity of PPTN

In our experiment, the monkeys performed VGST in which they were required to make saccades toward an ST after extinction of the FT (FIG. 1A). To control the level of motivation (reward expectation or prediction) for the task,

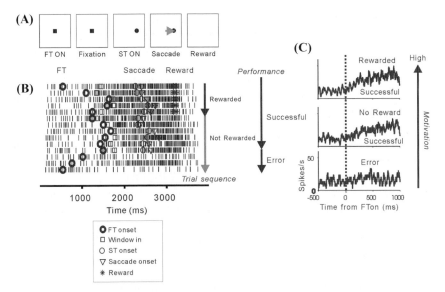

FIGURE 1. Neuronal activity of PPTN during VGST. **(A)** The temporal sequence of the visually guided saccade task (VGST) showing the screen in front of the monkey during successive epochs of a single trial for the VGST. A fixation target (FT) appears at the center of the screen and the monkey achieves foveal fixation. After a fixation period (400–1,000 msec), the FT is turned off and a peripheral saccade target (ST) appears at the left or the right with 10° eccentricity. The monkey is required to make a saccade directly to the target. The *arrow* indicates the direction of eye movement. After maintaining fixation on the target, the target stimulus is turned off, and the monkey receives the juice reward. **(B)** Each horizontal raster represents one trial. Trials are shown in order of presentation from top to bottom and each tick represents a neuronal impulse. Trials were changed from rewarded to reward omission. **(C)** Neural discrimination of task performance. Averages of responses to onset of FT for rewarded, reward omission, and error trials are shown. The firing rates are aligned on FT appearance.

the reward size was changed across a block of trials (i.e., initially rewarded trials, and then zero reward trials, shown in FIG. 1B). As a result of reward control, the decreased level of reward prediction increased the error rate (FIG. 1B). This indicates that we could control the level of motivation (i.e., the prediction of reward for the task) of the monkeys by a reward control.

Next, we will show that the neuronal activity of the PPTN varied with motivational level of the animal. Many of the recorded neurons responded reliably to the onset of the FT, and half of these neurons exhibited reliable sustained activity during execution of the task as reported previously.[33] As shown in FIGURE 1B, for the representative neuron, responses to the FT were stronger when the motivation of the monkeys was higher. During rewarded trials (the monkey could get a predicted reward on successful trials), the discharge increased gradually after onset of the FT, remained elevated for the duration of the fixation period, and was sustained until reward delivery (FIG. 1B, upper rasters).

On the other hand, during reward omission trials (reward was not delivered for the successful trial), there was a decrease in the spike rate after the FT, compared with rewarded trials (FIG. 1B, top and middle rastergrams). Yet, despite the decrease, the sustained activity remained above baseline until the end of trial, consistent with the fact that the monkey successfully completed the reward omission trials (low motivation) (FIG. 1B, middle). In this sense, we can expect that error trials where monkeys canceled the trial with the least motivation would show the least activity. In FIGURE 1C, activities after FT onset were plotted for rewarded, reward omission, and error trials. As we predicted, the activity was the lowest for the error trials. Thus, higher success rate (high motivation or a high level of prediction of upcoming reward for the task) was associated with enhanced neural activity of the PPTN in response to the FT (initial target to direct gaze for executing the VGST).[34]

Tonic and Phasic Performance-Related Activity to the FT

To investigate the temporal dynamics of the output of the PPTN neurons elicited by FT appearance, we examined the time course of this effect. The two illustrated neurons in FIGURE 2 show a higher FT response for successful trials (FIG. 2A, B) than for error trials (FIG. 2C, D), as shown in the previous section. The temporal activation pattern for FT appearance varied among neurons. FIGURE 2A, B shows the sustained tonic type response for one representative neuron, and FIGURE 2B, D shows the transient phasic type response for another neuron. Most of the transient phasic response disappears within 100 msec of its abrupt activation. For the sustained tonic responses, most of the sustained activity was maintained until the reward was delivered.

Interestingly, a substantial number of short latency responses to FT (possibly the most salient stimulus in the task) was observed, compared to the cue stimulus response of DA neurons.[35] For these neurons, the activity latency for FT was less than 100 msec (FIG. 2A, B).

Neural Activity Elicited by the Reward Outcome in the Fully Learned Condition

In addition to the motivational or reward prediction activity to FT appearance, we examined responses to the reward outcome. The PPTN neurons showed an abrupt increase in their firing in response to an unpredicted free reward delivered outside the task (FIG. 3A). This response was consistent with our previous observation.[33] The latency of the reward response (about 100 msec) was slightly shorter than that reported for DA neurons (mean 113 msec[36]).

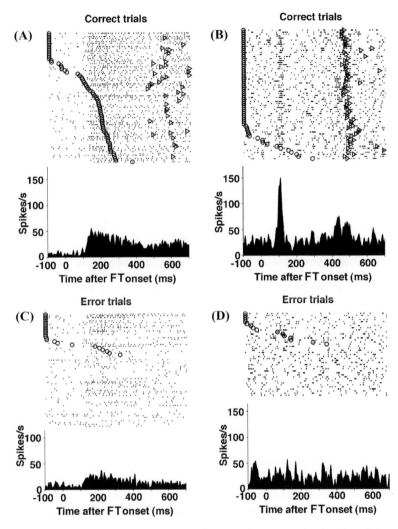

FIGURE 2. Tonic and phasic response to fixation target. Sustained tonic activity to the appearance of the fixation target (FT) in a representative neuron for successful trials (**A**) and error trials (**C**). Phasic activity to FT of a representative neuron for successful trials (**B**) and error trials (**D**). Each raster and histogram is aligned on FT onset. *Circles* indicate the time when the monkey began to fixate the FT. *Triangles* indicate saccade onset.

Most of the neurons that exhibited this reward response to free reward showed responses to reward during the fully learned task trials (FIG. 3B). These results are also consistent with our previous findings.[33] During the continuous saccade task, we did not observe adaptive suppression,[33] which is a remarkable feature of the reward response of the DA neuron.

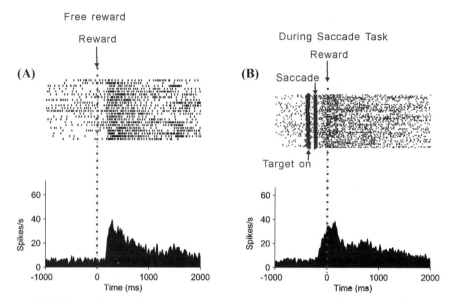

FIGURE 3. Neuronal activity for actual reward. Rastergrams and histograms of activity from a single PPTN neuron when (**A**) a free reward delivery outside task and (**B**) when the reward was delivered during VGST. Monkeys could predict upcoming reward during the saccade task. This figure is modified from Kobayashi et al.[33]

DISCUSSION

Possible Reward Prediction Signal in PPTN

In this article, we demonstrated neuronal activity of PPTN in response to FT appearance, and the responses are predictive of an animal's performance on the task (FIG. 1).[33] It is possible that the activity of PPTN to FT appearance is related to motivation level, reward prediction for the task and the sensory response to be conditioned. The result is consistent with a previous study of cats showing that activity in PPTN neurons was elicited during classical conditioning tasks in response to the conditioned stimulus.[11] Our result further suggests that the salience of the conditioned stimulus in this task (i.e., FT onset in the VGST) was influenced by motivation for the task. Thus, PPTN neurons may comprise a substrate whose role is to transform a sensory cue into a behavioral action. If this hypothesis is correct, it is quite reasonable to expect that the cue responses of the same neuron in a cue–reward association task would be modulated by the magnitude of the expected reward.

Interestingly, similar to the cholinergic PPTN presented here, the noradrenergic LC system has been implicated in response to both salient and motivational sensory events. In that study, LC neurons were phasically activated prior to behavioral responses on both correct and incorrect trials but were not activated by stimuli that failed to elicit lever responses or by lever movements outside the task.[37] In contrast to the LC neurons, we observed a sustained tonic activity during the task in the PPTN. For the next step, we should compare activity of neurons in dorsal raphe nucleus (DRN), LC, and PPTN while controlling arousal, motivation, and learning.

From where does the PPTN receive this motivational or reward prediction signal? First, we propose that the excitatory signals travel via the ventral striatum–ventral pallidum pathway, which receives input mainly from the limbic cortex.[38] This pathway quite possibly includes limbic information, but it uses inhibitory connections from the GABAergic (γ-amino butyric acid) pallidum pathway. There, excitation of PPTN is elicited by a double inhibition mechanism. Second, other possible excitatory sources may include the amygdala and STN.[12] Third, the activity may originate from the cerebral cortex. Recently, Matsumura has emphasized the functional role of cortical input to the PPTN in the integration mechanism of limbic-motor control.[16]

The Short Latency Phasic Response to Salient Stimuli in the PPTN

DA neurons respond to expected, unexpected, and salient sensory events with short latency, but little is known about the sensory systems underlying this response.[39] Studies of rats indicate that the superior colliculus (SC) makes direct synaptic contacts with DA neurons in SNc,[40,41] and there, SC relays visual information to DA neurons. In addition to the SC–SNc connection, the PPTN strongly innervates DA neurons as described above. Further, PPTN receives input from SC.[42–44] In this article, we propose that PPTN may also relay visual information to DA neurons. We showed that PPTN neurons exhibited responses to FT (a salient visual stimulus) that varied with performance of the task (FIG. 2). For some of these neurons, responses occurred with shorter latency (less than 100 msec) than the reported latency to the cue signal of DA neurons (50–120 msec[15,36]) in a rather phasic manner. There have been only a few studies examining visual responses of PPTN neurons. Pan and Hyland reported visual responses of PPTN neurons in rats, and examined response latency to the onset of a light stimulus (mean 70 msec), and they observed no variable visual responses for reward prediction.[45] Contradicting Pan and Hyland's results, a population of our recorded PPTN neurons in primates responded differently to a visual stimulus with motivational state. Our results may be closer to another study of PPTN in cats whose conditioned cue response occurred with a short latency.[11] We should further examine the reward prediction effect on the short latency response to salient stimulus in the PPTN.

Possible Primary Reward Signal in the PPTN

In the PPTN, we observed phasic reward responses for free reward and reward during the task (FIG. 3). It is possible that the reward signal comes from the lateral hypothalamus.[46] This pathway directly excites the PPTN,[12] which fires a brief burst and then accommodates or habituates.[11,47] This brief burst directly excites the SNc via cholinergic and glutamatergic projections and thereby causes a phasic burst in DA neurons projecting to the striatum for unexpected reward.[48,49] In future work, we will examine whether the response properties of the PPTN fulfill the necessary features of a primary reward signal; that is, the activity is related to reward occurrence, to value coding, and the activity has no adaptation under a fully leaned condition.

Computation of Reward Prediction Error Signal in DA Neurons

As described above, DA neurons have unique firing patterns related to the predicted volume and actual times of reward.[15,36,39,50–52] Recent computational models of DA firing have noted similarities between the response pattern of DA neurons and well-known learning algorithms,[53–58] especially temporal differ-ence (TD) models.[53,54,58] In the TD model, a sustained tonic reward prediction pulse originating from the striatum is temporally differentiated to produce an onset burst followed by an offset suppression. The TD model uses fast-sustained excitatory reward prediction and delayed slow-sustained inhibitory pulse signals in DA neurons. In that model the neurons in the striatum (the striosome) provide a significant source of GABAergic inhibition to DA neu-rons,[49] and the fast excitatory reward-predicting signals are derived via a double inhibition mechanism to DA neurons (matriosome–pallidum–DA neuron path-way[56]). Thus, the polysynaptic double inhibition pathway and monosynaptic direct inhibition may provide temporal differentiation of reward prediction in DA neurons. However, the model may not be realistic because it is assumed that (1) the polysynaptic net excitatory signal is faster than the inhibitory signal via the monosynaptic pathway to DA neuron and (2) the double inhibition pathway is required to strongly excite burst activity of DA neuron to a conditioned cue.

A significant difference between our new model, derived from our present findings, and the previous model is the source of excitation for DA neurons.[57] It is possible that the excitatory PPTN neurons may send both a tonic reward prediction signal (FIGS. 1 and 2) and a phasic reward signal (FIG. 3) to DA neurons. Thus, interestingly, the PPTN may relay both predictive and actual reward information to DA neurons.

On the basis of our finding that the PPTN maintains a sustained tonic ex-citatory input to DA neurons, we predict that it is a sustained inhibitory input from the striatum to these neurons, which prevents their tonic activation by this excitatory input after learning. With regard to the delayed, slow inhibition of DA neurons, Houk and colleagues proposed that the striosome–DA neuron connection provides a source of prolonged inhibition of these neurons, which persists from the time of the reward prediction signal to the time when the reward occurs.[56] Here, computation between a fast-sustained excitatory input

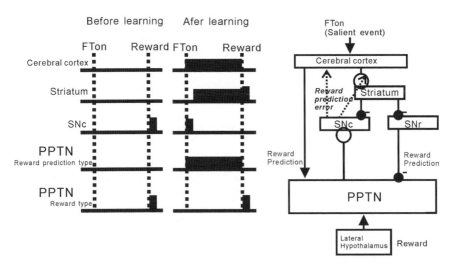

FIGURE 4. Temporal sequence of possible neuronal activity for reinforcement learning and its circuit diagram.

and a delayed slow inhibition occurs in DA neurons. If the fast-sustained excitatory signals from the PPTN appear shortly before the slow-sustained inhibitory inputs from the striosome, the time lag results in a phasic activation in DA neuronal activity. The fast excitatory and slow inhibitory activation maintains an equivalent level of activity in DA neurons from after cue-related phasic activation to just before reward delivery. If the excitatory signals from the PPTN disappear shortly before reward delivery, then the remaining inhibitory inputs should suppress the DA neuron at the time of reward. Thus, when expected rewards are not received, sustained-slow striosomal inhibition of DA neurons, unopposed by excitation, results in a phasic drop in DA neuronal activity. We need to further test PPTN neuronal involvement in the prediction of the timing of reward delivery with a classical conditioning task, as conducted by Schultz's group.[15]

Finally, we illustrated where and how to compute the reward prediction error signal in FIGURE 4. In contrast to the previous TD models that compute temporal derivatives prior to the DA neuron based on a double inhibition mechanism and a specific delay, our PPTN model uses two distinct pathways to the DA neuron: the PPTN for fast excitatory reward prediction and striatal neurons in the striosome for delayed, inhibitory reward prediction. The fast excitation and delayed inhibition are thus computed by separate structures and meet at the DA neuron, rather than by a single temporal differentiator. Further, the strong cholinergic excitatory input source allows short latency burst firing of DA neurons to the cue stimulus, rather than a double inhibition mechanism. Thus, the PPTN is an important center providing information of both reward prediction and the outcome of reward to DA neurons.

ACKNOWLEDGMENT

This study was supported by a Grant-in-Aid for Scientific Research on Priority Areas-System study on higher-order brain functions from the Ministry of Education, Culture, Sports, Science and Technology of Japan (17022027, 18020019).

REFERENCES

1. LINDSLEY, D.B. 1958. The reticular system and perceptual discrimination. *In* The Reticular Formation of the Brain. 513–534. Little, Brown & Co. Boston.
2. STECKLER, T. *et al.* 1994. The pedunculopontine tegmental nucleus: a role in cognitive processes? Brain Res. Brain Res. Rev. **19:** 298–318.
3. HALLANGER, A.E. & B.H. WAINER. 1988. Ascending projections from the pedunculopontine tegmental nucleus and the adjacent mesopontine tegmentum in the rat. J. Comp. Neurol. **274:** 483–515.
4. PERRY, E. *et al.* 1999. Acetylcholine in mind: a neurotransmitter correlate of consciousness? Trends Neurosci. **22:** 273–280.
5. LAVOIE, B. & A. PARENT. 1994. Pedunculopontine nucleus in the squirrel monkey: projections to the basal ganglia as revealed by anterograde tract-tracing methods. J. Comp. Neurol. **344:** 210–231.
6. EDLEY, S.M. & A.M. GRAYBIEL. 1983. The afferent and efferent connections of the feline nucleus tegmenti pedunculopontinus, pars compacta. J. Comp. Neurol. **217:** 187–215.
7. MENA-SEGOVIA, J., J.P. BOLAM & P.J. MAGILL. 2004. Pedunculopontine nucleus and basal ganglia: distant relatives or part of the same family? Trends Neurosci. **27:** 585–588.
8. KOYAMA, Y. & Y. KAYAMA. 1993. Mutual interactions among cholinergic, noradrenergic and serotonergic neurons studied by ionophoresis of these transmitters in rat brainstem nuclei. Neuroscience **55:** 1117–1126.
9. GARCIA-RILL, E. 1991. The pedunculopontine nucleus. Prog. Neurobiol. **36:** 363–389.
10. ALDERSON, H.L. & P. WINN. 2005. The pedunculopontine and reinforcement. Basal Ganglia VIII **56:** 523–532.
11. DORMONT, J.F., H. CONDE & D. FARIN. 1998. The role of the pedunculopontine tegmental nucleus in relation to conditioned motor performance in the cat. I. Context-dependent and reinforcement-related single unit activity. Exp. Brain Res. **121:** 401–410.
12. SEMBA, K. & H.C. FIBIGER. 1992. Afferent connections of the laterodorsal and the pedunculopontine tegmental nuclei in the rat: a retro- and antero-grade transport and immunohistochemical study. J. Comp. Neurol. **323:** 387–410.
13. CHIBA, T., T. KAYAHARA & K. NAKANO. 2001. Efferent projections of infralimbic and prelimbic areas of the medial prefrontal cortex in the Japanese monkey, *Macaca fuscata*. Brain Res. **888:** 83–101.
14. BENINATO, M. & R.F. SPENCER. 1988. The cholinergic innervation of the rat substantia nigra: a light and electron microscopic immunohistochemical study. Exp. Brain Res. **72:** 178–184.
15. SCHULTZ, W. 1998. Predictive reward signal of dopamine neurons. J. Neurophysiol. **80:** 1–27.

16. MATSUMURA, M. 2005. The pedunculopontine tegmental nucleus and experimental parkinsonism. A review. J. Neurol. **252**(Suppl 4): iv5–iv12.

17. FUTAMI, T., K. TAKAKUSAKI & S.T. KITAI. 1995. Glutamatergic and cholinergic inputs from the pedunculopontine tegmental nucleus to dopamine neurons in the substantia nigra pars compacta. Neurosci. Res. **21**: 331–342.

18. TAKAKUSAKI, K. *et al.* 1996. Cholinergic and noncholinergic tegmental pedunculopontine projection neurons in rats revealed by intracellular labeling. J. Comp. Neurol. **371**: 345–361.

19. LOKWAN, S.J. *et al.* 1999. Stimulation of the pedunculopontine tegmental nucleus in the rat produces burst firing in A9 dopaminergic neurons. Neuroscience **92**: 245–254.

20. SCROGGS, R.S. *et al.* 2001. Muscarine reduces calcium-dependent electrical activity in substantia nigra dopaminergic neurons. J. Neurophysiol. **86**: 2966–2972.

21. YAMASHITA, T. & T. ISA. 2003. Fulfenamic acid sensitive, Ca(2+)-dependent inward current induced by nicotinic acetylcholine receptors in dopamine neurons. Neurosci. Res. **46**: 463–473.

22. BLAHA, C.D. & P. WINN. 1993. Modulation of dopamine efflux in the striatum following cholinergic stimulation of the substantia nigra in intact and pedunculopontine tegmental nucleus-lesioned rats. J. Neurosci. **13**: 1035–1044.

23. INGLIS, W.L., M.C. OLMSTEAD & T.W. ROBBINS. 2000. Pedunculopontine tegmental nucleus lesions impair stimulus-reward learning in autoshaping and conditioned reinforcement paradigms. Behav. Neurosci. **114**: 285–294.

24. INGLIS, W.L., J.S. DUNBAR & P. WINN. 1994. Outflow from the nucleus accumbens to the pedunculopontine tegmental nucleus: a dissociation between locomotor activity and the acquisition of responding for conditioned reinforcement stimulated by d-amphetamine. Neuroscience **62**: 51–64.

25. ALDERSON, H.L. *et al.* 2002. The effect of excitotoxic lesions of the pedunculopontine tegmental nucleus on performance of a progressive ratio schedule of reinforcement. Neuroscience **112**: 417–425.

26. HIKOSAKA, O., K. NAKAMURA & H. NAKAHARA. 2006. Basal ganglia orient eyes to reward. J. Neurophysiol. **95**: 567–584.

27. SATOH, T. *et al.* 2003. Correlated coding of motivation and outcome of decision by dopamine neurons. J. Neurosci. **23**: 9913–9923.

28. WICKENS, J.R., R. KOTTER & M.E. ALEXANDER. 1995. Effects of local connectivity on striatal function: stimulation and analysis of a model. Synapse **20**: 281–298.

29. DOYA, K. 2002. Metalearning and neuromodulation. Neural Netw. **15**: 495–506.

30. BOLAM, J.P., C.M. FRANCIS & Z. HENDERSON. 1991. Cholinergic input to dopaminergic neurons in the substantia nigra: a double immunocytochemical study. Neuroscience **41**: 483–494.

31. BLAHA, C.D. *et al.* 1996. Modulation of dopamine efflux in the nucleus accumbens after cholinergic stimulation of the ventral tegmental area in intact, pedunculopontine tegmental nucleus-lesioned, and laterodorsal tegmental nucleus-lesioned rats. J. Neurosci. **16**: 714–722.

32. BROWN, J., D. BULLOCK & S. GROSSBERG. 1999. How the basal ganglia use parallel excitatory and inhibitory learning pathways to selectively respond to unexpected rewarding cues. J. Neurosci. **19**: 10502–10511.

33. KOBAYASHI, Y. *et al.* 2002. Contribution of pedunculopontine tegmental nucleus neurons to performance of visually guided saccade tasks in monkeys. J. Neurophysiol. **88**: 715–731.

34. KOBAYASHI, Y. *et al.* 2005. Reward-predicting activity of pedunculopontine tegmental nucleus neurons during visually guided saccade tasks. Soc. Neurosci. Abstract.

35. ROMO, R. & W. SCHULTZ. 1992. Role of primate basal ganglia and frontal cortex in the internal generation of movements. III. Neuronal activity in the supplementary motor area. Exp. Brain Res. **91:** 396–407.

36. MIRENOWICZ, J. & W. SCHULTZ. 1994. Importance of unpredictability for reward responses in primate dopamine neurons. J. Neurophysiol. **72:** 1024–1027.

37. CLAYTON, E.C. *et al.* 2004. Phasic activation of monkey locus ceruleus neurons by simple decisions in a forced-choice task. J. Neurosci. **24:** 9914–9920.

38. SCHULTZ, W. *et al.* 1992. Neuronal activity in monkey ventral striatum related to the expectation of reward. J. Neurosci. **12:** 4595–4610.

39. LJUNGBERG, T., P. APICELLA & W. SCHULTZ. 1992. Responses of monkey dopamine neurons during learning of behavioral reactions. J. Neurophysiol. **67:** 145–163.

40. COMOLI, E. *et al.* 2003. A direct projection from superior colliculus to substantia nigra for detecting salient visual events. Nat. Neurosci. **6:** 974–980.

41. MCHAFFIE, J.G. *et al.* 2006. A direct projection from superior colliculus to substantia nigra pars compacta in the cat. Neuroscience **138:** 221–234.

42. REDGRAVE, P., I.J. MITCHELL & P. DEAN. 1987. Descending projections from the superior colliculus in rat: a study using orthograde transport of wheatgerm-agglutinin conjugated horseradish peroxidase. Exp. Brain Res. **68:** 147–167.

43. HUERTA, M.F. & J.K. HARTING. 1982. Tectal control of spinal cord activity: neuroanatomical demonstration of pathways connecting the superior colliculus with the cervical spinal cord grey. Prog. Brain Res. **57:** 293–328.

44. MAY, P.J. & J.D. PORTER. 1992. The laminar distribution of macaque tectobulbar and tectospinal neurons. Vis Neurosci. **8:** 257–276.

45. PAN, W.X. & B.I. HYLAND. 2005. Pedunculopontine tegmental nucleus controls conditioned responses of midbrain dopamine neurons in behaving rats. J. Neurosci. **25:** 4725–4732.

46. NAKAMURA, K. & T. ONO. 1986. Lateral hypothalamus neuron involvement in integration of natural and artificial rewards and cue signals. J. Neurophysiol. **55:** 163–181.

47. TAKAKUSAKI, K., T. SHIROYAMA & S.T. KITAI. 1997. Two types of cholinergic neurons in the rat tegmental pedunculopontine nucleus: electrophysiological and morphological characterization. Neuroscience **79:** 1089–1109.

48. CONDE, H. 1992. Organization and physiology of the substantia nigra. Exp. Brain Res. **88:** 233–248.

49. GERFEN, C.R. 1992. The neostriatal mosaic: multiple levels of compartmental organization. Trends Neurosci. **15:** 133–139.

50. SCHULTZ, W., P. APICELLA & T. LJUNGBERG. 1993. Responses of monkey dopamine neurons to reward and conditioned stimuli during successive steps of learning a delayed response task. J. Neurosci. **13:** 900–913.

51. SCHULTZ, W. *et al.* 1995. Reward-related signals carried by dopamine neurons. *In* Models of Information Processing in the Basal Ganglia. J. Houk, J. Davis & D. Beiser, Eds.: 233–248. MIT Press. Cambridge, MA.

52. HOLLERMAN, J.R. & W. SCHULTZ. 1998. Dopamine neurons report an error in the temporal prediction of reward during learning. Nat. Neurosci. **1:** 304–309.

53. SCHULTZ, W., P. DAYAN & P.R. MONTAGUE. 1997. A neural substrate of prediction and reward. Science **275:** 1593–1599.

54. MONTAGUE, P.R., P. DAYAN & T.J. SEJNOWSKI. 1996. A framework for mesencephalic dopamine systems based on predictive Hebbian learning. J. Neurosci. **16:** 1936–1947.

55. BERNS, G.S. & T.J. SEJNOWSKI. 1998. A computational model of how the basal ganglia produce sequences. J. Cogn. Neurosci. **10:** 108–121.

56. HOUK, J.C., J.L. ADAMS & A.G. BARTO, Eds. 1995. A model of how the basal ganglia generate and use neural signals that predict reinforcement. *In* Models of Information Processing in the Basal Ganglia. 249–270. MIT Press. Cambridge, MA.

57. CONTRERAS-VIDAL, J.L., S. GROSSBERG & D. BULLOCK. 1997. A neural model of cerebellar learning for arm movement control: cortico- spino-cerebellar dynamics. Learn. Mem. **3:** 475–502.

58. SURI, R.E. & W. SCHULTZ. 1998. Learning of sequential movements by neural network model with dopamine-like reinforcement signal. Exp. Brain Res. **121:** 350–354.

A Computational Model of Craving and Obsession

A. DAVID REDISH[a] AND ADAM JOHNSON[b]

[a]Department of Neuroscience, University of Minnesota, Minneapolis, Minnesota, USA

[b]Graduate Program in Neuroscience and the Center for Cognitive Sciences, University of Minnesota, Minneapolis, Minnesota, USA

ABSTRACT: If addictions and problematic behaviors arise from interactions between drugs, reward sequences, and natural learning sytems, then an explanation of clinically problematic conditions (such as the self-administration of drugs or problem gambling) requires an understanding of the neural systems that have evolved to allow an agent to make decisions. We hypothesize a unified decision-making system consisting of three components—a situation recognition system, a flexible, planning-capable system, and an inflexible, habit-like system. In this article, we present a model of the planning-capable system based on a planning process arising from experimentally observed look-ahead dynamics in the hippocampus enabling a forward search of possibilities and an evaluation process in the nucleus accumbens. Based on evidence that opioid signaling can provide hedonic evalutation of an achieved outcome, we hypothesize that similar opioid-signaling processes evaluate the value of expected outcomes. This leads to a model of craving, based on the recognition of a path to a high-value outcome, and obsession, based on a value-induced limitation of the search process. This theory can explain why opioid antagonists reduce both hedonic responses and craving.

KEYWORDS: craving; obsession; addiction; hippocampus; nucleus accumbens; opioid signaling; opiates; dopamine

INTRODUCTION

We start from the assumption that neural systems have evolved to allow an agent to make decisions that will allow it to survive and procreate. This means that if we want to understand action-selection processes that lead to clinically problematic situations, such as self-administration of drugs[1-3] or the continued pursuit of problematic behaviors such as gambling,[4-7] we need to

Address for correspondence: A. David Redish, Department of Neuroscience, University of Minnesota, Minneapolis, MN 55455, USA. Voice: 612-626-3738; fax: 651-626-5009.
    redish@ahc.umn.edu

Ann. N.Y. Acad. Sci. 1104: 324–339 (2007). © 2007 New York Academy of Sciences.
doi: 10.1196/annals.1390.014

first understand that natural learning system and how those addictive processes access it. Making optimal decisions requires calculations of the *expected utility* or *value* of taking specific actions in specific situations. Expected utility (or value) can be defined as the expected reward, taking into account the expected magnitude of the reward, the expected probability of receiving the reward, and the expected delay before receiving that reward.[8,9]

To predict the expected reward and the appropriate action to achieve that reward, the agent must first recognize the situation it is in. This recognition process is fundamentally a classification problem—this situation is like these and not like those. For example, if one is deciding whether or not to put a dollar in a soda machine, to predict the consequences of putting the dollar in and pushing the soda button, one needs to correctly recognize that one is in front of a soda machine, not in front of a bank ATM. In the psychology literature, this is referred to as accessing the correct *schema*. Importantly, one needs to recognize not only the general soda machine schema, but also to determine whether there are any specific situation cues available. For example, is this a Coke or a Pepsi machine? Is this machine more or less reliable than other machines?

Once one has identified the situation one is in, calculating the value of the actions to be taken requires some combination of cached memory and search of the possibilities.[10] In the computer science literature, this has been termed *depth of search*, and is a fundamental basis of heuristic reasoning. Interestingly, there is strong behavioral evidence that there are two systems in the mammalian brain with differing levels of search: (1) a flexible system, which is capable of being learned quickly, but is computationally expensive to use, and (2) an inflexible system, which can act quickly, but must be learned slowly. The flexible system allows the planning of multiple paths to achieve a goal and takes the expectation of that goal into account in its decision making. In contrast, the inflexible system simply retrieves the remembered action for a given situation.[11–14] The flexible system can be learned quickly because of its flexibility—knowing the existence of a potential path to a goal does not commit one to taking that path. However, the complexity of planning through those potential paths makes the flexible system computationally expensive. In contrast, the inflexible system must be learned slowly because it would be dangerous to commit to always taking an action in a situation until one knows that that action is the correct one. However, the limited search done in the inflexible (habit) system allows it to work quickly requiring only limited computational resources.

The existence of these two systems has been proposed in both the animal navigation (*cognitive map* vs. *route* strategies,[11,12] *place* vs. *response* strategies[13]) and learning theory literatures (*situation–outcome* (*S–O*) vs. *situation–action* (*S–A*) associations[14,15]).

In the navigation literature, the interaction of multiple navigation systems can be seen in how rats solve the classic single-T maze task.[13,16–19] Limited training leads to a place strategy in which animals return to the same goal location when started from multiple starting points, even through this may require different actions. In contrast, extended training leads to a response strategy in which animals perform the same actions on entering the maze, even if that leads them to different goals. The place strategy depends on the integrity of the hippocampus and ventromedial striatum, whereas the response strategy depends on the integrity of the dorsolateral striatum.[13,18,19]

In the learning theory literature, the interaction of multiple learning systems can be seen in how rats respond to devaluation.[14,15,20] Classically, these differences are measured by first training an animal to take an action sequence leading to reward, and then, changing the value of the reward to the animal, usually in a different context. The value of a reward can be changed by providing excess amounts of the reward (satiation[14]) or by pairing the reward with an aversive stimulus, such as LiCl (devaluation[20,21]). Finally, the animal is provided the chance to take the action. If the action selection process takes into account the current value of the reward, then the animal will not respond, but if the action selection process is an association between the situation and the action (thus does not take into account the value of the reward), the animal will continue to respond. With extended training of a reliable association, animals switch from a devaluation-capable system to a devaluation-independent system.[14,22] The devaluation-capable system (S–O) is dependent on the integrity of the ventral striatum,[23,24] the prelimbic medial prefrontal cortex,[22] and the orbitofrontal cortex,[20,25] whereas the devaluation-independent system (S–A) is dependent on the integrity of the dorsal striatum[19,26,27] and the infralimbic cortex.[22,28]

This leads us to hypothesize a unified system incorporating three subsystems, a situation recognition system and two contrasting decision systems—a flexible, planning-capable system that accommodates multiple paths to goals and takes into account the value of potential outcomes, and an inflexible, habit-like system, which reacts with a single action to each situation and does not take into account the value of potential outcomes (see FIG. 1).

Both the planning-capable and habit-like systems require a recognition of the agent's situation. This recognition system entails a categorization process, which is likely to arise in cortical systems through competitive learning,[29–32] using content-addressable memory mechanisms.[33–35]

The first (flexible, planning) decision-making system requires recognition of a situation S, recognition of a means of achieving outcome O from situation S,

$$S \cdots \overset{(a)}{\rightarrow} O \qquad (1)$$

as well as the evaluation of the value of achieving outcome O, which will depend on the agent's current needs N

$$E(V) = V(E(O), N) \qquad (2)$$

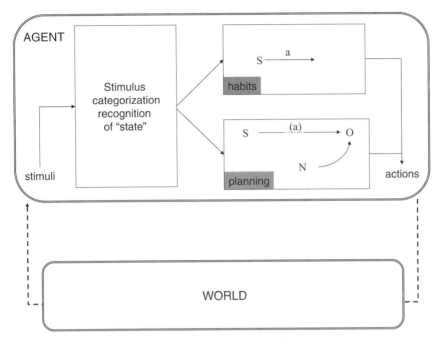

FIGURE 1. Three systems involved in decision making.

where $E(V)$ is the expectation of the value of taking action a in situation S, which is a function of the expected outcome[*] $E(O)$ and the needs of the agent N. Because the value of the outcome is calculated on the fly (online), that calculation can take into account the needs (N) of the agent.

The second (inflexible, habit) decision-making system entails a simple association between situation and action. Thus, the habit system requires recognition of a situation S, and a single, identified action to take within that situation. We describe this system with the simple formulation

$$S \xrightarrow{a} \tag{3}$$

Evaluation in the second (inflexible, habit) system entails a memory recall of the learned associated (cached[10]) value of taking action a in situation S,

$$E(V) = V(S, a) \tag{4}$$

most likely learned through temporal-difference reinforcement learning mechanisms.[8] These two systems, along with a situation-recognition component (S), form a unified theory of decision-making processes.

---

[*]Computationally, the outcome O is simply a future state S', but we refer to it as the "outcome" to emphasize the importance of the "completion of needs" that the new situation $O = S'$ will achieve.

While the $S \xrightarrow{a}$ system has been well modeled through the TDRL algorithms,[9,36–39] the mechanisms that underlie the $S \cdots\xrightarrow{(a)} O$ system are more controversial.[40] Following a recent suggestion that one possible difference between $S \xrightarrow{a}$ and $S \cdots\xrightarrow{(a)} O$ systems is the depth of search,[10] we propose a model of the $S \cdots\xrightarrow{(a)} O$ system based on a *consideration of possibilities* signal provided by the hippocampus.

A COMPUTATIONAL MODEL OF THE DEVALUATION-CAPABLE/MAP-NAVIGATION SYSTEM

The key to both devaluation and map navigation is the ability to consider the possible consequences of one's actions. This hypothesized mechanism needs three components: (1) a recognition of the situation at hand (S), (2) a process by which the system can calculate the expected consequences of taking available actions (retrieval of the $S \cdots\xrightarrow{(a)} O$ relationship), and (3) evaluation of the expected outcome ($E(V) = V(E(O), N)$). We hypothesize that the planning component is instantiated through hippocampal dynamics and the evaluation component is instantiated through processing in orbitofrontal cortex and through opioid signaling in the nucleus accumbens.

Planning

We have recently observed look-ahead dynamics in the hippocampal neural ensemble recordings of rats facing a high-cost choice.[41,42] Briefly, rats were trained to run a choice task in which they made choices to receive food. Rats, particularly early in the session, paused at high-cost choices and showed behavior reminiscent of vicarious trial and error (VTE[43–45]). Because of the spatial tuning of hippocampal pyramidal cells,[12,46] it is possible to reconstruct the position of the animal x from the firing pattern F using Bayesian reconstruction techniques.[47–50] The reconstructed distribution $P(x|F)$ tracked the animal well as the animal ran through the central path. When the animal paused, the reconstructed distribution moved out along one choice, and then the other, alternating a few times before the rat began moving again.

It is not known what effect these nonlocal planning signals seen in the hippocampus have on downstream structures, but it is known that other hippocampal processes representing nonlocal information (i.e., sharp wave ripple complexes occurring during slow-wave sleep in which replay of recent experiences is known to occur[51–54]) do have effects on downstream structures, such as nucleus accumbens.[55] Thus, it is likely that these representations could also be translated downstream, providing a potential planning signal (recognition of a potential $S \cdots\xrightarrow{(a)} O$ path) for decision making.

Alternative Structures Involved in Planning

Historically, planning and expectation of outcome have been associated with prefrontal structures,[56–58] and Daw *et al.*[10] have suggested the prefrontal cortex as the site of the $S \overset{(a)}{\cdots\to} O$ process. Hippocampus projects to medial prefrontal cortex,[59] and prefrontal structures have been observed to contain goal-related processes.[60,61] The relative roles of hippocampus and prefrontal cortex in planning remain to be elucidated.

In the motor control fields, the cerebellum has been hypothesized to be the site of "forward models" predicting the consequences of one's actions.[62–64] While the cerebellum has been identified in cognitive processes as well as motor,[65] the processes controlled by the cerebellum tend to be those with tightly controlled timing, likely controlled by highly specialized cerebellar circuits,[62,63,66,67] which once learned become inflexible. In contrast, the planning processes addressed above require flexible circuits capable of evaluating consequences over variable and longer time periods. Neither devaluation nor map navigation have been found to be dependent on cerebellar integrity.

EVALUATION

To evaluate the value of an outcome, the system needs a signal that recognizes hedonic value. Two structures that have been suggested to be involved in the evalutation of an outcome are the orbitofrontal cortex[20,68–72] and the ventral striatum.[9,73–76] Neurons in the ventral striatum show reward correlates,[75,77–81] and anticipate predicted reward.[77,78,82,83] The hippocampus projects to ventral striatum,[84–86] and ventral striatal firing patterns reflect hippocampal neural activity.[55,87] Neurons in the orbitofrontal cortex encode parameters relating to the value of potential choices.[68,69] Both fMRI,[70,71] and lesion[20,56,88,89] data have also implicated the orbitofrontal cortex in the evaluation of value. Anticipatory neural firing of goal-related information in orbitofrontal cortex is dependent on hippocampal integrity.[90]

Berridge and Robinson[91,92] suggest that hedonic signals ("liking") are carried by opioid signaling, as evidenced by the effect of opioid agonists and antagonists on taste reactivity. Consistent with these ideas, Levine and colleagues[93,94] report that opioid antagonists directly interfere with the reported qualia of hedonic pleasure associated with eating sweet, without interfering in taste discrimination.

There are multiple opioid receptor types in the mammalian brain (μ, κ, δ,[95–97]). Whereas μ-receptor agonists are rewarding, euphorigenic, and support self-administration, κ-receptor agonists are aversive, dysphoric, and interfere with self-administration.[95–103]† μ-receptor antagonists block

---

†The role of δ receptors is more controversial.[96,102,104]

self-administration and conditioned approach to drug cues, but blocking the other opioid receptors (κ, δ) do not.[95,102,103] Each receptor type is associated with a preferential endogenous opioid signaling peptide (μ: β-endorphin, the endomorphins; κ: dynorphin, δ: the enkephalins).[96,99,105] These data suggest that the opioid system is well situated to provide a direct evaluation of an event: rewarding signals via μ receptors and aversive signals via κ receptors.

It is important to differentiate hedonic rewards and costs from reinforcement and aversion.[37,91,92,106] Reinforcement and aversion entail changes reflecting changes in expectation (i.e., the *value prediction error* term in temporal difference learning[8,9]). If one correctly predicts the hedonic pleasure provided by a reward, then one's value prediction error signal is zero, even though one presumably still feels that hedonic pleasure on achieving the reward. While euphoria and dysphoria have been associated with opioid signals,[95-103] reinforcement signals have been associated with dopamine.[9,91,92,106,107]

If endogenous opioids signal the actual hedonic evaluation of an achieved outcome, then when faced with potential outcome signals arriving from the hippocampus, one might expect similar proccesses to evaluate the value of expected outcomes. This predicts that the effect of hippocampal planning signals on accumbens structures will be to trigger evaluative processes similar to those that occur in response to actual achieved outcomes. This has immediate consequences for craving and obsession.

IMPLICATIONS

Craving

Craving is the intense desire for something. It is, fundamentally, a subjective, internal feeling, and may or may not always be reflected in external actions. In the terminology presented above, craving is the recognition that there is a pathway to a high-value outcome. This expectation can only occur in the $S \cdots \xrightarrow{(a)} O$ (planning) system; the $S \xrightarrow{a}$ (habit) system does not include a recognition of the expected outcome. Because the flexible (planning) system only entails the recognition that an action can lead to a potential path to a goal and does not entail a commitment to action, craving is not necessarily going to produce action selection. In the $S \cdots \xrightarrow{(a)} O$ planning system, when the forward planning (hippocampal) component reaches a goal that is evaluated to have a high value, this will produce a strong desire to achieve that goal. We suggest that the psychological effect of that recognition is to produce "craving."

Obsession

It is important to remember that the forward search component of the $S \cdots \xrightarrow{(a)} O$ system requires a memory retrieval process. This search process entails the exploration of multiple consequences from situation S. Oversensitization of a single $S \cdots \xrightarrow{(a)} O$ relationship is likely to limit the exploration of possibilities, which would appear as a cognitive blinding to alternatives. Sensitization of an $S \cdots \xrightarrow{(a)} O$ relation would also mean that when an animal is returned to situation S, it is more likely to remember that it can reach outcome O, which would make it more likely to remember the existence of outcome O, thus more likely to experience craving in situation S. Craving would then lead to a recurring search of the same $S \cdots \xrightarrow{(a)} O$ path, which would appear as cognitive blinding or obsession.

DISCUSSION: PREDICTIONS AND OPEN QUESTIONS

In this article, we have proposed a model of craving based on a computational theory of planning processes,[10] which we have suggested arise from an interaction between a consideration-of-possibilities process involving hippocampus and an evaluative process involving nucleus accumbens or orbitofrontal cortex. Essentially, this produces an outcome-expectancy[108,109] model of craving:[110,112] craving entails recognition that there is a means of achieving a highly charged positive outcome (or of relieving a highly charged negative outcome). This model is consistent with new interpretations of Pavlovian conditioning as a memory-of-expectations process.[113] This process is fundamentally an associative memory process in that it requires the memory that there is a path to outcome O from situation S. Thus, it suggests that craving should involve structures involved in memory, particularly working memory, such as frontal cortex[114,115] and hippocampus.[12,116] Craving should also involve structures involved in the evaluation of future rewards, such as orbitofrontal cortex (OFC)[68,117] and nucleus accumbens.[118–120] Evidence from cue-induced craving responses in addicts supports these hypotheses.[112,121–123] The theory also provides immediate explanations for why opioid antagonists can be used to block craving, and makes predictions about a hippocampal role in devaluation.

Competitive opioid antagonists have been used clinically to reduce craving.[98,124–126] The model of the planning system laid out above provides an immediate explanation for this effect: when the predictive component of the planning system identifies the completion of an $S \cdots \xrightarrow{(a)} O$ pathway and a potential means of achieving an outcome, the evaluative component will release reward signals (i.e., endogenous opioids), identifying the value of that outcome for evaluative purposes. As noted above, the identification of a pathway to high reward leads to craving for that reward. The hypothesis that reward signals are released on recognition of a pathway to a high-value outcome implies that blocking those reward signals would not only dampen the subjective hedonic value of receiving reward, but would also dampen craving for those rewards. If that reward signal is based on opioid signaling, then this may explain why opioid antagonists such as naltrexone or nalmefene can reduce craving.

Addiction has been proposed to entail a transition from exploratory use, to (in some users) the development of strong desires (craving), followed in some users by a strong, habitual use in which the user loses control of the drug use.[127–130] This sequence follows the sequence of normal learning. Flexible, map-based, devaluation-capable strategies are learned first;[12,13,17] but with repeatable, regular experience, animals switch to automated, inflexible, route-based, devaluation-resistant strategies.[12–14,17,28,131] In animals, drug-seeking also first involves more ventromedial aspects of striatum[132,133] and later involves the more dorsolateral aspects.[133,134] This theory predicts that drug addiction should progress through a flexible strategy based on intense craving to an inflexible, habit-based strategy, which is independent of craving.

This unified hypothesis leads to important open questions and predictions. An important, but as yet unresolved question is: How well does the map/route differentiation in the navigation literature[11,12] translate to the devaluation/nondevaluation distinction?[14] In the navigation literature, the key difference between map- and route-based strategies is flexibility. Map strategies are highly flexible, allowing paths around obstacles,[11,12] and journeys to the same location from different starting points.[13,16] In contrast, route strategies are highly inflexible, requiring the same paths under the same conditions.[11–13,135] In early maze experiments, overtrained rats were found to run full speed into novel obstacles[136,137] or off shortened tracks.[138] In the devaluation literature, the key difference lies in the inclusion of the outcome in action selection. $S-O$ strategies entail a consideration of the outcome, while $S-A$ strategies do not. Anatomically, map learning is critically dependent on the hippocampus.[11,12] However, Corbit and Balleine[139] found that hippocampal lesions had no effect on devaluation. Importantly, these lesions were partial and occurred before training. Similarly sized partial lesions that occurred before training have little or no effect on place finding in the Morris water maze,[140] which is the classic hippocampal-dependent navigation task. Ostlund and Balleine[141] report that hippocampal lesions after training devastate devaluation learning, as it does place finding in the Morris water maze.[140,142]

The crucial test of this hypothesis, however, is the prediction that similar opiate signaling will occur in response to both veridical inputs (reflecting real receipt of reward/punishment, leading to euphoria/dysphoria) and to hypothetical inputs (reflecting planning, leading to craving/dread). These hypotheses could be tested with simultaneous recordings of hippocampus and ventral striatum.

ACKNOWLEDGMENTS

This work was supported by a Career Development Award from the University of Minnesota TTURC (to ADR, NCI/NIDA P50 DA01333), by a graduate fellowship from the Center for Cognitive Sciences at the University of Minnesota (to AJ, T32HD007151), by a Fulbright scholarship (to AJ), as well as by NIMH R01-MH06829 and by the Land Grant Professorship program at the University of Minnesota. We thank Daniel Smith, Carolyn Fairbanks, Jadin Jackson, Zeb Kurth-Nelson, Suck-Won Kim, Steve Jensen, and Paul Schrater for helpful discussions.

REFERENCES

1. VOLKOW, N. & T.-K. LI. 2005. The neuroscience of addiction. Nature Neurosci. **8:** 1429–1430.
2. KALIVAS, P.W. & N.D. VOLKOW. 2005. The neural basis of addiction: a pathology of motivation and choice. Am. J. Psychiatry **162:** 1403–1413.
3. O'BRIEN, C.P., N. VOLKOW & T.-K. LI. 2006. What's in a word? Addiction versus dependence in DSM-V. Am. J. Psychiatry **163:** 764–765.
4. CUSTER, R.L. 1984. Profile of the pathological gambler. J. Clin. Psychiatry **45:** 35–38.
5. WAGENAAR, W.A. 1988. Paradoxes of Gambling Behaviour. Hillsdale, NJ.
6. PETRY, N.M. 2006. Should the scope of addictive behaviors be broadened to include pathological gambling? Addiction **101:** 152–159.
7. POTENZA, M.N. 2006. Should addictive disorders include non-substance-related conditions? Addiction **101:** 142–151.
8. SUTTON, R.S. & A.G. BARTO. 1998. Reinforcement Learning: An Introduction. MIT Press. Cambridge, MA.
9. DAW, N.D. 2003. Reinforcement learning models of the dopamine system and their behavioral implications. Ph.D. thesis, Carnegie Mellon University. Pittsburgh, PA.
10. DAW, N.D., Y. NIV & P. DAYAN. 2005. Uncertainty-based competition between prefrontal and dorsolateral striatal systems for behavioral control. Nature Neurosci. **8:** 1704–1711.
11. O'KEEFE, J. & L. NADEL. 1978. The Hippocampus as a Cognitive Map. Clarendon Press. Oxford.
12. REDISH, A.D. 1999. Beyond the Cognitive Map: From Place Cells to Episodic Memory. MIT Press. Cambridge, MA.
13. PACKARD, M.G. & J.L. McGAUGH. 1996. Inactivation of hippocampus or caudate nucleus with lidocaine differentially affects expression of place and response learning. Neurobiol. Learn. Mem. **65:** 65–72.
14. BALLEINE, B.W. & A. DICKINSON. 1998. Goal-directed instrumental action: contingency and incentive learning and their cortical substrates. Neuropharmacology **37:** 407–419.
15. DICKINSON, A. 1980. Contemporary Animal Learning Theory. Cambridge University Press. New York.
16. TOLMAN, E.C., B.F. RITCHIE & D. KALISH. 1946. Studies in spatial learning. II. Place learning versus response learning. J. Exp. Psychol. **36:** 221–229.
17. RESTLE, F. 1957. Discrimination of cues in mazes: a resolution of the 'place-vs-response' question. Psychol. Rev. **64:** 217–228.
18. BARNES, C.A., L. NADEL & W.K. HONIG. 1980. Spatial memory deficit in senescent rats. Can. J. Psychol. **34:** 29–39.

19. YIN, H.H. & B.J. KNOWLTON. 2004. Contributions of striatal subregions to place
     and response learning. Learn. Mem. **11:** 459–463.
20. SCHOENBAUM, G., M. ROESCH & T.A. STALNAKER. 2006. Orbitofrontal cortex,
     decision making, and drug addiction. Trends Neurosci. **29:** 116–124.
21. NELSON, A. & S. KILLCROSS. 2006. Amphetamine exposure enhances habit for-
     mation. J. Neurosci. **26:** 3805–3812.
22. KILLCROSS, S. & E. COUTUREAU. 2003. Coordination of actions and habits in the
     medial prefrontal cortex of rats. Cerebral Cortex **13:** 400–408.
23. SCHOENBAUM, G., T.A. STALNAKER & M.R. ROESCH. 2006. Ventral striatum fails to
     represent bad outcomes after cocaine exposure. Soc. Neurosci. Abstr. Program
     No. 485.16. 2006 Neuroscience Meeting Planner. Atlanta, GA: Society for
     Neuroscience, 2006. Online.
24. CORBIT, L.H., J.L. MUIR & B.W. BALLEINE. 2001. The role of the nucleus ac-
     cumbens in instrumental conditioning: evidence of a functional dissociation
     between accumbens core and shell. J. Neurosci. **21:** 3251–3260.
25. SCHOENBAUM, G. *et al.* 2006. Encoding changes in orbitofrontal cortex in reversal-
     impaired aged rats. J. Neurophysiol **95:** 1509–1517.
26. YIN, H.H., B. KNOWLTON & B.W. BALLEINE. 2004. Lesions of dorsolateral stria-
     tum preserve outcome expectancy but disrupt habit formation in instrumental
     learning. Eur. J. Neurosci. **19:** 181–189.
27. YIN, H.H., B.J. KNOWLTON & B.W. BALLEINE. 2006. Inactivation of dorsolateral
     striatum enhances sensitivity to changes in the action-outcome contingency in
     instrumental conditioning. Behav. Brain Res. **166:** 189–196.
28. COUTUREAU, E. & S. KILLCROSS. 2003. Inactivation of the infralimbic prefrontal
     cortex reinstates goal-directed responding in overtrained rats. Behav. Brain Res.
     146: 167–174.
29. GROSSBERG, S. 1976. Adaptive pattern classification and universal recoding: I.
     parallel development and coding of neural feature detectors. Biol. Cyber. **23:**
     121–134.
30. RUMELHART, D.E. & J.L. MCCLELLAND, Eds. 1986. PDP: Explorations in the
     Microstructures of Cognition. Vol. 1. Foundations. MIT Press. Cambridge, MA.
31. ARBIB, M., Ed. 1995. The Handbook of Brain Theory and Neural Networks. MIT
     Press. Cambridge, MA.
32. REDISH, A.D. 2005. Implications of the temporal difference reinforcement learn-
     ing model for addiction and relapse. Neuropsychopharmacology **30:** S27–S28.
33. HEBB, D.O. 1949. The Organization of Behavior. Wiley, New York. Reissued
     2002. LEA.
34. HOPFIELD, J.J. 1982. Neural networks and physical systems with emergent col-
     lective computational abilities. Proc. Natl. Acad. Sci. USA **79:** 2554–2558.
35. HERTZ, J., A. KROGH & R.G. PALMER. 1991. Introduction to the Theory of Neural
     Computation. Addison-Wesley. Reading, MA.
36. MONTAGUE, P. R., P. DAYAN & T.J. SEJNOWSKI. 1996. A framework for mesen-
     cephalic dopamine systems based on predictive Hebbian learning. J. Neurosci.
     16: 1936–1947.
37. REDISH, A.D. 2004. Addiction as a computational process gone awry. Science
     306: 1944–1947.
38. DOYA, K. 2000. Reinforcement learning in continuous time and space. Neur.
     Comput. **12:** 219–245.
39. DAW, N.D., A.C. COURVILLE & D.S. TOURETZKY. 2006. Representation and timing
     in theories of the dopamine system. Neur. Comput. **18:** 1637–1677.

40. DAYAN, P. & B.W. BALLEINE. 2002. Reward, motivation, and reinforcement learning. Neuron **36:** 285–298.
41. JOHNSON, A. & A.D. REDISH. 2005. Observation of transient neural dynamics in the rodent hippocampus during behavior of a sequential decision task using predictive filter methods. Acta Neurobiol. Exp. **65:** 103.
42. JOHNSON, A. & A.D. REDISH. 2006. Neural ensembles in CA3 transiently encode paths forward of the animal at a decision point: a possible mechanism for the consideration of alternatives. Program No. 574.2. 2006 Neuroscience Meeting Planner. Atlanta, GA: Society for Neuroscience, 2006. Online.
43. MEUNZINGER, K.F. 1938. Vicarious trial and error at a point of choice I. a general survey of its relation to learning efficiency. J. Genet. Psychol. **53:** 75–86.
44. TOLMAN, E.C. 1939. Prediction of vicarious trial and error by means of the schematic sowbug. Psychol. Rev. **46:** 318–336.
45. HU, D. & A. AMSEL. 1995. A simple test of the vicarious trial-and-error hypothesis of hippocampal function. Proc. Natl. Acad. Sci. **92:** 5506–5509.
46. O'KEEFE, J. 1976. Place units in the hippocampus of the freely moving rat. Exp. Neurol. **51:** 78–109.
47. RIEKE, F. *et al.* 1997. Spikes. MIT Press. Cambridge, MA.
48. ZHANG, K. *et al.* 1998. Interpreting neuronal population activity by reconstruction: Unified framework with application to hippocampal place cells. J. Neurophysiol. **79:** 1017–1044.
49. BROWN, E.N. *et al.* 1998. A statistical paradigm for neural spike train decoding applied to position prediction from ensemble firing patterns of rat hippocampal place cells. J. Neurosci. **18:** 7411–7425.
50. JOHNSON, A., J. JACKSON & A.D. REDISH. In press. Measuring distributed properties of neural representations beyond the decoding of local variables— implications for cognition. *In* Mechanisms of Information Processing in the Brain: Encoding of Information in Neural Populations and Networks. C. HÖLSCHER & M.H.J. MUNK, Eds. Cambridge University Press. Cambridge, UK.
51. WILSON, M.A. & B.L. MCNAUGHTON. 1994. Reactivation of hippocampal ensemble memories during sleep. Science **265:** 676–679.
52. KUDRIMOTI, H.S., C.A. BARNES & B.L. MCNAUGHTON. 1999. Reactivation of hippocampal cell assemblies: effects of behavioral state, experience, and EEG dynamics. J. Neurosci. **19:** 4090–4101.
53. NÁDASDY, Z. *et al.* 1999. Replay and time compression of recurring spike sequences in the hippocampus. J. Neurosci. **19:** 9497–9507.
54. LEE, A.K. & M.A. WILSON. 2002. Memory of sequential experience in the hippocampus during slow wave sleep. Neuron **36:** 1183–1194.
55. PENNARTZ, C.M.A. *et al.* 2004. The ventral striatum in off-line processing: ensemble reactivation during sleep and modulation by hippocampal ripples. J. Neurosci. **24:** 6446–6456.
56. BECHARA, A. 2005. Decision making, impulse control and loss of willpower to resist drugs: a neurocognitive perspective. Nature Neurosci. **8:** 1458–1463.
57. ZERMATTEN, A. *et al.* 2005. Impulsivity and decision making. J. Nerv. Ment. Dis. **193:** 647–650.
58. OWEN, A.M. 1997. Cognitive planning in humans: neuropsychological, neuroanatomical and neuropharmacological perspectives. Progr. Neurobiol. **53:** 431–450.
59. JONES, M.W. & M.A. WILSON. 2005. Theta rhythms coordinate hippocampal-prefrontal interactions in a spatial memory task. PLOS Biol. **3:** e402.

60. JUNG, M.W. *et al.* 1998. Firing characteristics of deep layer neurons in prefrontal cortex in rats performing spatial working memory tasks. Cerebral Cortex **8:** 437–450.

61. HOK, V. *et al.* 2005. Coding for spatial goals in the prelimbic/infralimbic area of the rat frontal cortex. Proc. Natl. Acad. Sci. USA **102:** 4602–4607.

62. HIKOSAKA, O. *et al.* 1998. Differential roles of the frontal cortex, basal ganglia, and cerebellum in visuomotor sequence learning. Neurobiol. Learn. Mem. **70:** 137–149.

63. DOYA, K. 1999. What are the computations of the cerebellum, the basal ganglia, and the cerebral cortex? Neur. Networks **12:** 961–974.

64. MIALL, R.C. 1998. The cerebellum, predictive control and motor coordination. Novartis Found. Symp. **218:** 272–284.

65. SEIDLER, R.D. *et al.* 2002. Cerebellum activation associated with performance change but not motor learning. Science **296:** 2043–2046.

66. DOYON, J. *et al.* 1998. Role of the striatum, cerebellum and frontal lobes in the automatization of a repeated visuomotor sequence of movements. Neuropsychologia **36:** 625–641.

67. ITO, M. 2000. Mechanisms of motor learning in the cerebellum. Brain Res. **886:** 237–245.

68. PADOA-SCHIOPPA, C. & J.A. ASSAD. 2006. Neurons in the orbitofrontal cortex encode economic value. Nature **441:** 223–226.

69. SCHOENBAUM, G. & M. ROESCH. 2005. Orbitofrontal cortex, associative learning, and expectancies. Neuron **47:** 633–636.

70. VOLKOW, N.D. & J.S. FOWLER. 2000. Addiction, a disease of compulsion and drive: Involvement of the orbitofrontal cortex. Cerebral Cortex **10:** 318–325.

71. O'DOHERTY, J. *et al.* 2001. Abstract reward and punishment representations in the human orbitofrontal cortex. Nature Neurosci. **4:** 95–102.

72. FEIERSTEIN, C.E. *et al.* 2006. Representation of spatial goals in rat orbitofrontal cortex. Neuron **60:** 495–507.

73. MOGENSON, G.J., D.L. JONES & C.Y. YIM. 1980. From motivation to action: Functional interface between the limbic system and the motor system. Progr. Neurobiol. **14:** 69–97.

74. MOGENSON, G.J. & M. NIELSEN. 1984. Neuropharmacological evidence to suggest that the nucleus accumbens and subpallidal region contribute to exploratory locomotion. Behav. Neural Biol. **42:** 52–60.

75. LAVOIE, A.M. & S.J.Y. MIZUMORI. 1994. Spatial-, movement- and reward-sensitive discharge by medial ventral striatum neurons in rats. Brain Res. **638:** 157–168.

76. O'DOHERTY, J. *et al.* 2004. Dissociable roles of ventral and dorsal striatum in instrumental conditioning. Science **304:** 452–454.

77. MARTIN, P.D. & T. ONO. 2000. Effects of reward anticipation, reward presentation, and spatial parameters on the firing of single neurons recorded in the subiculum and nucleus accumbens of freely moving rats. Behav. Brain Res. **116:** 23–38.

78. MIYAZAKI, K. *et al.* 1998. Reward-quality dependent anticipation in rat nucleus accumbens. NeuroReport **9:** 3943–3948.

79. CARELLI, R.M., S.G. IJAMES & A.J. CRUMLING. 2000. Evidence that separate neural circuits in the nucleus accumbens encode cocaine versus "natural" (water and food) reward. J. Neurosci. **20:** 4255–4266.

80. CARELLI, R.M. 2002. Nucleus accumbens cell firing during goal-directed behaviors for cocaine vs. 'natural' reinforcement. Physiol. Behav. **76:** 379–387.

81. CARELLI, R.M. & J. WONDOLOWSKI. 2003. Selective encoding of cocaine versus natural rewards by nucleus accumbens neurons is not related to chronic drug exposure. J. Neurosci. **23:** 11214–11223.
82. YUN, I.A. *et al.* 2004. The ventral tegmental area is required for the behavioral and nucleus accumbens neuronal firing responses to incentive cues. J. Neurosci. **24:** 2923–2933.
83. SCHULTZ, W. *et al.* 1992. Neuronal activity in monkey ventral striatum related to the expectation of reward. J. Neurosci. **12:** 4595–4610.
84. MCGEORGE, A.J. & R.L. FAULL. 1989. The organization of the projection from the cerebral cortex to the striatum in the rat. Neuroscience **29:** 503–537.
85. FINCH, D.M. 1996. Neurophysiology of converging synaptic inputs from rat prefrontal cortex, amygdala, midline thalamus, and hippocampal formation onto single neurons of the caudate/putamen and nucleus accumbens. Hippocampus **6:** 495–512.
86. SWANSON, L.W. 2000. Cerebral hemisphere regulation of motivated behavior. Brain Res. **886:** 113–164.
87. MARTIN, P.D. 2001. Locomotion towards a goal alters the synchronous firing of neurons recorded simultaneously in the subiculum and nucleus accumbens of rats. Behav. Brain Res. **124:** 19–28.
88. SCHOENBAUM, G., A.A. CHIBA & M. GALLAGHER. 1999. Neural encoding in orbitofrontal cortex and basolateral amygdala during olfactory discrimination learning. J. Neurosci. **19:** 1876–1884.
89. CHUDASAMA, Y. *et al.* 2003. Dissociable aspects of performance on the 5-choice serial reaction time task following lesions of the dorsal anterior cingulate, infralimbic and orbitofrontal cortex in the rat: differential effects on selectivity, impulsivity and compulsivity. Behav. Brain Res. **146:** 105–119.
90. DAVIS, J.B. *et al.* 2006. Hippocampal dependence of anticipatory neuronal firing in the orbitofrontal cortex of rats learning an odor-sequence memory task. Soc. Neurosci. Abstr. Program No. 66.7. 2006 Neuroscience Meeting Planner. Atlanta. GA: Society for Neuroscience, 2006. Online.
91. BERRIDGE, K.C. & T.E. ROBINSON. 1998. What is the role of dopamine in reward: hedonic impact, reward learning, or incentive salience? Brain Res. Rev. **28:** 309–369.
92. BERRIDGE, K.C. & T.E. ROBINSON. 2003. Parsing reward. Trends Neurosci. **26:** 507–513.
93. ARBISI, P.A., C.J. BILLINGTON & A.S. LEVINE. 1999. The effect of naltrexone on taste detection and recognition threshold. Appetite **32:** 241–249.
94. LEVINE, A.S. & C.J. BILLINGTON. 2004. Opioids as agents of reward-related feeding: a consideration of the evidence. Physiol. Behav. **82:** 57–61.
95. DE VRIES, T.J. & T.S. SHIPPENBERG, 2002. Neural systems underlying opiate addiction. J. Neurosci. **22:** 3321–3325.
96. HERZ, A. 1997. Endogenous opioid systems and alcohol addiction. Psychopharmacology **129:** 99–111.
97. HERZ, A. 1998. Opioid reward mechanisms: a key role in drug abuse? Can. J. Physiol. Pharmacol. **76:** 252–258.
98. MEYER, R. & S. MIRIN. 1979. The Heroin Stimulus. Plenum. New York.
99. CHAVKIN, C., I.F. JAMES & A. GOLDSTEIN. 1982. Dynorphin is a specific endogenous ligand of the kappa opioid receptor. Science **215:** 413–415.
100. MUCHA, R.F. & A. HERZ. 1985. Motivational properties of kappa and mu opioid receptor agonists studied with place and taste preference conditioning. Psychopharmacology **86:** 274–280.

101. BALS-KUBIK, R., A. HERZ & T. SHIPPENBERG. 1989. Evidence that the aversive effects of opioid antagonists and κ-agonists are centrally mediated. Psychopharmacology **98:** 203–206.

102. MATTHES, H.W.D. *et al.* 1996. Loss of morphine-induced analgesia, reward effect, and withdrawal symptoms in mice lacking the μ-opioid-receptor gene. Nature **383:** 819–823.

103. KIEFFER, B.L. 1999. Opioids: first lessons from knockout mice. Trends Pharmacol. Sci. **20:** 19–26.

104. BROOM, D.C. *et al.* 2002. Nonpeptidic δ-opioid receptor agonists reduce immobility in the forced swim assay in rats. Neuropsychopharmacology **26:** 744–755.

105. ZADINA, J.E. *et al.* 1997. A potent and selective endogenous agonist for the μ-opiate receptor. Nature **386:** 499–502.

106. SCHULTZ, W. 2002. Getting formal with dopamine and reward. Neuron **36:** 241–263.

107. MONTAGUE, P.R. *et al.* 1995. Bee foraging in uncertain environments using predictive hebbian learning. Nature **377:** 725–728.

108. TOLMAN, E.C. 1948. Cognitive maps in rats and men. Psychol. Rev. **55:** 189–208.

109. BOLLES, R.C. 1972. Reinforcement, expectancy, and learning. Psychol. Rev. **79:** 394–409.

110. MARLATT, G.A. 1985. Cognitive factors in the relapse process. *In* Relapse Prevention. G.A. MARLATT & J.R. GORDON, Eds.: 128–200. Guilford. New York.

111. GOLDMAN, M.S., S.A. BROWN & B.A. CHRISTIANSEN. 1987. Expectancy theory: Thinking about drinking. *In* Psychological Theories of Drinking and Alcoholism. H.T. BLAINE & K.E. LEONARD, Eds.: 181–226. Guilford. New York.

112. TIFFANY, S.T. 1999. Cognitive concepts of craving. Alcohol Res. Health **23:** 215–224.

113. RESCORLA, R.A. 1988. Pavlovian conditioning: It's not what you think it is. Am. Psychol. **43:** 151–160.

114. GOLDMAN-RAKIC, P.S., S. FUNAHASHI & C.J. BRUCE. 1990. Neocortical memory circuits. Cold Spring Harbor Symposia on Quant. Biol. LV: 1025–1038.

115. FUSTER, J.M. 1997. The Prefrontal Cortex: Anatomy, Physiology, and Neuropsychology of the Frontal Lobe, 3rd ed. Lippincot-Raven. Philadelphia, PA.

116. OLTON, D.S. & R.J. SAMUELSON. 1976. Remembrance of places passed: spatial memory in rats. J. Exp. Psych.: Anim. Behav. Processes **2:** 97–116.

117. ROESCH, M.R., A.R. TAYLOR & G. SCHOENBAUM. 2006. Encoding of time-discounted rewards in orbitofrontal cortex is independent of value representation. Neuron **51:** 509–520.

118. KALIVAS, P.W., N. VOLKOW & J. SEAMANS. 2005. Unmanageable motivation in addiction: A pathology in prefrontal-accumbens glutamate transmission. Neuron **45:** 647–650.

119. ANAGNOSTARAS, S.G., T. SCHALLERT & T.E. ROBINSON. 2002. Memory processes governing amphetamine-induced psychomotor sensitization. Neuropsychopharmacology **26:** 703–715.

120. LI, Y., M.J. ACERBO & T.E. ROBINSON. 2004. The induction of behavioural sensitization is associated with cocaine-induced structural plasticity in the core (but not shell) of the nucleus accumbens. Eur. J. Neurosci. **20:** 1647–1654.

121. CHILDRESS, A.R. *et al.* 1993. Cue reactivity and cue reactivity interventions in drug dependence. NIDA Res. Monogr. **137:** 73–94.

122. GRANT, S. *et al.* 1996. Activation of memory circuits during cue-elicited cocaine craving. Proc. Natl. Acad. Sci. **93:** 12040–12045.

123. HOMMER, D.W. 1999. Functional imaging of craving. Alcohol Res. Health **23:** 187–196.
124. KIEFER, F. & K. MANN. 2005. New achievements and pharmacotherapeutic approaches in the treatment of alcohol dependence. Eur. J. of Pharmacol. **526:** 163–171.
125. O'BRIEN, C.P. 2005. Anticraving medications for relapse prevention: a possible new class of psychoactive medications. Am. J. Psychiatry **162:** 1423–1431.
126. GRANT, J.E. *et al.* 2006. Multicenter investigation of the opioid antagonist nalmefene in the treatment of pathological gambling. Am. J. Psychiatry **163:** 303–312.
127. ALTMAN, J. *et al.* 1996. The biological, social and clinical bases of drug addiction: commentary and debate. Psychopharmacology **125:** 285–345.
128. LOWINSON, J.H. *et al.*, Eds. 1997. Substance Abuse: A Comprehensive Textbook 3rd ed. Williams and Wilkins. Baltimore.
129. ROBBINS, T.W. & B.J. EVERITT. 1999. Drug addiction: bad habits add up. Nature **398:** 567–570.
130. EVERITT, B.J. & T.W. ROBBINS. 2005. Neural systems of reinforcement for drug addiction: from actions to habits to compulsion. Nature Neurosci. **8:** 1481–1489. Corrected online after print.
131. SCHMITZER-TORBERT, N.C. & A.D. REDISH. 2002. Development of path stereotypy in a single day in rats on a multiple-T maze. Archives Italiennes de Biologie **140:** 295–301.
132. ITO, R. *et al.* 2000. Dissociation in conditioned dopamine release in the nucleus accumbens core and shell in response to cocaine cues and during cocaine-seeking behavior in rats. J. Neurosci. **20:** 7489–7495.
133. LETCHWORTH, S.R. *et al.* 2001. Progression of changes in dopamine transporter binding site density as a result of cocaine self-administration in rhesus monkeys. J. Neurosci. **21:** 2799–2807.
134. ITO, R. *et al.* 2002. Dopamine release in the dorsal striatum during cocaine-seeking behavior under the control of a drug-associated cue. J. Neurosci. **22:** 6247–6253.
135. REDISH, A.D. & D.S. TOURETZKY. 1998. The role of the hippocampus in solving the Morris water maze. Neural Computation **10:** 73–111.
136. WATSON, J.B. 1907. Kinaesthetic and organic sensations: their role in the reactions of the white rat to the maze. Psychol. Rev. **8:** 43–100.
137. CARR, H. & J.B. WATSON. 1908. Orientation in the white rat. J. Comp. Neurol. Psychol. **18:** 27–44.
138. DENNIS, W. 1932. Multiple visual discrimination in the block elevated maze. J. Comp. Physiol. Psychol. **13:** 391–396.
139. CORBIT, L.H. & B.W. BALLEINE. 2000. The role of the hippocampus in instrumental conditioning. J. Neurosci. **20:** 4233–4239.
140. MOSER, M.-B. & E.I. MOSER. 1998. Distributed encoding and retrieval of spatial memory in the hippocampus. J. Neurosci. **18:** 7535–7542.
141. OSTLUND, S.B. & B.W. BALLEINE. 2004. Post-training neurotoxic lesions of the dorsal hippocampus disrupt goal-directed actions. Soc. Neurosci. Abstr. Program no. 897.17. 2004 Abstract Viewer/Itinerary Planner. Washington, DC: Society for Neuroscience, 2004. Online.
142. MORRIS, R.G.M. 1981. Spatial localization does not require the presence of local cues. Learn. Motiv. **12:** 239–260.

Calculating the Cost of Acting in Frontal Cortex

MARK E. WALTON, PETER H. RUDEBECK, DAVID M. BANNERMAN, AND MATTHEW F. S. RUSHWORTH

Department of Experimental Psychology, University of Oxford, Oxford, United Kingdom

ABSTRACT: To make informed and successful decisions, it is vital to be able to evaluate whether the expected benefits of a course of action make it worth tolerating the costs incurred to obtain them. The frontal lobe has been implicated in several aspects of goal-directed action selection, social interaction, and optimal choice behavior. However, its exact contribution has remained elusive. Here, we discuss a series of studies in rats and primates examining the effect of discrete lesions on different aspects of cost-benefit decision making. Rats with excitotoxic lesions of the anterior cingulate cortex became less willing to invest effort for reward but showed no change when having to tolerate delays. Orbitofrontal cortex-lesioned rats, by contrast, became more impulsive, yet were just as prepared as normal animals to expend energy to obtain reward. The sulcal region of primate anterior cingulate cortex was also shown to be essential for dynamically integrating over time the recent history of choices and outcomes. Selecting a particular course of action may also come at the expense of gathering important information about other individuals. Evaluating social information when deciding whether to respond was demonstrated to be a function of the anterior cingulate gyrus. Taken together, this indicates that there may be dissociable pathways in the frontal lobe for managing different types of response cost and for gathering social information.

KEYWORDS: anterior cingulate cortex; orbitofrontal cortex; decision making; effort; delay; risk; social

INTRODUCTION

Over the past decades, significant strides have been made in our understanding of the mechanisms underlying goal-directed actions. From the background

Address for correspondence: Mark Walton, Department of Experimental Psychology, University of Oxford, South Parks Road, Oxford OX1 3UD, UK. Voice: +44 (0)1865 271315; fax: +44 (0)1865 3103447.
mark.walton@psy.ox.ac.uk

Ann. N.Y. Acad. Sci. 1104: 340–356 (2007). © 2007 New York Academy of Sciences.
doi: 10.1196/annals.1390.009

of a stringent theoretical framework, systematic research is starting to uncover the brain regions and neural mechanisms involved in learning and representing the connection between responses and their consequences.[1] Obviously, the availability of such information is essential to make beneficial decisions when confronted with multiple options as it allows organisms to begin to construct a notion of value of the available alternatives given their current motivation.

However, one failing with applying such an approach to decision making is that it has tended to assume that actions lead directly to reward with few intervening costs or negative events in between the choice and the outcome. As has been appreciated for many years in other behavioral sciences, such as behavioral ecology and economics, animals, including humans, do not just make decisions and select actions on the basis of an expected reward but also weigh up the potential costs of the different courses of action that might be pursued.[2–5] Such costs may involve the investment of time or effort, a willingness to tolerate a risk that a reward may not be forthcoming, or to endure pain in the pursuit of a goal. Similarly, selecting one particular option may also come at the expense of gathering other significant information, such as the proximity and intentions of other competing animals. To date, though, it is far from clear how the brain incorporates calculations of such costs with expected reward to guide response selection.

Neuropsychological studies have demonstrated that parts of the frontal lobe are integral for goal-directed action selection, strategy implementation, and social behavior, all of which are essential components of optimal decision making.[6–9] More recently, the direct contribution of these regions to cost-benefit choices has been investigated, but there remains a high degree of confusion over which regions are critical and what their exact contributions are. Patients with damage to parts of the prefrontal cortex can simultaneously exhibit prolonged deliberation about choices accompanied by subsequent irresponsible, risky behavior.[10–13] Similarly, several studies have shown that such lesions can cause both symptoms of apathy and indifference as well as poor impulse control.[14–16]

Such combinations of apparently contradictory deficits may seem surprising given behavioral evidence for potential partial dissociations between the processing of different response costs.[17] However, the damage in most prefrontal patients encompasses several neuroanatomically separable areas. By contrast, by making experimental lesions in animals, it is possible to examine the direct contribution of discrete regions to aspects of cost-benefit decision making. Here, we discuss recent evidence from our laboratory that indicates that different parts of the frontal lobe—in particular, the orbitofrontal cortex (OFC) and anterior cingulate cortex (ACC)—play dissociable roles in action selection depending on the type of response cost and environmental information that is encountered.

THE ACC AND EFFORT-BASED VALUATION

If an animal is faced with a choice between two courses of action, one of which leads to a larger reward than the other, then it is simple to work out which is more valuable and therefore which it should choose. However, if the same choice is presented except that now the larger reward is only obtained after a period of time has elapsed or an amount of effort has been expended, then the optimal decision is no longer immediately self-evident. This diminution of the value of a reward by the cost incurred to achieve it is known as discounting and has been shown consistently to influence the way in which animals and humans make choices.[3,17,18] If we assume that the basic machinery of motivation is designed to bias animals toward courses of action that result in more certain, easily obtained and immediate reward,[19] then the question arises that what mechanisms exist to resist such temptation when more taxing options may result in greater overall utility.

Such situations can be modeled in animals by presenting them with a choice between an easily obtained low reward (LR) option and a high reward (HR) option attained only after overcoming some type of additional response cost. Several studies in rodents have implicated parts of the frontal lobe in allowing animals to overcome response costs to earn greater reward. Using a cost-benefit T-maze paradigm originally designed by Salamone and colleagues,[20] Walton and colleagues demonstrated that the medial frontal cortex is essential for allowing an animal to put in extra work for greater reward.[21] Rats chose between investing effort by surmounting a large barrier to obtain the HR or selecting the LR, which did not incur any additional response cost. Following excitotoxic lesions of this region, animals became profoundly cost averse, switching from selecting to climb the barrier for the HR on the majority of trials presurgery to choosing the more easily obtained LR option on almost all occasions. However, when the response costs were equated by adding an identical barrier into the LR arm, meaning that the rats were now required to put in the same amount of effort to obtain either size of reward, all animals returned to choosing the HR. This implies that the deficit was not caused by any gross motor deficits or spatial impairments, but is instead primarily one concerned with making optimal decisions.

Subsequent experiments have localized this effect to the ACC, with lesions of this region (including pre- and perigenual Cg1 and Cg2, and ACd FIG. 1A) causing a comparable switch to low effort responses whereas lesions of adjacent prelimbic cortex or to the OFC having no effect on the allocation of responses in this task (FIG. 1B).[22,23] However, the ACC lesions do not affect response selection or make animals cost averse in all decision-making situations. When a rule has to be learned and applied to work out what action to make, such as in a delayed matching-to-position task, animals with ACC lesions perform comparably to control animals.[22,24] It had previously been shown in an operant cost-benefit decision-making task where rats chose between an immediately

(A) **(B)**

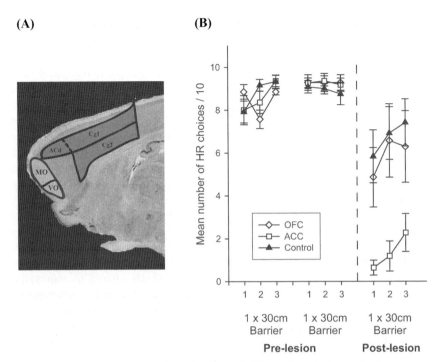

FIGURE 1. (A) Representation of ACC and OFC regions in the rat brain that were lesioned in the study by Rudebeck and colleagues.[26] ACC includes pre- and perigenual Cg1 and Cg2 and dorsal ACC (ACd). OFC includes medial (MO), ventral (VO), and lateral orbital regions (latter not depicted). **(B)** Choice performance of ACC- and OFC-lesioned animals and controls in two preoperative testing sessions and a single postoperative testing session of the effort-based decision-making task. In each block, rats chose between climbing a 30-cm barrier for the HR (4 food pellets) or selecting the unoccupied arm for the LR (2 pellets) (denoted in the figure as "1 × 30 cm Barrier"). Adapted from Rudebeck *et al.*[26]

available LR or a HR presented after an increasing delay that neither ACC nor prelimbic cortex lesions caused any change in the pattern of responses.[25] Similarly, in a T-maze analogue of the barrier task described above, Rudebeck and colleagues also found no change in delay-based decision making following identical ACC lesions to those which affect effort-related choices.[26] The role of rodent ACC in guiding decisions in uncertain situations or in tasks with probabilistic reinforcement has yet to be tested.

THE OFC AND DELAY-BASED VALUATION

Although OFC lesions did not make animals cost averse on the T-maze barrier task, several neuropsychological studies in humans and animals have implicated the OFC in mediating certain types of cost-benefit decision making, in particular in aspects of impulsivity and foresight.[11,12,27] Recent studies have found cells in the OFC that respond to the amount of time before reward presentation, with activity appearing to reflect a discounted value of the delayed reward.[28,29] However, just as the human literature can appear contradictory in highlighting both impulsive choices and long deliberation times of patients with damage to this region, the results from studies of delay-based decision making in rodents are similarly conflicting, with some showing an increase in impulsive choices following OFC lesions[30,31] and another the opposite effect.[32] Moreover, all of these studies were performed in operant boxes that have markedly different response selection and cue-control elements to a T-maze, making direct comparison with the effort-based decision-making findings tricky.

To investigate this inconsistency, Rudebeck and colleagues also tested animals with excitotoxic lesions of the OFC on a T-maze delay-based decision-making task where they could choose between a delayed HR or an immediately available LR option.[26] Prior to surgery, all animals preferred to wait for the HR. However, postoperatively, in contrast to the animals with ACC lesions, the OFC-lesioned rats became cost averse, switching to the LR option in the majority of trials (FIG. 2). This was not caused by a general hyperactivity as the same animals showed no increase in spontaneous locomotor activity. Moreover, as with the ACC-lesioned animals on the T-maze barrier task, when the costs were equated by requiring the animals to wait for an identical amount of time for both the HR and LR options, the OFC group returned to choosing the HR. This again implies that the increase in impulsive choices was a consequence of an alteration in the way the costs and benefits of the options were processed rather than being a simple impairment in spatial or reward magnitude processing.

WHAT ARE THE CONTRIBUTIONS OF THE ACC AND OFC TO OVERCOMING RESPONSE COSTS?

The studies discussed above clearly demonstrate that both the ACC and OFC are crucial to allow animals to make optimal cost-benefit decisions. More importantly, they also illustrate that there is anatomical separation in choice behavior depending on whether animals are required to integrate predicted energetic expenditure or delay information with expected reward magnitude. However, their respective roles in guiding different types of cost-benefit decision making are not unique or categorical. Inactivating the basolateral amygdala (BLA) with bupivacaine or lesions of the dopamine terminals in the nucleus accumbens, for example, makes rats comparably cost averse when choosing whether or not to invest effort for greater reward.[20,33] Similarly, excitotoxic lesions of the BLA or nucleus accumbens also render them impulsive.[25,32] This indicates that extended interconnected, but partially dissociable, frontal-subcortical circuits are required to calculate the value of the available options and, where

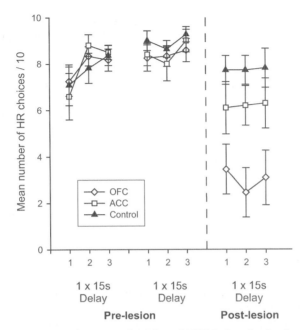

FIGURE 2. Choice performance of ACC- and OFC-lesioned animals and controls in two preoperative testing sessions and a single postoperative testing session of the delay-based decision-making task. In each block, rats chose between an immediately available LR (1 pellet) or waiting 15 sec for the HR (10 pellets) (denoted in the figure as "1 × 15 s Delay"). Adapted from Rudebeck *et al.*[26]

appropriate, to resist the temptation of the most easily available reward. Moreover, the fact that either ACC- or OFC-lesioned animals were able to return to choosing the HR option when either the effort or delay, respectively, was equated for both response options demonstrates that the surgery, rather than rendering them completely unable to tolerate response costs, caused them instead to be biased away from the high cost option.

Interestingly, even this bias does not appear to be permanent. In Rudebeck and colleagues' study,[26] the animals on each cost-benefit decision-making task were retested after the equal cost manipulation with the original choice contingencies of a high-cost HR and a low-cost LR restored. On both the effort- and delay-based tasks, the group that had previously been impaired immediately after surgery now performed comparably to the control animals. This meant that the ACC-lesioned rats were now just as likely to choose to surmount the barrier for the HR as the sham or OFC-lesioned groups (FIG. 3A) and the OFC-lesioned rats opted to tolerate the delay for the HR at comparable levels to the other two groups (FIG. 3B). Further analysis and testing revealed that this was not caused by simple recovery from the lesion as there was no correlation between rats' choice performance and the length of time from

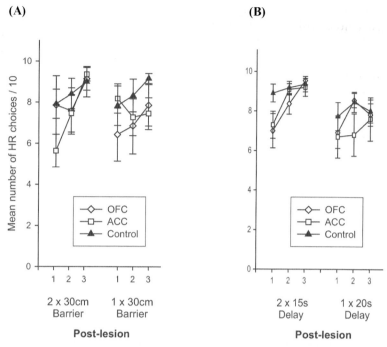

FIGURE 3. Postoperative choice performance of ACC- and OFC-lesioned animals and controls in the equal cost condition ("2 × 30 cm Barrier" or "2 × 15 s Delay") and a retest of the original cost-benefit choice contingency on either **(A)** the effort-based ("1 × 30 cm Barrier") or **(B)** delay-based decision-making task ("1 × 20 s Delay"). Adapted from Rudebeck *et al.*[26]

surgery, and the same animals were impaired on further social and locomotor experiments conducted several weeks later.

The question therefore arises as to what is the exact contribution of the rat ACC and OFC that allows each to bias animals to tolerate investing effort or time, respectively, for greater reward. It is known that the OFC, in particular, plays a crucial role with interconnected structures such as the BLA in learning associations between stimuli and rewards, signaling expected outcomes and representing incentive value.[34–37] Moreover, it has been demonstrated that lesions to the OFC cause a reduction, though not abolition, in expected reward-related neuronal activity in the BLA.[38] Primate ACC too has been shown to contain cells that track the progress toward an expected reward across a sequence of actions,[39] though no such evidence yet exists of how lesions to this region affect the responses in other interconnected structures. Based on this evidence, it may be that the ACC and OFC play crucial and dissociable roles in representing the anticipated goal of a course of action across effort- and delay-based costs, especially in the absence of mediating cues and extensive preoperative training. Lesions to these regions could degrade this representation, pushing animals away from being sufficiently motivated to select the costly HR in the face of the easily obtainable LR. However, following the equal cost condition in which animals are repeatedly exposed to the HR option, animals are able to reevaluate their decision reference, perhaps using their intact BLA and nucleus accumbens, meaning that they can continue to make optimal choices when reexposed to the original cost-benefit choice contingencies. Further experiments will be required to validate these hypotheses.

WEIGHING UP WHAT TO DO IN AN UNCERTAIN WORLD

All of the experiments described above deliberately use paradigms in which the contingencies are static and fully learned. However, in more naturalistic settings, outcomes are often uncertain and vary as a function of the patterns of choices made by both the agent in question and other organisms in the environment. Being able to dynamically integrate the expected magnitude of a reward with the uncertainty of its availability is vitally important to maintain up-to-date and accurate representations of the value of available options. This is particularly the case for a foraging animal having to choose whether to stay with what is a depleting patch of food or to move away to try to explore other potentially more fruitful sources of nourishment.[3,4,40]

There is an increasingly voluminous literature implicating ACC, particularly in primates the sulcal region (ACCs) that has direct projections to the motor system[41,42] (FIG. 4A), in learning and representing the value of available actions.[43–45] To investigate the role of this region in allocating responses in an uncertain environment, Kennerley and colleagues[46] taught monkeys a version of the "matching" task, originally devised by Herrnstein[47] and used more recently in macaque monkeys by Sugrue and colleagues,[48] which captures several of the key features faced by foraging animals described above. In this task, animals can choose between one of two joystick movements—a lift or a turn response—which are rewarded with unequal probabilities. Importantly, these rewards are assigned independently to each action in each trial and remain available until captured. As only one of the two actions is possible in each trial, this means that the cumulative probability of the ignored option increases the more trials in which it is not chosen. Therefore, it is not optimal for an animal continually simply to select the option with a higher probability as there will come a point when the likelihood of a reward being available on the less profitable option is actually greater than for the more profitable response.

Normal monkeys discover that they need to sample both options to develop a sense of the yield of each alternative and usually learn within 150 trials how often it is advantageous (within 97% of the optimal) for them to switch away

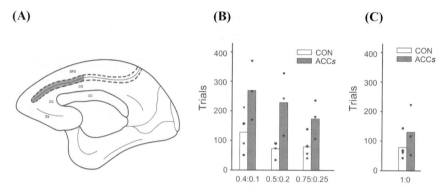

FIGURE 4. (**A**) Schematic of the location of the ACC*s* lesion. (**B**) Number of trials required to reach and sustain performance within 97% of the optimal rate when the outcome for each choice was probabilistic. Control and ACC*s* lesion data are shown by white and gray bars, respectively. All data come from the postoperative testing period. The optimal rate was defined as

$$r_{opt} = \frac{p - pq}{p + q - 2pq}$$

where p and q are the probabilities of a new reward being assigned to the lift or turn responses, respectively. (**C**) Number of trials required to reach and sustain performance within 97% of the optimal rate when the outcome for each choice was deterministic. Adapted from Kennerley et al.[46]

from the more profitable option to the one that normally leads to a poorer revenue of reward. However, by contrast, macaques with discrete lesions to the ACC*s* took significantly more trials to approach the optimum response ratio with each of the three sets of probabilistically rewarded contingencies (FIG. 4B). This was not caused by an inability to work out which response was better or to sustain a response as the lesioned animals were just as good as the controls when the outcomes were deterministic (correct or incorrect; FIG. 4C). Instead, coupled with a previous error-guided switching experiment by Kennerley et al.,[46] this suggests that the ACC*s* may be one crucial cortical component for integrating the extended history of an animal's choices and payoffs to guide action selection.

THE VALUE OF SOCIAL INFORMATION

Although it is more common to associate the term reward with homeostatic necessities, such as food and liquid, other information that can guide decision making can also be intrinsically valuable. For a social, competitive animal, for instance, choices outside the laboratory will need to take account of the presence and position of other individuals who may either guide it to a fruitful source of food or challenge it for its provisions. In humans, a whole branch of mathematics has been devoted to describing the optimal way of making decisions in a world populated with intelligent competitors.[49,50] There is also a range of evidence to suggest that monkeys find the acquisition of social information rewarding and scale their valuation depending on how important the information would be to them, even going so far as to sacrifice fluid to gain access to a picture of a high-status male or female perineum.[51,52]

It has long been known that damage to ventromedial parts of prefrontal cortex in humans, encompassing both the ACC and OFC that result in decision-making impairments, can also cause profound changes in personality and disrupt normal patterns of social interaction.[6] To investigate the involvement of these regions in evaluating social information, Rudebeck and colleagues[53] gave monkeys a task in which they were presented with an appetizing food item at the same time as either a neutral or a socially interesting stimulus. Their latencies to pick up the food item served as an index of how much they valued obtaining further information about the stimulus against the rewarding incentive of the food. Normal animals delayed reaching for a foodstuff in the presence of images of male or female monkeys, becoming increasingly retarded as the stimuli became more socially interesting (FIG. 5A). However, animals with selective lesions to the ACC gyrus (ACCg, FIG. 5B) were seemingly not interested by the social information, retrieving the food rapidly regardless of whether an image of a monkey or a neutral picture was presented. By contrast, no such alterations in performance were seen in animals with either lesions encompassing lateral orbital and ventral prefrontal cortex or to ACCs, with these monkeys showing an analogous scaling as controls with the social stimuli. The lack of concern shown by the ACCg group was not simply caused by a basic impairment in

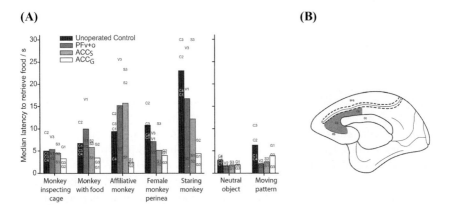

FIGURE 5. (**A**) Response latencies to pick up a food item in the presence of either a social stimulus (5 left-hand columns) or two neutral stimuli (2 right-hand columns). Symbols indicate scores for each individual. Redrawn, based on Rudebeck *et al.*[53] (**B**) Schematic of the location of the ACCg lesion.

flexibly assigning value as they were just as proficient as controls on a stimulus reversal learning task. This finding is consistent with several recent imaging studies requiring complex social interaction or use of social knowledge that has shown activations in a potentially homologous region of the ACC.[54–57] However, Rudebeck and colleagues' data indicate that the crucial role of this area may be in integrating the value of the social information with the other incentives in the environment to decide whether to act. How these social functions interact with other cognitive and autonomic roles of the ACC in decision making is presently unexplored.

CONCLUSIONS AND OUTSTANDING QUESTIONS

Making advantageous decisions involves a complex process of weighing up the potential benefits of a particular course of action given the current motivational state against the costs of achieving that goal and the value of gathering information from other possible sources. A series of neuropsychological studies in rats and monkeys has indicated that parts of the frontal lobe play essential and dissociable roles in integrating the costs and benefits of available alternatives. Rodent ACC is necessary to allow an animal to invest effort for greater reward whereas the OFC is needed to avoid impulsive choices. This separation in the processing of costs may map onto differences seen in human patients, where dysfunction in an ACC subcortical circuit can result in apathy whereas damage to the OFC subcortical circuit may cause disinhibition and agitation.[15] Neither decision-making deficit was insurmountable, however, with the lesioned rats changing their behavior after extensive experience of choosing to tolerate the cost to obtain the reward. This suggests that neuropsychiatric patients may also be aided in their decision making with practice and by providing mediating cues to help them make favorable choices. Primate ACCs was also shown to be vital for learning about the extended history of choices and outcomes and guiding behavior appropriately. Finally, the ACCg appears crucial for evaluating social information relating to other individuals in the environment when deciding whether to respond.

Although the ACC and OFC are clearly vital for guiding optimal decision making, their exact contributions in concert with interconnected regions is as yet far from clear. As previously discussed, the BLA, nucleus accumbens, and monoamines, such as dopamine and serotonin, all play roles in allowing animals to avoid being cost averse, with some having dissociable functions depending on the cost the animal has to overcome.[58] Similarly, cells in dorsolateral prefrontal cortex and the caudate, both of which are connected with the ACCs,[59–62] also appear to represent the current action value based on the history of outcomes.[63,64] Other regions, such as posterior cingulate cortex and parts of parietal cortex, are also sensitive to reward probability and value.[48,65] However, it is also the case that even invertebrates, such as gastropods and locusts, with much more simple nervous systems can be shown to integrate costs and benefits and develop state-based valuation systems.[66,67] It is a priority for future research to try to discover what the specific contributions of the ACC and OFC to the process of valuation and deciding are.

Part of this will involve improved definition of what constitutes a response cost and how animals learn to mediate them. There have been descriptions that partially equate aspects of risk taking and impulsivity, with the former being aided by myopia for potentially negative consequences in the future.[68,69] Similarly, it is possible to conceive of an effort task with multiple unrewarded steps leading to an eventual predicted goal as having a probabilistic component (a fixed ratio [FR] 4 schedule and a reward probability of 0.25 should result in a comparable rate of rewarded and unrewarded responses). Moreover, in an uncertain and changeable environment, it will be vital to learn when it is worth persisting with a mode of responding even if it is temporarily unfruitful and when it would be better to switch to a different course of action and explore other possibilities.

Although the dissociations described in the present article are striking, it will also be important to try to find whether there are also commonalities in the way such costs are dealt with. One possibility is that this will require a closer connection to be made between the types of cost-benefit decision making discussed here and other types of goal-directed action. It is known, for instance, that both the ACC and OFC play roles in types of associative learning.[35,70] Behavioral ecologists have advanced an associative learning hypothesis to explain how rats discount the value of outcomes as a function of the cost, with the strength of the attribution between the choice and its consequences being scaled as a function of the cost between the two (i.e., the longer the delay between the choice and the outcome, the weaker the causal connection between the two, which can in turn lead to the propensity to discount delayed rewards).[71] Similarly, frontal-striatal circuits have also long been associated with aspects of rule-based learning,[72,73] which may provide a way of generating heuristics for the contexts in which it is worth investing time or effort, taking risks, or exploring the environment. Further experiments will be required that explicitly manipulate the instrumental and Pavlovian elements of the tasks and that examine whether animals are devising strategies or are making choices based more simply on their history of reinforcement.

It is also an intriguing question how the changes in social behavior observed following ACC lesions relate to the other types of cost-benefit decision making discussed above. In monkeys, there appears to be an anatomical dissociation between the parts of the ACC concerned more closely with integrating reward history to guide action selection (ACCs) and those involved in social valuation (ACCg).[46,53] However, there is recent evidence that the same ACC lesions in rats that cause alterations in effort-related decision making also disrupt the acquisition and retention of social information (Rudebeck *et al.*, in preparation). Although there might also be finer-scale anatomical distinctions in the rat ACC, this latter finding raises the possibility that parts of the ACC might play a common role in integrating many different factors to decide whether one course of action is worth choosing over the other possibilities. A similar connection can be made between the OFC's role in processing fearful stimuli and in aspects of reversal learning,[74–76] with both potentially related to an underlying representation of the reinforcement value of stimuli.[35] By investigating how such crucial determinants of decisions as the availability of social information interact with goal-directed action selection, it should be possible to gain a better understanding of how exactly frontal regions guide animals to make optimal choices.

ACKNOWLEDGMENTS

This work was supported by the Medical Research Council, the Royal Society, and the Wellcome Trust. Thanks are extended to B.B. Herman for advice and encouragement on this manuscript.

REFERENCES

1. BALLEINE, B.W. 2005. Neural bases of food-seeking: affect, arousal and reward in corticostriatolimbic circuits. Physiol. Behav. **86:** 717–730.
2. KACELNIK, A. 1997. Normative and descriptive models of decision making: time discounting and risk sensitivity. Ciba. Found. Symp. **208:** 51–67; discussion 67–70.
3. STEPHENS, D.W. & J.R. KREBS. 1986. Foraging Theory. Princeton University Press. Princeton, NJ.
4. CHARNOV, E.L. 1976. Optimal foraging: the marginal value theorem. Theor. Pop. Biol. **9:** 129–136.
5. KAHNEMAN, D. & A. TVERSKY. 1979. Prospect theory: an analysis of decision under risk. Econometrica **47:** 263–291.
6. BECHARA, A., H. DAMASIO & A.R. DAMASIO. 2000. Emotion, decision making and the orbitofrontal cortex. Cereb. Cortex **10:** 295–307.
7. BURGESS, P.W., et al. 2000. The cognitive and neuroanatomical correlates of multitasking. Neuropsychologia **38:** 848–863.
8. PASSINGHAM, R.E. 1993. The Frontal Lobes and Voluntary Action. Oxford University Press. London.
9. MILLER, E.K. & J.D. COHEN. 2001. An integrative theory of prefrontal cortex function. Annu. Rev. Neurosci. **24:** 167–202.
10. MANES, F., et al. 2002. Decision-making processes following damage to the prefrontal cortex. Brain **125:** 624–639.
11. ROGERS, R.D., et al. 1999. Dissociable deficits in the decision-making cognition of chronic amphetamine abusers, opiate abusers, patients with focal damage to prefrontal cortex, and tryptophan-depleted normal volunteers: evidence for monoaminergic mechanisms. Neuropsychopharmacology **20:** 322–339.

12. BECHARA, A., D. TRANEL & H. DAMASIO. 2000. Characterization of the decision-making deficit of patients with ventromedial prefrontal cortex lesions. Brain **123**(Pt 11): 2189–2202.
13. FELLOWS, L.K. & M.J. FARAH. 2005. Different underlying impairments in decision-making following ventromedial and dorsolateral frontal lobe damage in humans. Cereb. Cortex **15**: 58–63.
14. BARRASH, J., D. TRANEL & S.W. ANDERSON. 2000. Acquired personality disturbances associated with bilateral damage to the ventromedial prefrontal region. Dev. Neuropsychol. **18**: 355–381.
15. CUMMINGS, J.L. 1993. Frontal-subcortical circuits and human behavior. Arch. Neurol. **50**: 873–880.
16. LEVY, R. & B. DUBOIS. 2006. Apathy and the functional anatomy of the prefrontal cortex—basal ganglia circuits. Cereb. Cortex **16**: 916–928.
17. STEVENS, J.R., *et al.* 2005. Will travel for food: spatial discounting in two new world monkeys. Curr. Biol. **15**: 1855–1860.
18. BAUTISTA, L.M., J. TINBERGEN & A. KACELNIK. 2001. To walk or to fly? How birds choose among foraging modes. Proc. Natl. Acad. Sci. USA **98**: 1089–1094.
19. MONTEROSSO, J. & G. AINSLIE. 1999. Beyond discounting: possible experimental models of impulse control. Psychopharmacology (Berl) **146**: 339–347.
20. SALAMONE, J.D., M.S. COUSINS & S. BUCHER. 1994. Anhedonia or anergia? Effects of haloperidol and nucleus accumbens dopamine depletion on instrumental response selection in a T-maze cost/benefit procedure. Behav. Brain Res. **65**: 221–229.
21. WALTON, M.E., D.M. BANNERMAN & M.F. RUSHWORTH. 2002. The role of rat medial frontal cortex in effort-based decision making. J. Neurosci. **22**: 10996–11003.
22. WALTON, M.E., *et al.* 2003. Functional specialization within medial frontal cortex of the anterior cingulate for evaluating effort-related decisions. J. Neurosci. **23**: 6475–6479.
23. SCHWEIMER, J. & W. HAUBER. 2005. Involvement of the rat anterior cingulate cortex in control of instrumental responses guided by reward expectancy. Learn. Mem. **12**: 334–342.
24. DIAS, R. & J.P. AGGLETON. 2000. Effects of selective excitotoxic prefrontal lesions on acquisition of nonmatching- and matching-to-place in the T-maze in the rat: differential involvement of the prelimbic-infralimbic and anterior cingulate cortices in providing behavioural flexibility. Eur. J. Neurosci. **12**: 4457–4466.
25. CARDINAL, R.N., *et al.* 2001. Impulsive choice induced in rats by lesions of the nucleus accumbens core. Science **292**: 2499–2501.
26. RUDEBECK, P.H., *et al.* 2006. Separate neural pathways process different decision costs. Nat. Neurosci. **9**: 1161–1168.
27. FELLOWS, L.K. & M.J. FARAH. 2005. Dissociable elements of human foresight: a role for the ventromedial frontal lobes in framing the future, but not in discounting future rewards. Neuropsychologia **43**: 1214–1221.
28. ROESCH, M.R. & C.R. OLSON. 2004. Neuronal activity related to reward value and motivation in primate frontal cortex. Science **304**: 307–310.
29. ROESCH, M.R., A.R. TAYLOR & G. SCHOENBAUM. 2006. Encoding of time-discounted rewards in orbitofrontal cortex is independent of value representation. Neuron **51**: 509–520.

30. MOBINI, S., *et al.* 2002. Effects of lesions of the orbitofrontal cortex on sensitivity to delayed and probabilistic reinforcement. Psychopharmacology (Berl) **160:** 290–298.
31. KHERAMIN, S., *et al.* 2002. Effects of quinolinic acid-induced lesions of the orbital prefrontal cortex on inter-temporal choice: a quantitative analysis. Psychopharmacology (Berl) **165:** 9–17.
32. WINSTANLEY, C.A., *et al.* 2004. Contrasting roles of basolateral amygdala and orbitofrontal cortex in impulsive choice. J. Neurosci. **24:** 4718–4722.
33. FLORESCO, S.B. & S. GHODS-SHARIFI. 2007. Amygdala-prefrontal cortical circuitry regulates effort-based decision making. Cereb. Cortex **17:** 251–260.
34. BAXTER, M.G., *et al.* 2000. Control of response selection by reinforcer value requires interaction of amygdala and orbital prefrontal cortex. J. Neurosci. **20:** 4311–4319.
35. SCHOENBAUM, G. & M. ROESCH. 2005. Orbitofrontal cortex, associative learning, and expectancies. Neuron **47:** 633–636.
36. CARDINAL, R.N. 2006. Neural systems implicated in delayed and probabilistic reinforcement. Neural Netw. **19:**1277–1301.
37. GALLAGHER, M., R.W. MCMAHAN & G. SCHOENBAUM. 1999. Orbitofrontal cortex and representation of incentive value in associative learning. J. Neurosci. **19:** 6610–6614.
38. SADDORIS, M.P., M. GALLAGHER & G. SCHOENBAUM. 2005. Rapid associative encoding in basolateral amygdala depends on connections with orbitofrontal cortex. Neuron **46:** 321–331.
39. SHIDARA, M. & B.J. RICHMOND. 2002. Anterior cingulate: single neuronal signals related to degree of reward expectancy. Science **296:** 1709–1711.
40. DAW, N.D., *et al.* 2006. Cortical substrates for exploratory decisions in humans. Nature **441:** 876–879.
41. HE, S.Q., R.P. DUM & P.L. STRICK. 1995. Topographic organization of corticospinal projections from the frontal lobe: motor areas on the medial surface of the hemisphere. J. Neurosci. **15:** 3284–3306.
42. WANG, Y., *et al.* 2001. Spatial distribution of cingulate cells projecting to the primary, supplementary, and pre-supplementary motor areas: a retrograde multiple labeling study in the macaque monkey. Neurosci. Res. **39:** 39–49.
43. RUSHWORTH, M.F., *et al.* 2004. Action sets and decisions in the medial frontal cortex. Trends Cogn. Sci. **8:** 410–417.
44. AMIEZ, C., J.P. JOSEPH & E. PROCYK. 2006. Reward encoding in the monkey anterior cingulate cortex. Cereb. Cortex **16:** 1040–1055.
45. MATSUMOTO, K. & K. TANAKA. 2004. The role of the medial prefrontal cortex in achieving goals. Curr. Opin. Neurobiol. **14:** 178–185.
46. KENNERLEY, S.W., *et al.* 2006. Optimal decision making and the anterior cingulate cortex. Nat. Neurosci. **9:** 940–947.
47. HERRNSTEIN, R.-J. 1997. The Matching Law: Papers in Psychology and Economics. H. Rachlin & D.I. Laibson, Eds. Harvard University Press. Cambridge, MA.
48. SUGRUE, L.P., G.S. CORRADO & W.T. NEWSOME. 2004. Matching behavior and the representation of value in the parietal cortex. Science **304:** 1782–1787.
49. VON NEUMANN, J. & O. MORGENSTERN. 1944. The Theory of Games and Economic Behaviour. Princeton University Press. Princeton, NJ.
50. NASH, J.F. 1950. Equilibrium points in n-person games. Proc. Natl. Acad. Sci. USA **36:** 48–49.

51. DEANER, R.O., A.V. KHERA & M.L. PLATT. 2005. Monkeys pay per view: adaptive valuation of social images by rhesus macaques. Curr. Biol. **15:** 543–548.
52. ANDERSON, J.-R. 1998. Social stimuli and social rewards in primate learning and cognition. Behav. Processes **42:** 159–175.
53. RUDEBECK, P.H., *et al.* 2006. A role for the macaque anterior cingulate gyrus in social valuation. Science **313:** 1310–1312.
54. FRITH, C.D. & U. FRITH. 1999. Interacting minds—a biological basis. Science **286:** 1692–1695.
55. TOMLIN, D., *et al.* 2006. Agent-specific responses in the cingulate cortex during economic exchanges. Science **312:** 1047–1050.
56. RILLING, J., *et al.* 2002. A neural basis for social cooperation. Neuron **35:** 395–405.
57. RILLING, J.K., *et al.* 2004. The neural correlates of theory of mind within interpersonal interactions. Neuroimage **22:** 1694–1703.
58. DENK, F., *et al.* 2005. Differential involvement of serotonin and dopamine systems in cost-benefit decisions about delay or effort. Psychopharmacology (Berl) **179:** 587–596.
59. KUNISHIO, K. & S.N. HABER. 1994. Primate cingulostriatal projection: limbic striatal versus sensorimotor striatal input. J. Comp. Neurol. **350:** 337–356.
60. BATES, J.F. & P.S. GOLDMAN-RAKIC. 1993. Prefrontal connections of medial motor areas in the rhesus monkey. J. Comp. Neurol. **336:** 211–228.
61. TAKADA, M., *et al.* 2004. Organization of prefrontal outflow toward frontal motor-related areas in macaque monkeys. Eur. J. Neurosci. **19:** 3328–3342.
62. TAKADA, M., *et al.* 2001. Organization of inputs from cingulate motor areas to basal ganglia in macaque monkey. Eur. J. Neurosci. **14:** 1633–1650.
63. BARRACLOUGH, D.J., M.L. CONROY & D. LEE. 2004. Prefrontal cortex and decision making in a mixed-strategy game. Nat. Neurosci. **7:** 404–410.
64. SAMEJIMA, K., *et al.* 2005. Representation of action-specific reward values in the striatum. Science **310:** 1337–1340.
65. McCOY, A.N. & M.L. PLATT. 2005. Risk-sensitive neurons in macaque posterior cingulate cortex. Nat. Neurosci. **8:** 1220–1227.
66. POMPILIO, L., A. KACELNIK & S.T. BEHMER. 2006. State-dependent learned valuation drives choice in an invertebrate. Science **311:** 1613–1615.
67. GILLETTE, R., *et al.* 2000. Cost-benefit analysis potential in feeding behavior of a predatory snail by integration of hunger, taste, and pain. Proc. Natl. Acad. Sci. USA **97:** 3585–3590.
68. LOGUE, A.W. 1988. Research on self-control: an integrated framework. Behav. Brain Sci. **11:** 665–709.
69. EVENDEN, J.L. 1999. Varieties of impulsivity. Psychopharmacology (Berl) **146:** 348–361.
70. GABRIEL, M. 1990. Functions of anterior and posterior cingulate cortex during avoidance learning in rabbits. Prog. Brain Res. **85:** 467–482; discussion 482–483.
71. KACELNIK, A. 2003. The economics of patience: hyperbolic discounting as rate maximising. *In*: Time and Decision: Economic and Psychological Perspectives on Intertemporal Choice. G. Loewenstein, D. Read & R. Baumeister, Eds. Russell Sage Foundation. New York.
72. BUNGE, S.A. 2004. How we use rules to select actions: a review of evidence from cognitive neuroscience. Cogn. Affect. Behav. Neurosci. **4:** 564–579.

73. BUNGE, S.A., *et al.* 2005. Neural circuitry underlying rule use in humans and nonhuman primates. J. Neurosci. **25:** 10347–10350.
74. IZQUIERDO, A., R.K. SUDA & E.A. MURRAY. 2004. Bilateral orbital prefrontal cortex lesions in rhesus monkeys disrupt choices guided by both reward value and reward contingency. J. Neurosci. **24:** 7540–7548.
75. FELLOWS, L.K. & M.J. FARAH. 2003. Ventromedial frontal cortex mediates affective shifting in humans: evidence from a reversal learning paradigm. Brain **126:** 1830–1837.
76. SCHOENBAUM, G., *et al.* 2003. Lesions of orbitofrontal cortex and basolateral amygdala complex disrupt acquisition of odor-guided discriminations and reversals. Learn. Mem. **10:** 129–140.

Cost, Benefit, Tonic, Phasic

What Do Response Rates Tell Us about Dopamine and Motivation?

YAEL NIV[a,b]

[a] *Gatsby Computational Neuroscience Unit, UCL, London, United Kingdom*

[b] *Interdisciplinary Center for Neural Computation, The Hebrew University of Jerusalem, Jerusalem, Israel*

ABSTRACT: The role of dopamine in decision making has received much attention from both the experimental and computational communities. However, because reinforcement learning models concentrate on discrete action selection and on phasic dopamine signals, they are silent as to how animals decide upon the rate of their actions, and they fail to account for the prominent effects of dopamine on response rates. We suggest an extension to reinforcement learning models in which response rates are optimally determined by balancing the tradeoff between the cost of fast responding and the benefit of rapid reward acquisition. The resulting behavior conforms well with numerous characteristics of free-operant responding. More importantly, this framework highlights a role for a tonic signal corresponding to the net rate of rewards, in determining the optimal rate of responding. We hypothesize that this critical quantity is conveyed by tonic levels of dopamine, explaining why dopaminergic manipulations exert a global affect on response rates. We further suggest that the effects of motivation on instrumental rates of responding are mediated through its influence on the net reward rate, implying a tight coupling between motivational states and tonic dopamine. The relationships between phasic and tonic dopamine signaling, and between directing and energizing effects of motivation, as well as the implications for motivational control of habitual and goal-directed instrumental action selection, are discussed.

KEYWORDS: tonic dopamine; phasic dopamine; motivation; response rate; energizing; reinforcement learning; free operant; cost/benefit; generalized drive

Address for correspondence: Yael Niv, Center for the Study of Brain, Mind and Behavior, Green Hall, Princeton University, Princeton, NJ 08544, USA. Voice: +1-609-258-7511; fax: +1-609-258-2574.
yael@princeton.edu

Ann. N.Y. Acad. Sci. 1104: 357–376 (2007). © 2007 New York Academy of Sciences.
doi: 10.1196/annals.1390.018

INTRODUCTION

Browsing through any random selection of experimental psychology papers will reveal that the dependent variable most commonly used to study animal behavior is response rate.[1] The effects of experimental manipulations as diverse as changes in the amount of reward that an animal can earn, alterations of the requirements or conditions under which rewards or punishments are delivered, lesions of neural structures, or the administration of drugs, are commonly discerned through changes in response rates. In terms of decision making and action selection, response rates are, in fact, inseparable from responding itself: accompanying any choice of which action to perform is a choice of how fast (or at what instantaneous rate) to perform this action. It may come as a surprise, then, that *normative* models of responding, such as reinforcement learning, which have done much to explain *why* it is appropriate for animals to choose actions the way they do, have completely ignored the choice of response rates.

Response rates have played a more prominent role in descriptive models. These aim to quantify the relationships between experimental variables and response rates (e.g., the Matching Law[2]) but not why, or in what sense these relationships are appropriate in different scenarios. In the absence of normative models (which deal exactly with these latter aspects), questions, such as why does motivation influence response rates, and how should dopamine affect rate selection, are left unanswered. In previous work,[3–6] on which we focus in this review, we proposed to remedy this by extending the framework of reinforcement learning to the optimal selection of response rates.

In our model,[3,6] animals choose with what latency (i.e., how fast, or with what instantaneous rate) to perform actions, by optimally balancing the costs of fast performance and the benefits of rapid reward acquisition. Focusing on this tradeoff, the model highlights the *net expected rate of rewards* as the important determinant of the cost of delaying future rewards and the optimal rate of responding. We marshal evidence suggesting that this quantity is signaled by tonic levels of dopamine, and argue that this explains why higher levels of dopamine are associated with faster performance, while low levels of dopamine induce lethargy. We further leverage the normative framework to argue that motivation and dopamine are tightly linked in controlling response vigor, as the effect of motivation on response rates is mediated by a change in the expected net rate of rewards.

In the following, we first detail the basic characteristics of response rates, which we expect our model to reproduce. We then describe the new model emphasizing the tradeoffs that must be negotiated optimally to maximize reward intake. In particular, we focus on the role of the expected rate of reward in determining the opportunity cost of time and the optimal rate of responding. The following section relates this signal to tonic levels of dopamine, and discusses the implications for understanding the role of dopamine in action selection. In the next section, we use this normative model of response rates to analyze the effects of motivation on responding. We first discuss how both the directing and energizing effects of motivation are manifest in the model. The results suggest a parcellation of motivational effects into outcome-specific and outcome-general effects, leading to a new understanding of the susceptibility of goal-directed behavior on the one hand, and habitual behavior on the other, to motivational manipulations. Finally, we argue that the outcome-general energizing effects of motivation on response rates are mediated through changes in the expected net rate of rewards, implying a strong link between tonic dopamine and motivation. In the last section, we discuss some open questions, such as the extension of the model to Pavlovian behavior, the relationship between phasic dopaminergic signals and motivation, and the neural locus of cost/benefit tradeoff computations.

WHAT DO WE KNOW ABOUT RESPONSE RATES?

Action selection has most frequently been studied in instrumental conditioning paradigms, on which we will focus here. In the commonly used *free-operant* form of these,[7] animals (typically rats, mice, or pigeons) perform an action (e.g., pressing a lever, pecking a key) to obtain some coveted reinforcement (such as food for a hungry animal). Importantly, rather than performing actions at discrete, predefined time points (as is typically modeled in reinforcement learning[8]), free-operant responding is *self-paced*, and animals are free to choose their rate of responding.

Numerous experiments have shown that the schedule of reinforcement (e.g., ratio or interval), the nature or amount of the rewards used, and the motivational state of the animal profoundly affect the rate of instrumental responding. In general, responding is slower the longer the interval duration or ratio requirement,[9–12] and faster for higher magnitude rewards or more desirable rewards.[13,14] More refined characteristics of free-operant behavior include the observation of higher response rates on ratio schedules compared to yoked interval schedules[15–17] and response allocation that matches payoff rates when two interval schedules are concurrently available.[2,18,19]

The fact that response rates are affected by manipulations of the schedule of reinforcement suggests that animals choose with which rate to perform different actions as an adaptation to the specifics of the task they are solving. Furthermore, in most cases behavior in such schedules is well below ceiling rates, evidence that response latencies are not constrained by decision times, or motor or perceptual requirements, but rather the particular response rate was selected as appropriate for the task at hand. In the following discussion we will assume that the choice of response rate is the result of an *optimization process* that is influenced by two opposing goals: the desire to acquire rewards rapidly on the one hand, and to minimize effort costs on the other hand.

OPTIMAL RESPONDING: COST/BENEFIT TRADEOFFS

Consider a situation in which a rat can choose between several actions: it can poke its nose into a (possibly empty) food well, it can press a lever that may cause food to fall into the food well, it can pull a chain that may cause water to pour into the food well, and so forth (FIG. 1A). The choice of which sequence of actions to take, and at what rate (or with what latency) to take each action, can be seen as an optimization problem, if we assume that the goal of the rat is to harvest rewards at as high a rate as possible, while incurring minimal effort costs. Because for free-operant tasks the problem can be defined computationally as a (semi-)Markov decision process, the optimal solution can be derived as a series of optimal decisions: the rat should first choose the currently optimal action and execute it with the optimal latency, and then, based on the consequences of this action (the resulting "state" of the world, e.g., whether the action resulted in food falling into the food well or not), choose the next optimal action and latency, and so forth. The optimal policy of which actions to choose in the different states,[a] and with what latency to perform the chosen actions, can be found using reinforcement learning methods, such as "value iteration,"[8] or online "temporal difference" learning[8,20] (for a full computational exposition of the model equations and solution, see Refs. 3, 5). To gain insight into the optimal policy, we will now analyze the factors that affect a single decision within the series of actions. In our model this consists of two parts: the rat must choose which action to perform, and *how fast* (or with what latency) to perform it. It turns out that these two subdecisions depend on different characteristics of the task.

The choice of which action to perform depends on the utility of the rewards potentially available for each of the actions, the probability that the action will indeed be rewarded, and the effort cost of performing the action. For instance, if pressing the lever is rewarded with food with a probability of 20%, this would be preferable to an action that leads to the same outcome but with only 10% chance. What about a choice between actions that lead to different rewards? When comparing the worth of qualitatively different outcomes, such as food and water, the motivational state of the animal must come into consideration, as it determines the utility of each outcome to the animal.[4] A hungry rat may prefer to press the lever for food, while a thirsty one might choose to pull the chain to obtain water. The choice of which action to perform also depends on how costly the action itself is, in terms of effort: for instance, if pulling the chain necessitates much effort to jump and reach it, the benefit of a small amount of water may not be worth this effort. To summarize, the optimal choice of

---

[a] The states we refer to here are states of the environment, such as whether there is food in the magazine, whether the lever is extended and available for pressing, etc. These should not be confused with the motivational state of the animal, which we will discuss later. For modeling simplicity, we assume that the animal's motivational state is constant during the experimental session.

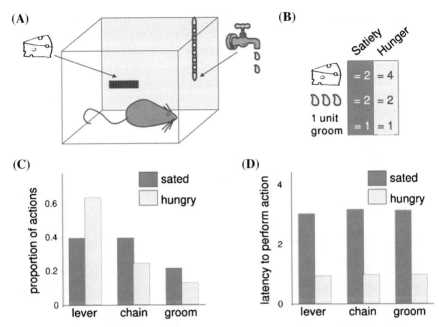

FIGURE 1. Two behavioral consequences of a motivational shift. (**A**) A simulated rat, trained in an operant chamber, can perform several actions: it can press a lever for a 20% chance of obtaining cheese, it can pull a chain for a 20% chance of obtaining water, or it can groom itself (with possibly an internal reward). (**B**) Even when relatively sated, the cheese and water have slightly higher utilities than grooming. A shift to hunger, however, markedly enhances the utility of cheese, compared to the other utilities that are left unchanged. (**C**) One effect of the shift from satiety to hunger is to "direct" the rat to choose to press the lever (to obtain cheese) more often, at the expense of either grooming or chain pulling (which are still performed, albeit less often). (**D**) A second orthogonal consequence of the motivational shift is that all actions are now performed faster. Measurements of the latencies to perform individual actions in the simulation reveal that not only is the rate of lever pressing enhanced, but, when performed, grooming and chain pulling are also executed faster. This "energizing" effect of the motivational shift is thus not specific to the action leading to the favored outcome, and can be regarded an outcome-independent effect. Figure modified from Ref. 4.

which action to perform can be determined by comparing the available actions in terms of the worth and probability of the potential reward for each action, and the effort cost of performing the action.

The optimal choice of *how fast to perform the chosen* action is determined by an altogether different cost/benefit tradeoff. First, we must assume that it is more costly for the rat to perform an action quickly rather than slowly (otherwise rats would always perform actions at the fastest possible rate, which is clearly not the case). Against what should this cost of fast performance be weighed, when deciding on the optimal speed of an action? Completing the chosen action slowly, for instance, moving toward the lever and pressing it without haste, will of course delay the availability of the possible reward for this action. But, more important, all future actions and rewards will be delayed. So the effort cost of behaving quickly should be weighed against the cost of delaying *all* future rewards. Note that while the choice of which action to perform is affected by parameters local to the different actions and their (potentially long-term) outcomes, the choice of response rate influences the timing of all future actions, and is thus affected by global considerations.

How can the rat estimate the cost of delaying all future rewards? Average reward reinforcement learning techniques reveal a simple solution.[21,22] A specific policy of action choices and response latencies will lead to an average rate of rewards obtained per unit time, at an average effort cost per unit time.[23] The rate of rewards minus costs—the influx of net benefit per unit time, which we will refer to as the net reward rate—is exactly the worth of time under this policy,[24-26] or the *opportunity cost* of wasted time.[3,5] That is, in every second in which the current policy of responding will not be performed, on average this amount of net benefit will be lost. This means that when selecting a rate of performance, or a speed of execution for each individual action, the lower cost of performing the action more slowly should be weighed against the opportunity cost of the extra execution time, that is, the net reward rate, which could have been obtained during this time. The formal average-reward reinforcement learning solution ensures that such a choice of actions and latencies will indeed lead to the highest possible net influx of benefit, and so will be the truly optimal solution.

Simulating a wide variety of free-operant experiments using this model of optimal behavioral choice showed that the well-known characteristics of free-operant behavior indeed qualitatively match the optimal solution: simulated rats showed a higher response rate when the magnitude of reward was larger or the schedule was more rewarding (lower interval or ratio requirement), response rates were lower on interval schedules compared to yoked ratio schedules, and when tested on two concurrent interval schedules the simulated rats matched their response rates on each lever to their payoff rates.[5]

OPPORTUNITY COSTS AND TONIC DOPAMINE

Optimal action selection based on online learning of the values of different actions has previously been suggested as a model of action selection in the basal ganglia.[27-30] In one version of these, called actor–critic models, it has been suggested[29] that ventral striatal areas (the so-called "critic") learn to evaluate situations or states of the world, by using a reward prediction error signal provided by dopaminergic neurons in the ventral tegmental area. The dorsal striatum (the "actor"), in turn, learns the values of different actions in these states, based on a similar dopaminergic prediction error signal originating in the substantia nigra pars compacta (for a review of the underlying neural data see Ref. 31). These models have emphasized the role of phasic dopaminergic firing patterns, which signal temporally local errors in the prediction of future outcomes, in providing the basis of optimal learning of long-term values of actions and states.[28,32]

In addition to requiring this phasic prediction error signal to determine the optimal selection of actions and rates, our model highlights the importance of a new signal, which should indicate the expected net rate of rewards, that is, the opportunity cost of time. In a certain class of problems to which this model is applicable, the net reward rate is a global, slowly changing term, common to all the states and to all actions and rates evaluated.[23] That is, whether deciding how fast to perform the next lever press, or the next nose-poke, and regardless of whether a reward is currently available in the food well or not, the opportunity cost of time is the same—the long-term average reward rate forfeited in that time.

What could be the neural bearer of such a global, slowly changing signal? We hypothesize this to be the *tonic level of dopamine* in basal ganglia and pre-frontal areas.[5] The tonic level of dopamine is suitable to indicate the net rate of rewards on computational, neural, and psychological grounds. Computationally, resulting from the very definition of temporal difference reward prediction errors, averaging of the phasic dopaminergic prediction errors over time, will exactly result in the correct average rate of reward. Neurally, dopamine concentrations in target areas, such as the striatum, are relatively homogeneous,[33] and a recent investigation using fast scan cyclic voltammetry indeed showed that time averaging of phasic dopaminergic activity in target areas results in a stable tonic level,[34] well within the range expected from microdialysis measurements.[35]

Finally, psychological theories of dopamine function have long focused on a putative role for dopamine in modulating the vigor of behavior.[36–43] The identification of tonic dopamine levels with the opportunity cost of time explains, for the first time, *why* dopaminergic manipulations affect response rates as they do. According to our theory, artificially elevating the tonic level of dopamine increases the opportunity cost of time, with the effect of making the optimal response rates for all actions higher. Suppressing dopamine levels will lead to a reduced cost of time, and slothful behavior. Indeed, the most prominent effect of dopaminergic interventions is an enhancement or reduction of overall response rates as a result of increased or decreased dopaminergic transmission, respectively.[42,44–52] Modeling dopamine manipulations as changes in the effective net reward rate, we can simulate and replicate many of these results.[5]

THE EFFECTS OF MOTIVATION ON RESPONDING

Using this model of response rates, we can now analyze the effects of motivation on response selection. Understanding how motivation influences behavior is complicated by the fact that animals use a number of different action selection systems, which are differentially sensitive to motivation, and with which we will separately deal below. But first, let us consider in general how motivation can affect the optimal cost/benefit tradeoff we have discussed above. One way to define motivational states is as a *mapping* between outcomes (or significant events in the world) and the utility they confer to the animal.[4] For instance, food holds high utility for a hungry rat, but a low utility for a sated or thirsty rat (FIG. 1B). Using this simple definition, a straightforward means by which motivation can affect action selection, is through the determination of the utility of the outcomes of the different available actions. This corresponds to the traditional "directing" role ascribed to motivation, because by determining which are the most valuable outcomes, motivation can direct action selection toward those actions that will lead to these outcomes.

But this is not the only way that the motivational mapping can affect responding: the outcome utilities will also affect the net rate of rewards (which is measured in units of utility per time). Because the net reward rate serves as the opportunity cost of time, motivation will affect the optimal response rates of all chosen actions. For instance, consider a rat pressing a lever for food pellets on a random interval 30-sec schedule. On average, the net rate of reward is equal to the utility of two pellets per minute, minus the costs per minute of the actions emitted to obtain and harvest these pellets. If the rat is now made hungrier, the utility of each of the pellets increases and with it the net reward rate, thus increasing the opportunity cost of time and favoring faster responding. In this way, higher motivational states cause higher response rates, while lower motivational states, such as satiety, decrease the rate of responding. This corresponds to the "energizing" role of motivation, and the much debated notion of "generalized drive."[53-55]

In sum, motivation can exert a twofold influence on responding in our model: a "directing" effect on the choice of which action to perform, and an "energizing" effect on the rates with which all actions are performed.[3,5] FIGURE 1C, D illustrates these two effects, and their qualitative differences. The choice of action depends on a comparison of the local utilities of the outcomes of different actions, and so the "directing" effect of motivation is *outcome-specific* (i.e., motivation differentially affects different actions, based on their consequent outcomes; FIG. 1C). In contrast, the choice of response rate depends on the global opportunity cost of time, thus motivation exerts a similar "energizing" effect on *all* prepotent actions, regardless of their specific outcome (FIG. 1D). This explains some hitherto paradoxical observations of "generalized drive," such as the fact that hungrier rats will also work harder for water rewards.

Multiple Action Selection Mechanisms

Although motivation can potentially influence action selection in two ways, different action selection mechanisms may be differentially sensitive to the "directing" or "energizing" effects of motivation.[4] In addition to the traditional distinction between Pavlovian and instrumental mechanisms for action control, a recent series of sophisticated studies has teased apart two different types of instrumental control, namely, goal-directed and habitual behavior, based exactly on their susceptibility to motivational influences.[56] The evidence points to two neurally distinct[57] behavioral controllers, which employ different computational strategies to estimate what is the currently optimal behavior.[58] The goal-directed system uses a forward model (or action → outcome knowledge) to iterate forward to the expected consequences of a series of actions. As such, its decision-making process is directly sensitive to the utilities of the outcomes consequent on the different actions.[59–62] Conversely, habitual decision making eschews the online simulation of potential consequences of actions, relying instead on estimates of the long-term values of actions, which have been previously learned and stored. These value estimates summarize previous experience about the consequences of actions, but do not represent the outcomes themselves. As a result, habitual responding is not immediately sensitive to changes in action–outcome contingencies,[60,63–65] and similarly can not react to a change in outcome utilities without the relatively slow relearning of new values of actions.[58]

How do the two effects of motivation interact with the constraints of these instrumental action selection systems? We can expect goal-directed action selection, which chooses actions based on the utility of their consequent outcomes, to express the "directing" influence of motivation naturally, selecting those actions that lead to desired outcomes based on the current motivational state of the animal. This is, in fact, the characteristic hallmark of goal-directed behavior.[56] Moreover, the effects of motivational shifts on goal-directed responding have been shown to depend on a process of "incentive learning" (in which animals experience the utilities of different outcomes in different motivational states[66–71]), testifying that motivational states indeed affect action selection through outcome utilities.

The habitual controller, however, can choose actions that are optimal for the current motivational state, only if the animal has learned and stored the long-term values of different actions in this motivational state. This means that a rat that has been extensively trained (to the point of habitization), in a state of hunger, to press one lever for food and another for water, will not be able to adjust its behavior flexibly and will continue to predominantly press the food lever, even when shifted to a motivational state of thirst. Only through subsequent learning of the new values of the lever press actions in terms of the utility of their consequent outcomes in the new motivational state, will habitual behavior be sensitive to the "directing" effects of motivation.[58]

Does this mean that habitual behavior is initially totally insensitive to motivational manipulations? We argue to the contrary.[4] Because motivation also exerts a global effect on response rates, which is independent of the specific outcomes of the different actions, motivational states can "energize" both habitual and goal-directed behavior. Assuming that the animal can estimate at least the *direction* in which the net reward rate will change in the new motivational state (which depends on whether the current motivational state is lower or higher than previously, and whether the animal has reason to expect the availability of outcomes that are relevant to this state), the rate of responding, whether habitual or goal-directed, can be adjusted appropriately so as to approximate the optimal solution. Our model thus predicts that habitual behavior should be sensitive to the "energizing" aspects of motivation, while goal-directed behavior should be affected by both the "energizing" and the "directing" aspects.[4]

DISCUSSION

Building on and extending previous normative models of action selection, we have suggested a model of optimal selection of response rates in free-operant tasks. Our analysis focused on the critical tradeoffs that need to be negotiated to reap rewards at the highest possible rate and the lowest possible cost. This revealed that, different from the decision of which action to perform that is determined by outcome-specific considerations, decisions regarding response rates are determined by global considerations as the consequence of slow performance is to delay all future outcomes. This insight provided the basis for a novel outlook on the effects of motivation on the one hand, and of dopamine on the other, on instrumental responding.

In our model, the global quantity used to evaluate the cost of delaying all future rewards, that is, the opportunity cost of time, is the net rate of rewards. We suggest that this quantity is reported by the tonic level of dopamine, which explains why high levels of dopamine are associated with generally high response rates, and lower levels of dopamine induce lethargy. Consequently, dopamine has a dual effect on behavior: an effect on action choice through learning, based on phasic aspects of dopaminergic signaling, and an effect on rate selection, mediated by tonic levels. Different from other roles that have been suggested for tonic dopamine,[72–74] our analysis is the first to suggest a normative role, and to imply that the tonic level of dopamine is a quantity that represents specific aspects of the task and of the animal's performance in it. From this follow computationally specific predictions: our model predicts that tonic levels of dopamine will be higher when performing a more rewarding or a less costly task, and lower when working harder or for fewer rewards.

We have further argued that motivation also exerts a twofold effect on responding. By determining the mapping between outcomes and their utility, motivation "directs" action selection to those actions that are expected to yield the most valued outcomes, and "energizes" all ongoing behavior through affecting the overall reward rate. However, due to the computational limitations of the habitual system, only the goal-directed system is susceptible to the "directing" effect of motivation. The "energizing" effect, in contrast, can influence both habitual and goal-directed behavior. It is this latter effect that we hypothesize to be mediated by tonic levels of dopamine, suggesting a strong link between motivation and dopaminergic control.[75] The direct prediction, which has yet to be tested, is that higher motivational states will be associated with higher tonic levels of dopamine (providing the animal has reason to believe that motivation-relevant outcomes are forthcoming).

Incentive Motivation and Dopamine

In our model, response rates are determined based on the vigor cost of the action and the overall net reward rate, but importantly, without regard for the outcome contingent on the specific action. However, behavioral results from discrete trial experiments show that specific outcome expectancies do affect response latencies, with responding to cues predictive of higher reward being typically faster than responding to less valuable cues.[76-79] Furthermore, although in our model the speed of responding is generally associated with the tonic level of dopamine, dopaminergic recordings have shown a linear relationship between reaction times and phasic dopaminergic responding.[80,81]

If the tonic average reward signal is indeed computed by slow averaging of the phasic prediction error signals, then this result is perhaps not surprising. Cues associated with higher reward expectancies induce larger phasic reward prediction signals,[82,83] which would transiently elevate dopamine tone,[81,84,85] influencing vigor selection and resulting in faster responding. This explanation is a slightly different outlook on ideas about "incentive motivation," according to which different outcomes exert a motivational effect on responding by virtue of their incentive value.[56,86,87]

Pavlovian Responding

We have accounted for the role of dopamine, and that of motivation, in controlling habitual and goal-directed instrumental responding. But what about the third class of behavior, namely, Pavlovian responding? The answer to this is not straightforward. On the one hand, phasic dopamine reward prediction errors have been implicated in optimal learning of Pavlovian predictive values, as well as instrumental values. On the other hand, Pavlovian responding itself is not necessarily normative—rather than a flexible, optimal, adaptation to a task, it seems as if Pavlovian responding is adaptive only on an evolutionary timescale.

Within an animal's behavioral repertoire, Pavlovian responses are characterized by their inflexibility, and tasks can be constructed in which they are strictly suboptimal. For instance, Pavlovian behavior persists even in circumstances (such as omission schedules) in which the occurrence of the Pavlovian response *prevents* the delivery of a reward. It therefore seems that Pavlovian responses are an inevitable consequence of the predictive value of cues.[88] A normative model is thus limited in its applicability to Pavlovian responding.

There is another sense in which our model is ill-suited for Pavlovian behavior: a critical simplification of our model is that once a decision is made regarding the next optimal action and the latency with which to perform it, the validity of this decision does not change while the action is executed. That is, we have assumed that the state of the world (e.g., whether a reward is available in the food well or not) does not change while an animal is executing an action. Though this is true in free-operant schedules, our framework cannot be used without modification to model tasks in which this assumption is invalid, such as instrumental avoidance conditioning (in which an aversive outcome occurs if a response is not performed fast enough). More generally, the model cannot incorporate Pavlovian state changes, for example, stimuli appearing and disappearing, and rewards that are given regardless of the animal's actions.

Having said this, we can still derive some insight from the model as to the effect Pavlovian cues or rewards *should* have on instrumental behavior in a simplified setting. Consider the case of a rat performing an appetitive free-operant task, to which we now add a "free" reward that is delivered independent of the animal's actions, and does not require any harvesting actions (for instance, brain stimulation reward delivered with some fixed probability at every second). Extending our framework to this special case is straightforward, and we can analyze the effect of this free reward on ongoing instrumental behavior. According to the optimal solution, and consistent with common sense, such a reward should *have no effect* on any ongoing instrumental behavior: any action and rate of responding that were optimal in the original task, are still optimal in the modified setting. This implies that the effective net reward rate used to determine the optimal rate of instrumental responding should be the same in both tasks, that is, that the net rate of rewards controlling instrumental behavior should be comprised of only those rewards that are instrumentally earned.

However, to infer which rewards are earned instrumentally and which would have been delivered regardless of one's actions is not at all a trivial problem, especially when behavior is habitual. Indeed, although animals show sensitivity to the contingencies between actions and rewards and reduce responding on a lever if rewards are offered at the same rate whether the lever is or is not pressed (a "contingency degradation" treatment[59, 61, 89, 90]), responding in such cases is not completely eliminated, evidence for some confusion on the part of the animal. As a result of such overestimation of agency in obtaining Pavlovian rewards, the net instrumental reward rate would be overestimated, leading to instrumental response rates that are higher than is optimal.

A more obvious example is the phenomenon of Pavlovian to instrumental transfer (PIT) in which the onset of a cue that has been associated previously with Pavlovian rewards, enhances the rate of ongoing instrumental behavior. This is clearly not optimal: the Pavlovian cue does nothing to change the tradeoff determining the optimal response rate. Nonetheless, PIT has been demonstrated in a wide host of settings.[91–95] It seems, then, that similar to the suboptimality of Pavlovian responding in general, Pavlovian effects on instrumental responding are suboptimal. Our model suggests that this is the result of erroneous inclusion of Pavlovian rewards in the expected net rate of instrumental rewards. Interestingly, there is an outcome-specific and an outcome-nonspecific component to PIT.[95] Based on our model and some suggestive experimental results,[40] it is tempting to propose that, like effects of motivation on behavior, the outcome-nonspecific effect of Pavlovian cues is indeed mediated by the tonic level of dopamine.

Where Is the Tradeoff Resolved?

Finally, where in the brain is the tradeoff controlling response rate resolved, is currently an open question. As this computation can be shared by both habitual and goal-directed controllers of instrumental behavior, it might not reside in either of these two neural systems. One potential candidate is the anterior cingulate cortex (ACC), and its projections to the nucleus accumbens, and to midbrain dopaminergic neurons.[96,97] The ACC has been implicated in monitoring conflict in cognitive tasks, specifically at the level of response selection, possibly as an index of task difficulty as part of a cost/benefit analysis underlying action selection.[98] Recent investigations using tasks specifically designed to probe cost/benefit tradeoffs,[48,99] confirmed that animals do indeed weigh the amount of effort required for obtaining a reward on each of the available options to decide which course of action to take.[96] In these same tasks, lesions to the ACC (but not to other medial frontal areas) affected animals' cost/benefit tradeoff, and caused them to prefer a low-effort/low-reward option to the high-effort/high-reward option preferred by nonlesioned rats.[96,97,100,101] Although a similar effect is seen with 6-hydroxydopamine lesions of the nucleus accumbens,[48,99] there are differences between the effects of ACC and accumbal dopaminergic lesions,[96] suggesting that the ACC and nucleus accumbens dopamine may fulfill different roles in the decision-making process, with nucleus accumbens dopamine computing and signaling the opportunity cost of time, and the ACC integrating this with expected immediate costs and benefits to determine the tradeoff for or against each possible action. Results to the opposite direction, showing excessive nose-poke responding in a go/no-go task after ACC lesions,[102] indeed suggest that ACC lesions do not merely tilt the balance toward less effortful options (as is suggested for accumbal dopamine depletions), but rather disrupt the instrumental cost/benefit analysis such that a less sophisticated Pavlovian default response pattern is chosen. That is, in a lever-pressing task in which the lever-press action is not the Pavlovian default, ACC lesions cause the animal to cease pressing, while in an appetitive approach task the Pavlovian default of approaching the food port dominates as a result of the lesion.

CONCLUSIONS

To conclude, from a detailed analysis of the factors affecting response rates we have gained not only a normative understanding of free-operant behavior, but also a new outlook on the effects of dopamine and motivation on responding. The tight coupling we suggest between motivation and dopamine is perhaps surprising: dopamine had been related to motivation in early theories, only to be dissociated from signaling reward motivation *per se* in contemporary normative models. However, we are not advocating to abandon ideas about reward prediction errors, and relapse to the "anhedonia hypothesis" of dopamine. Rather, we suggest to take normative models of dopamine one step forward, to account for tonic as well as phasic signaling, two distinct modes of transmission that can carry separate computational roles.

ACKNOWLEDGMENTS

This work was funded by a Hebrew University Rector Fellowship, and the Gatsby Charitable Foundation. The author is grateful to the organizers and participants of the "Reward and decision making in cortico-basal-ganglia networks" meeting for much stimulating discussion and feedback, and to Rui Costa, Nathaniel Daw, Peter Dayan, Daphna Joel, and Geoffrey Schoenbaum for helpful comments on the article.

REFERENCES

1. WILLIAMS, B.A. 1994. Reinforcement and choice. *In* Animal Learning and Cognition, ch. 4: 81–108. Academic Press. San Diego.
2. HERRNSTEIN, R.J. 1997. The Matching Law: Papers in Psychology and Economics. Harvard University Press. London.
3. NIV, Y., N.D. DAW & P. DAYAN. 2005. How fast to work: response vigor, motivation and tonic dopamine. *In* NIPS 18, Y. Weiss, B. Schölkopf & J. Platt, Eds.: 1019–1026. MIT Press. Cambridge, MA.
4. NIV, Y., D. JOEL & P. DAYAN. 2006. A normative perspective on motivation. Trends Cogn. Sci. **10:** 375–381.
5. NIV, Y., N.D. DAW, D. JOEL & P. DAYAN. 2006. Tonic dopamine: opportunity costs and the control of response vigor. Psychopharmacology (Berl.) **191**(3): 507–520.

6. NIV, Y. The effects of motivation on habitual instrumental behavior. PhD thesis, The Hebrew University of Jerusalem. Submitted 2007.
7. DOMJAN, M. 2003. The Principles of Learning and Behavior. Fifth edition.Thomson/Wadsworth. Belmont, CA.
8. SUTTON, R.S. & A.G. BARTO. 1998. Reinforcement Learning: An Introduction. MIT Press.
9. MAZUR, J.A. 1983. Steady-state performance on fixed-, mixed-, and random-ratio schedules. J. Exp. Anal. Behav. **39:** 293–307.
10. BAUM, W.M. 1993. Performances on ratio and interval schedules of reinforcement: data and theory. J. Exp. Anal. Behav. **59:** 245–264.
11. KILLEEN, P.R. 1995. Economics, ecologies and mechanics: the dynamics of responding under conditions of varying motivation. J. Exp. Anal. Behav. **64:** 405–431.
12. FOSTER, T.M., K.A. BLACKMAN & W. TEMPLE. 1997. Open versus closed economies: performance of domestic hens under fixed-ratio schedules. J. Exp. Anal. Behav. **67:** 67–89.
13. BRADSHAW, C.M., E. SZABADI & P. BEVAN. 1978. Relationship between response rate and reinforcement frequency in variable-interval schedules: the effect of concentration of sucrose reinforcement. J. Exp. Anal. Behav. **29:** 447–452.
14. BRADSHAW, C.M., H.V. RUDDLE & E. SZABADI. 1981. Relationship between response rate and reinforcement frequency in variable interval schedules: II. Effect of the volume of sucrose reinforcement. J. Exp. Anal. Behav. **35:** 263–270.
15. ZURIFF, G.E. 1970. A comparison of variable-ratio and variable-interval schedules of reinforcement. J. Exp. Anal. Behav. **13:** 369–374.
16. CATANIA, A.C., T.J. MATTHEWS, P.J. SILVERMAN & R. YOHALEM. 1977. Yoked variable-ratio and variable-interval responding in pigeons. J. Exp. Anal. Behav. **28:** 155–161.
17. DAWSON, G.R. & A. DICKINSON. 1990. Performance on ratio and interval schedules with matched reinforcement rates. Q. J. Exp. Psych. B **42:** 225–239.
18. SUGRUE, L.P., G.S. CORRADO & W.T. NEWSOME. 2004. Matching behavior and the representation of value in the parietal cortex. Science **304:** 1782–1787.
19. LAU, B. & P.W. GLIMCHER. 2005. Dynamic response-by-response models of matching behavior in rhesus monkeys. J. Exp. Anal. Behav. **84:** 555–579.
20. WATKINS, C.J.C.H. 1989. Learning with Delayed Rewards. PhD thesis, Cambridge University. Cambridge, UK.
21. SCHWARTZ, A. 1993. A reinforcement learning method for maximizing undiscounted rewards. *In* Proceedings of the Tenth International Conference on Machine Learning. 298–305. Morgan Kaufmann. San Francisco.
22. MAHADEVAN, S. 1996. Average reward reinforcement learning: foundations, algorithms and empirical results. Machine Learning **22:** 1–38.
23. BERTSEKAS, D.P. & J.N. TSITSIKLIS. 1996. Neuro-dynamic programming. Athena Scientific. Nashua, NH.
24. DAW, N.D. & D.S. TOURETZKY. 2002. Long-term reward prediction in TD models of the dopamine system. Neur. Comp. **14:** 2567–2583.
25. DAW, N.D. 2003. Reinforcement learning models of the dopamine system and their behavioral implications. PhD thesis, Carnegie Mellon University. Pittsburgh, PA.

26. DAW, N.D., A.C. COURVILLE & D.S. TOURETZKY. 2006. Representation and timing in theories of the dopamine system. Neur. Comp. **18:** 1637–1677.

27. BARTO, A.G. 1995. Adaptive critic and the basal ganglia. *In* Models of Information Processing in the Basal Ganglia, ch. 11. J.C. Houk, J.L. Davis & D.G. Beiser, Eds.: 215–232. MIT Press. Cambridge.

28. MONTAGUE, P.R., P. DAYAN & T.J. SEJNOWSKI. 1996. A framework for mesencephalic dopamine systems based on predictive hebbian learning. J. Neurosci. **16:** 1936–1947.

29. O'DOHERTY, J.P., P. DAYAN, J. SCHULTZ, *et al.* 2004. Dissociable roles of ventral and dorsal striatum in instrumental conditioning. Science **304:** 452–454.

30. DAW, N.D., Y. NIV & P. DAYAN. 2006. Actions, policies, values, and the basal ganglia. *In* Recent Breakthroughs in Basal Ganglia Research. E. Bezard, Ed.: 111–130. Nova Science Publishers. New York.

31. JOEL, D., Y. NIV & E. RUPPIN. 2002. Actor-critic models of the basal ganglia: New anatomical and computational perspectives. Neur. Netw. **15:** 535–547.

32. SCHULTZ, W., P. DAYAN & P.R. MONTAGUE. 1997. A neural substrate of prediction and reward. Science **275:** 1593–1599.

33. ARBUTHNOTT, G.W. & J. WICKENS. 2007. Space, time, and dopamine. Trends Neurosci. **30(2):** 62–69.

34. ROITMAN, M.F., A. SEIPEL, J.J. DAY, *et al.* 2006. Rapid onset, short-duration fluctuations in dopamine contribute to the tonic, steady-state level of dopamine concentration in the nucleus accumbens. In Society for Neuroscience Abstracts, **32:** 254.15.

35. C.J. WATSON, B.J. VENTON & R.T. KENNEDY. 2006. *In vivo* measurements of neurotransmitters by microdialysis sampling. Anal. Chem. **78:** 1391–1399.

36. BENINGER, R.J. 1983. The role of dopamine in locomotor activity and learning. Brain Res. Rev. **6:** 173–196.

37. BERRIDGE, K.C. & T.E. ROBINSON. 1998. What is the role of dopamine in reward: hedonic impact, reward learning, or incentive salience? Brain Res. Rev. **28:** 309–369.

38. SCHULTZ, W. 1998. Predictive reward signal of dopamine neurons. J. Neurophysiol. **80:** 1–27.

39. IKEMOTO, S. & J. PANKSEPP. 1999. The role of nucleus accumbens dopamine in motivated behavior: a unifying interpretation with special reference to reward-seeking. Brain Res. Rev. **31:** 6–41.

40. DICKINSON, A., J. SMITH & J. MIRENOWICZ. 2000. Dissociation of Pavlovian and instrumental incentive learning under dopamine agonists. Behav. Neurosci. **114:** 468–483.

41. WEINER, I. & D. JOEL. 2002. Dopamine in schizophrenia: dysfunctional information processing in basal ganglia-thalamocortical split circuits. *In* Handbook of Experimental Pharmacology Vol. 154/II, Dopamine in the CNS II. G. Di Chiara, Ed.: 417–472. Springer-Verlag. Berlin.

42. SALAMONE, J.D. & M. CORREA. 2002. Motivational views of reinforcement: implications for understanding the behavioral functions of nucleus accumbens dopamine. Behav. Brain Res. **137:** 3–25.

43. MURSCHALL, A. & W. HAUBER. 2006. Inactivation of the ventral tegmental area abolished the general excitatory influence of Pavlovian cues on instrumental performance. Learn. Mem. **13:** 123–126.

44. LYON, M. & T.W. ROBBINS. 1975. The action of central nervous system stimulant drugs: a general theory concerning amphetamine effects. *In* Current Developments in Psychopharmacology, Vol. 4. W. Essman & L. Valzelli, Eds.: 79–163. Spectrum. New York.

45. JACKSON, D.M., N. ANDEN & A. DAHLSTROM. 1975. A functional effect of dopamine in the nucleus accumbens and in some other dopamine-rich parts of the rat brain. Psychopharmacologia **45:** 139–149.

46. TAYLOR, J.R. & T.W. ROBBINS. 1986. 6-Hydroxydopamine lesions of the nucleus accumbens, but not of the caudate nucleus, attenuate enhanced responding with reward-related stimuli produced by intra-accumbens d-amphetamine. Psychopharmacology **90:** 390–397.

47. CARR, G.D. & N.M. WHITE. 1987. Effects of systemic and intracranial amphetamine injections on behavior in the open field: a detailed analysis. Pharmacol. Biochem. Behav. **27:** 113–122.

48. SOKOLOWSKI, J.D. & J.D. SALAMONE. 1998. The role of accumbens dopamine in lever pressing and response allocation: effects of 6-OHDA injected into core and dorsomedial shell. Pharmacol. Biochem. Behav. **59:** 557–566.

49. ABERMAN, J.E. & J.D. SALAMONE. 1999. Nucleus accumbens dopamine depletions make rats more sensitive to high ratio requirements but do not impair primary food reinforcement. Neuroscience **92:** 545–552.

50. SALAMONE, J.D., A. WISNIECKI, B.B. CARLSON & M. CORREA. 2001. Nucleus accumbens dopamine depletions make animals highly sensitive to high fixed ratio requirements but do not impair primary food reinforcement. Neuroscience **5:** 863–870.

51. M. CORREA, B.B. CARLSON, A. WISNIECKI & J.D. SALAMONE. 2002. Nucleus accumbens dopamine and work requirements on interval schedules. Behav. Brain Res. **137:** 179–187.

52. MINGOTE, S., S.M. WEBER, K. ISHIWARI, *et al.* 2005. Ratio and time requirements on operant schedules: effort-related effects of nucleus accumbens dopamine depletions. Eur. J. Neurosci. **21:** 1749–1757.

53. HULL, C.L. 1943. Principles of Behavior: An Introduction to Behavior Theory. Appleton-Century-Crofts. New York.

54. BROWN, J.S. 1961. The Motivation of Behavior. McGraw-Hill. New York.

55. BOLLES, R.C. 1967. Theory of Motivation. Harper & Row. New York.

56. DICKINSON, A. & B.W. BALLEINE. 2002. The role of learning in the operation of motivational systems. *In* Learning, Motivation and Emotion, volume 3 of Steven's Handbook of Experimental Psychology, ch. 12. C.R. Gallistel, Ed.: 497–533. John Wiley & Sons. New York.

57. BALLEINE, B.W. 2005. Neural bases of food-seeking: affect, arousal and reward in corticostriatolimbic circuits. Physiol. Behav. **86:** 717–730.

58. DAW, N.D., Y. NIV & P. DAYAN. 2005. Uncertainty-based competition between prefrontal and dorsolateral striatal systems for behavioral control. Nat. Neurosci. **8:** 1704–1711.

59. BALLEINE, B.W. & A. DICKINSON. 1998. Goal-directed instrumental action: contingency and incentive learning and their cortical substrates. Neuropharmacology **37:** 407–419.

60. KILLCROSS, S. & E. COUTUREAU. 2003. Coordination of actions and habits in the medial prefrontal cortex of rats. Cereb. Cortex **13:** 400–408.

61. YIN, H.H., S.B. OSTLUND, B.J. KNOWLTON & B.W. BALLEINE. 2005. The role of the dorsomedial striatum in instrumental conditioning. Eur. J. Neurosci. **22:** 513–523.

62. YIN, H.H., B.J. KNOWLTON & B.W. BALLEINE. 2005. Blockade of NMDA receptors in the dorsomedial striatum prevents action-outcome learning in instrumental conditioning. Eur. J. Neurosci. **22:** 505–512.
63. ADAMS, C.D. 1982. Variations in the sensitivity of instrumental responding to reinforcer devaluation. Quart. J. Exp. Psychol. **34B:** 77–98.
64. DICKINSON, A., B. BALLEINE, A. WATT, F. GONZALEZ & R.A. BOAKES. 1995. Motivational control after extended instrumental training. Anim. Learn. Behav. **23:** 197–206.
65. YIN, H.H., B.J. KNOWLTON & B.W. BALLEINE. 2004. Lesions of the dosolateral striatum preserve outcome expectancy but disrupt habit formation in instrumental learning. Eur. J. Neurosci. **19:** 181–189.
66. COLWILL, R.M. & R.A. RESCORLA. 1986. Associative structures in instrumental learning. Psychol. Learn. Motiv. **20:** 55–104.
67. DICKINSON, A. & G.R. DAWSON. 1988. Motivational control of instrumental performance: the role of prior experience with the reinforcer. Quart. J. Exp. Psychol. **40B:** 113–134.
68. DICKINSON, A. & G.R. DAWSON. 1989. Incentive learning and the motivational control of instrumental performance. Quart. J. Exp. Psychol. **41B:** 99–112.
69. BALLEINE, B.W. 1992. Instrumental performance following a shift in primary motivation depends on incentive learning. J. Exp. Psychol. Anim. Behav. Process. **18:** 236–250.
70. DICKINSON, A. & B.W. BALLEINE. 1994. Motivational control of goal-directed action. Anim. Learn. Behav. **22:** 1–18.
71. BALLEINE, B.W. 2000. Incentive processes in instrumental conditioning. In Handbook of Contemporary Learning Theories. R.R. Mowrer & S.B. Klein, Eds.: 307–366. Lawrence Erlbaum Associates. Mahwah, NJ.
72. DAW, N.D., S. KAKADE & P. DAYAN. 2002. Opponent interactions between serotonin and dopamine. Neur. Netw. **15:** 603–616.
73. COHEN, J.D., T.S. BRAVER & J.W. BROWN. 2002. Computational perspectives on dopamine function in prefrontal cortex. Curr. Opin. Neurobiol. **12:** 223–229.
74. REDISH, A.D. 2005. Implications of the temporal difference reinforcement learning model for addiction and relapse. Neuropsychopharmacology **30**(Suppl 1): S27–S28.
75. WILLNER, P., K. CHAWLA, D. SAMPSON, et al. 1988. Tests of functional equivalence between pimozide pretreatment, extinction and free feeding. Psychopharmacology **95:** 423–426.
76. WATANABE, M., H. CROMWELL, L. TREMBLAY, et al. 2001. Behavioral reactions reflecting differential reward expectations in monkeys. Exp. Brain Res. **140:** 511–518.
77. TAKIKAWA, Y.K., R.K. KAWAGOE, H.K. ITOH, et al. 2002. Modulation of saccadic eye movements by predicted reward outcome. Exp. Brain Res. **142:** 284–291.
78. LAUWEREYNS, J., K. WATANABE, B. COE & O. HIKOSAKA. 2002. A neural correlate of response bias in monkey caudate nucleus. Nature **418:** 413–417.
79. SCHOENBAUM, G., B. SETLOW, S.L. NUGENT, et al. 2003. Lesions of orbitofrontal cortex and basolateral amygdala complex disrupt acquisition of odor-guided discriminations and reversals. Learn. Mem. **10:** 129–140.

80. SATOH, T., S. NAKAI, T. SATO & M. KIMURA. 2003. Correlated coding of motivation and outcome of decision by dopamine neurons. J. Neurosci. **23:** 9913–9923.
81. ROITMAN, M.F., G.D. STUBER, P.E.M. PHILLIPS, *et al.* 2004. Dopamine operates as a subsecond modulator of food seeking. J. Neurosci. **24:** 1265–1271.
82. C.D. FIORILLO, P.N. TOBLER & W. SCHULTZ. 2003. Discrete coding of reward probability and uncertainty by dopamine neurons. Science **299:** 1898–1902.
83. TOBLER, P.N., C.D. FIORILLO & W. SCHULTZ. 2005. Adaptive coding of reward value by dopamine neurons. Science **307:** 1642–1645.
84. PHILLIPS, P.E.M. & R.M. WIGHTMAN. 2004. Extrasynaptic dopamine and phasic neuronal activity. Nat. Neurosci. **7:** 199.
85. WISE, R.A. 2004. Dopamine, learning and motivation. Nat. Rev. Neurosci. **5:** 483–495.
86. MCCLURE, S.M., N.D. DAW & P.R. MONTAGUE. 2003. A computational substrate for incentive salience. Trends Neurosci. **26:** 423–428.
87. BERRIDGE, K.C. 2004. Motivation concepts in behavioral neuroscience. Physiol. Behav. **81:** 179–209.
88. DAYAN, P., Y. NIV, B. SEYMOUR & N.D. DAW. 2006. The misbehavior of value and the discipline of the will. Neur. Netw. **19:** 1153–1160.
89. CORBIT, L.H. & B.W. BALLEINE. 2000. The role of the hippocampus in instrumental conditioning. J. Neurosci. **20:** 4233–4239.
90. CORBIT, L.H., S.B. OSTLUND & B.W. BALLEINE. 2002. Sensitivity to instrumental contingency degradation is mediated by the entorhinal cortex and its efferents via the dorsal hippocampus. J. Neurosci. **22:** 10976–10984.
91. DICKINSON, A. & B. BALLEINE. 1990. Motivational control of instrumental performance following a shift from thirst to hunger. Quart. J. Exp. Psychol. **24B:** 413–431.
92. COLWILL, R.M. & S.M. TRIOLA. 2002. Instrumental responding remains under the control of the consequent outcome after extended training. Behav. Process. **57:** 51–64.
93. CORBIT, L.H. & B.W. BALLEINE. 2003. Instrumental and Pavlovian incentive processes have dissociable effects on components of a heterogeneous instrumental chain. J. Exp. Psychol. Anim. Behav. Process. **29:** 99–106.
94. CORBIT, L.H. & B.W. BALLEINE. 2005. Double dissociation of basolateral and central amygdala lesions on the general and outcome-specific forms of Pavlovian-instrumental transfer. J. Neurosci. **25:** 962–970.
95. HOLLAND, P.C. 2004. Relations between Pavlovian-instrumental transfer and reinforcer devaluation. J. Exp. Psychol. Anim. Behav. Process. **30:** 104–117.
96. WALTON, M.E., S.W. KENNERLEY, D.M. BANNERMAN, *et al.* 2006. Weighing up the benefits of work: behavioral and neural analyses of effort-related decision making. Neur. Netw. **19:** 1302–1314.
97. WALTON, M.E., P.H. RUDEBECK, D.M. BANNERMAN & M.F.S. RUSHWORTH. 2007. Calculating the cost of acting in prefrontal cortex. Ann. N. Y. Acad. Sci. This issue.
98. BOTVINICK, M.M., J.D. COHEN & C.S. CARTER. 2004. Conflict monitoring and anterior cingulate cortex: an update. Trends Cogn. Sci. **8:** 539–546.
99. COUSINS, M.S., A. ATHERTON, L. TURNER & J.D. SALAMONE. 1996. Nucleus accumbens dopamine depletions alter relative response allocation in a T-maze cost/benefit task. Behav. Brain Res. **74:** 189–197.

100. WALTON, M.E., D.M. BANNERMAN, K. ALTERESCU & M.F.S. RUSHWORTH. 2003. Functional specialization within medial frontal cortex of the anterior cingulate for evaluating effort-related decisions. J. Neurosci. **23:** 6475–6479.
101. RUSHWORTH, M.F.S., M.E. WALTON, S.W. KENNERLEY & D.M. BANNERMAN. 2004. Action sets and decisions in the medial frontal cortex. Trends Cogn. Sci. **8:** 410–417.
102. JHOU, T.C., M.P. SADDORIS, J.N. MADDUX, *et al.* 2006. Lesions of the anterior cingulate cortex impair aversive response latencies in a go, no-go task. *In* Society for Neuroscience Abstracts, Vol. **32:** 463.2.

Erratum

In [1], the following error was published on page xii.

—ROGER C. YOUNG
University of Vermont, Burlington, Vermont, USA
Tel Aviv University, Tel Aviv, Israel

The text was incorrect and should have read:

—ROGER C. YOUNG
University of Vermont, Burlington, Vermont, USA

We apologize for this error.

REFERENCE

1. ELAD, D. *et al.* 2007. Preface. Ann. N.Y. Acad. Sci. **1101:** xi–xii.

Ann. N.Y. Acad. Sci. 1104: 377 (2007). © 2007 New York Academy of Sciences.
doi: 10.1196/annals.1390.040

Index of Contributors